ANNALS OF THE NEW YORK ACADEMY OF SCIENCES

The New York Academy of Sciences
2 East 63rd Street
New York, New York 10021

Volume 581

AMERICAN INSTITUTE OF PHYSICS CONFERENCE PROCEEDINGS

American Institute of Physics
Books Division
335 East 45th Street
New York, New York 10017

Number 213

EDITORIAL STAFF OF THE NEW YORK ACADEMY OF SCIENCES

Executive Editor
BILL BOLAND

Managing Editor
JUSTINE CULLINAN

Associate Editor
JOYCE HITCHCOCK

THE NEW YORK ACADEMY OF SCIENCES
(Founded in 1817)

Chairman of the Board
LEWIS THOMAS

President
CHARLES A. SANDERS

President-Elect
DENNIS D. KELLY

Honorary Life Governors
H. CHRISTINE REILLY
IRVING J. SELIKOFF

Vice Presidents
DAVID A. HAMBURG
CYRIL M. HARRIS
PETER D. LAX
CHARLES G. NICHOLSON

Secretary-Treasurer
HENRY A. LICHSTEIN

Elected Governors-at-Large
JOSEPH L. BIRMAN
FLORENCE L. DENMARK
LAWRENCE R. KLEIN
GERALD D. LAUBACH
LLOYD N. MORRISETT
GERARD PIEL

Past Chairman
WILLIAM T. GOLDEN

General Counsel
HELENE L. KAPLAN

Executive Director
OAKES AMES

AMERICAN INSTITUTE OF PHYSICS
(Founded in 1931)

Governing Board Chair
HANS FRAUENFELDER

Executive Director and CEO
KENNETH W. FORD

Member Societies
The American Physical Society
Optical Society of America
Acoustical Society of America
The Society of Rheology
American Association of Physics Teachers
American Crystallographic Association
American Astronomical Society
American Association of Physicists in Medicine
American Vacuum Society
American Geophysical Union

FRONTIERS IN CONDENSED MATTER THEORY

Proceedings of a US - USSR Conference

ANNALS OF THE NEW YORK ACADEMY OF SCIENCES

Volume 581

AMERICAN INSTITUTE OF PHYSICS CONFERENCE PROCEEDINGS

Number 213

FRONTIERS IN CONDENSED MATTER THEORY

PROCEEDINGS OF A US - USSR CONFERENCE

Edited by Melvin Lax, Lev P. Gor'kov, and Joseph L. Birman

The New York Academy of Sciences
New York, New York
1990

∞ *The paper used in this publication meets the minimum requirements of American National Standard for Information Sciences—Permanence of Paper for Printed Library Materials, ANSI Z39.48-1984.*

The art that appears on the softcover version of this volume is from the paper by D.J. Bishop *et al.* and appears on page 80.

Library of Congress Cataloging-in-Publication Data

Frontiers in condensed matter theory : proceedings of a US-USSR conference, held in New York City, at the Graduate Center of the City University of New York, December 4-8, 1989 / edited by Melvin Lax, Lev. P. Gor'kov & Joseph L. Birman.
p. cm. — (Annals of the New York Academy of Sciences : v. 581) (AIP conference proceedings : no. 213)
"First Binational US-USSR Conference on Frontiers in Condensed Matter Theory"—American editor's pref.
Includes bibliographical references and index.
ISBN 0-88318-771-X (cloth : AIP). — ISBN 0-88318-772-8 (paper : AIP)
1. Condensed matter—Congresses. 2. Superconductivity—Congresses. 3. Quantum Hall effect—Congresses. 4. Transport theory—Congresses. I. Lax, Melvin J. II. Gor'kov, L. P. (Lev Petrovich) III. Birman, Joseph Leon, 1927- . IV. Binational US-USSR Conference on Frontiers in Condensed Matter Theory (1st : 1989 : Graduate Center, City University of New York) V. Series. VI. Series: AIP conference proceedings ; no. 213.
Q11.N4 vol. 581
[QC173.4.C65]
500 s—dc20
[530.4'1] 90-6421
CIP

All orders for this volume should be sent to the American Institute of Physics, 335 East 45th Street, New York, New York 10017.

PCP
Printed in the United States of America
ISBN 0-88318-771-x (cloth)
ISBN 0-88318-772-8 (paper)

ANNALS OF THE NEW YORK ACADEMY OF SCIENCES

Volume 581

AMERICAN INSTITUTE OF PHYSICS CONFERENCE PROCEEDINGS

Number 213

July 2, 1990

Frontiers in Condensed Matter Theory

Proceedings of a US - USSR Conference [a]

Editors

MELVIN LAX, LEV P. GOR'KOV, AND JOSEPH L. BIRMAN

Superconductivity

Transport, Quantum Hall Effect, Localization, and Scattering in Random Systems

[a] This volume is the result of a conference entitled US-USSR Conference on Frontiers in Condensed Matter Theory held December 4-8, 1989 in New York City, and co-sponsored by The New York Academy of Sciences, The Academy of Sciences of the USSR, and City University of New York.

[b] The names in boldface are those of the paper presenters.

[c] The panel discussions do not appear in the book.

Superfluids

Dielectrics and Semiconductors

Two Dimensional Transport and the Quantum Hall Effect

Localization Effects

Poster Papers

Financial assistance was provided by:

- CCNY SCIENCE DIVISION AND PHYSICS DEPARTMENT
- CUNY GRADUATE CENTER
- CUNY RESEARCH FOUNDATION
- HAMAMATSU CORPORATION
- NEC RESEARCH LABORATORY
- NEW YORK STATE INSTITUTE ON SUPERCONDUCTIVITY
- NEW YORK STATE SCIENCE AND TECHNOLOGY FOUNDATION
- THE SOROS FOUNDATION—SOVIET UNION
- T. J. WATSON RESEARCH LABORATORIES OF IBM (YORKTOWN HEIGHTS)

FOREWORD AND DEDICATION

The concept of Binational US-USSR Conferences on Frontiers in Condensed Matter Theory and Frontiers in Mathematics originated in discussions between the late Dr. Heinz Pagels, then the Executive Director of the New York Academy of Sciences, and myself sometime in 1986-7. It was Heinz Pagels' idea that the experience of the New York Academy would be helpful in enhancing American-Soviet scientific cooperation: the Academy is a natural focus for high-level scientific activity in the Metropolitan New York area, and it wished to expand its activities in the physical and mathematical sciences. The *Annals of the New York Academy of Sciences* was available to publish the proceedings. As we formulated this idea, I grew more positive about the possibility of helping to bring additional first-rate science to the New York area. I took this plan to colleagues in the USA including Prof. M. Lax, and other members of the Metropolitan-New York USA Organizing Committee, and in the USSR, including Professors I.M. Khalatnikov, L.P. Gor'kov, L.D. Faddeev, S. Novikov, and others and found a very receptive, and positive reaction.

Owing to this positive reaction on both sides, the nascent idea was realized and colleagues on both sides—in condensed matter theory, and in mathematics—carried the plan still further—for example by getting needed funding from the two Academies and other sources—so that the Condensed Matter Theory Conference would be realized in December 1989—almost one year after the successful Mathematics Conference. Elsewhere thanks are expressed to many of those who work made these Frontier Conferences succeed.

Heinz Pagels did not see these "brain-children" of ours come to life owing to the tragic mountaineering accident in Aspen, Colorado which took his life in July 1987. I believe that he would have shared our satisfaction about these two successful Conferences which featured just that kind of scientific interchange and warm cooperation we hoped for. Let this volume of the Proceedings of the Condensed Matter Theory Conference—and the hoped-for companion volume of the Mathematics Conference—be a partial memento to one side of the many-faceted Heinz Pagels: namely to the importance of fostering scientific cooperation between all scientists!

JOSEPH L. BIRMAN

Physics Department
City College of
The City University of New York
New York, New York 10031

9 April 1990

AMERICAN EDITORS' PREFACE

The First Binational US-USSR Conference on Frontiers in Condensed Matter Theory was held at the Graduate Center of The City University of New York in New York City, December 4-9, 1989. This volume of the proceedings contains the full texts of the papers presented and prepared for presentation at this symposium. Participants in the conference included twelve Soviet physicists, coming to New York from various locales in the USSR—the majority were from the Landau Institute of Theoretical Physics, Moscow—and included one experimental physicist: Prof. V. Timofeev. On the American side, the speakers were drawn from major institutions of the Metropolitan-New York area, including sixteen theorists, and two experimentalists, from universities, industrial research laboratories, and the Brookhaven National Laboratory. Besides clearly identifiable American and Soviet physicists the Conference benefited from the presence of several Soviet scientists who are on long-term visits to American Institutions—these "semi" Soviet-Americans exemplify the material side of *glasnost* and *perestroika.* We hope the conference, and the wide distribution of these proceedings will further assist the developing processes of openness and cooperation between American and Soviet physicists. A memorable and unusual example of this was the award to Prof. Lev Petrovich Gor'kov of the degree of Doctor of Science *honoris causa* by the Graduate School of City University of New York in recognition of his scientific achievements.

Of course a conference of this magnitude could not have succeeded without the dedicated help of many individuals and the material and financial help of far-sighted individuals and institutions. First we want to thank the members of the American and Soviet Organizing Committees who helped to bring the speakers, session chairs, and other participants, to the conference. In this regard we want to single out Professor Olivier Martin of City College, and Professor Igor Dzyaloshinsky of the Landau Institute who gave special help to us in the organization and preorganization states. Professor Albert K. Levine helped with local organization at the Graduate Center of City University of New York. Manuscript preparation for the Soviet papers was in the capable hands of Dr. Wei Cai and Ting-Fang Zheng who put them into a uniform format. Great help in all phases of organization were graciously given by the staff of the New York Academy of Sciences, with the overall coordination by Dr. Maria Simpson, head of the Conference Department. In particular Ms. Renee Wilkerson and Ms. Sheila Kane were tirelessly and efficiently available to solve all problems, before, during and after the conference. Copublication of these proceedings had the essential editorial cooperation of Mr. Bill Boland and Ms. Joyce Hitchcock of the New York Academy of Sciences and Mr. Tim Taylor of the American Institute of Physics. Mrs. Elizabeth De Crescenzo, Mrs. Evelyne Rosenstock and others at City College also helped. To all, our hearty thanks—from all participants.

We next want to warmly thank those institutions which provided the essential support, financial and otherwise, for the conference. The New York Academy of Sciences—especially Dr. Oakes Ames and Mr. William Golden—and the Graduate Center of City University of New York—especially President Harold Proshansky—were mainstays of the conference. Important early and major support came from the Theoretical Department of the T.J. Watson Research Laboratories—for which thanks to Dr. Erling Pytte. The Research Foundation of CUNY, President Mathew Goldstein, The New York State Institute on Superconductivity (Professor David T. Shaw), The New York State Science and Technology Foundation (Mr. Graham Jones), The NEC Research Laboratory (Dr. Joseph Giordimaine), the Science Division of City College (Dean Michael Arons), the

Physics Department of City College (Prof. Philip Baumel, Chairman) and the CUNY Academy of Humanities and Sciences (Dr. Mitchell Gurfield) all provided support and we here record our thanks to all.

Now we want to record our special thanks to the G. Soros Foundation—Soviet Union (Mr. G. Soros) which generously supported our Soviet guests during their stay in the U.S.; and to Mr. B. Schoenfeld of the Schubert Organization which provided tickets for a memorable evening of theater entertainment for the Soviet scientists.

Finally, the American co-editors express our warm thanks to Prof. Lev Gor'kov, who, by methods both known and unknown to us, assembled the Soviet delegation and arranged for them to arrive at J.F. Kennedy Airport as planned, with a complete set of their manuscripts for this volume.

We hope this volume will be a useful and topical addition to the current literature and a long-time reminder of a productive and pleasant Conference.

JOSEPH L. BIRMAN MELVIN LAX

OPENING REMARKS

William T. Golden

When Joe Birman spoke of not starting quite on time, there came to mind, not entirely irrelevantly, a remark of the late Jimmy Walker, who was the Mayor of New York several decades ago and was distinguished in a number of ways. One of them was that he was habitually late. He once remarked "If you're there before it's over, you're on time." So, we're all always going to be on time. Now, Joe Birman also said that he had not asked in advance—or rather Professor Khalatnikov said that Joe Birman had not asked him in advance to prepare any remarks. I am not in that situation: Joe Birman instructed me *not* to prepare any remarks—not to make any speech. So everyone may now relax. But I am here, as Chairman of the Board of the New York Academy of Sciences, to welcome you and to express my pleasure that you are here—that this Second Joint Conference of the Soviet Academy of Sciences, the New York Academy of Sciences, and the City University of New York Graduate Center is underway. The first joint conference, on Frontiers of Higher Mathematics, held just about a year ago, was a great success, and I am confident that this one will be also.

At the New York Academy, and also elsewhere in scientific and cultural circles in New York City, we are looking forward to increasingly warm and intimate relationships, intellectually and culturally, between people of the United States and people of the Soviet Union. Certainly, developments of the past few years, and especially of the past year and of even more recent time, are all going in that direction. But most specifically, I hope, as a representative of the New York Academy of Sciences, that these conferences and mutual relationships of joint activity will prosper. Certainly, we are going to do everything we can at the New York Academy to work toward that end with you, our friends from the Soviet Union. I would like to add that just about a month ago I was in the Soviet Union as a guest of the Soviet Academy of Sciences. I was there as a representative of the American Association for the Advancement of Science, which most of you know because it is the publisher of *Science*—a journal with which you are no doubt familiar. The Soviet Academy invited us to send our officers, so four of us—I am the Treasurer and a member of the Board of Directors—went and were very well received. We were really fascinated with all that we saw and learned, and we enjoyed a great feeling of warmth and a great desire for closer relationships. We hope to have them.

Indeed, I'll mention, as I did to Professor Khalatnikov just now, that I am getting up a book that I have been working on for two years, which is entitled *Worldwide Science and Technology: Advice to the Highest Levels of Governments.* It will be published in New York and Oxford by Pergamon Press in mid-1990. In it, as editor, I have collected papers from 35 countries, all the major countries of the world *except* the Soviet Union. They were invited to contribute and have promised a paper, but it has not arrived and we are about to go to press. Perhaps, through Professor Khalatnikov's efforts, it will arrive. I mention this book because it is in the growing spirit of globalism and one-worldism. The advantages of globalism are evident in closer relationships in scientific channels and also in political and cultural channels.

With those remarks, I think that I have given my non-speech, except that I want to mention one more matter. At the New York Academy of Sciences we have a sub-group, the Science Policy Association, which holds breakfast meetings from 8:00-9:30 a.m. about once a month. At these meetings we have distinguished speakers and attract an invited audience of people of ability and prominence. These meetings have been a great success. When I was in the Soviet Union a month ago, I invited the members of the Academy whom we met, including President Marchuk and Academicians Velikhov, Sagdeyev, and Makarov, to address our Science Policy Association of the New York Academy. So that is another step that we are taking, and we are looking forward to their visit.

With that, I end where I began—with welcome to all of you. I am glad that you're here, and I am glad that I'm here.

ON THE SUBSTITUTION OF Zn FOR Cu ATOMS IN HIGH-T_c SUPERCONDUCTORS: EPR MEASUREMENTS OF $La_{1.82}Sr_{0.18}(Cu_{1-x}Zn_x)O_4$

A. M. Finkel'stein
Landau Institute for Theoretical Physics, Moscow, USSR

V. E. Kataev, E. F. Kukovitskii, G. B. Teitel'baum
Physical Technical Institute, Kazan', USSR

EPR measurements of $La_{1.82}Sr_{0.18}(Cu_{1-x}Zn_x)O_4$ compounds reveal that Zn doping induces formation of localized magnetic moments which, as has been found here, sit on Cu ions. At Zn concentration of 1 atomic percent, the quantity of these magnetic moments is close to the amount of substituted ions. This result is important for verifying diverse microscopic mechanisms of high-T_c superconductivity.

INTRODUCTION

As is known, shortly after the discovery high-T_c compounds [1], Anderson [2] made an assumption that we are dealing with substances for which the conventional Fermi-liquid description does not hold. Further, this trend in the study of high-T_c superconductivity has gained a strong impetus. Experimentally this trend was promoted by reports on the temperature-linear heat capacity in the superconducting phase at low temperatures and on the presence of a temperature-linear contribution to resistivity in the wide region above T_c . However, after the discovery of the Bi-Sr-Ca-Cu-O compounds the linear heat capacity was apparently ruled out. As to the temperature-linear resistivity, it is indeed the characteristic property of all high T_c superconductors. Initially it was considered as a fact confirming the RVB picture [3], but now it seems that this property can be accounted for by the traditional mechanism. Thus we come across an unpleasant situation; namely, rich experimental data have been accumulated which, nevertheless, coexist with various microscopic theories. One of the ways of investigating the nature of high-T_c superconductivity is to substitute Cu^{2+} ions by ions of other elements [4]. The most impressive has been the report [5] of the fact, that in lanthanium compounds doping with only 2.5 at.% Zn or Ga was enough to completely suppress superconductivity. Since Zn and Ga ions have a closed-shell $3d^{10}$ configuration in metal, this strong effect on superconductivity is striking. In terms of conventional superconductivity the role of such impurities should be minor. Therefore, with this drastic effect of T_c-suppression [5] taken into account, doping with Zn or Ga should be regarded as a way to probe the correlated electron liquid in high-T_c superconductors. Since substitution for Cu^{2+} in CuO_2 plane by an ion with a closed $3d^{10}$ shell leads to elimination of one of spins from the lattice, it seems quite probable that this will permit exploration of the role of spin correlations and may help us to exclude some versions of the theory and to narrow down the field of investigation.

Generally speaking there are diverse possibilities for suppressing superconductivity with doping, e.g.

a) superconducting pairing is nontrivial. Usually such types of pairing are not stable with respect to nonmagnetic impurities.

b) owing to spin lattice defects, spin correlations, which may be responsible for superconductivity, are destroyed.

c) owing to the Zn or Ga doping, localized magnetic moments (magnetic impurities) are created in spite of the fact that Zn or Ga ions are nonmagnetic. These moments can suppress superconductivity in a standard fashion. (Note that b) and c) may be combined)

etc). . .

Version a) seems not to be in agreement with the current experiment. Besides, it requires an explanation for why only the substitution defects proved to be effective in the T_c suppression, but not other types of disorder. The discussion of versions b) and c) needs experimental data on magnetic properties. Before discussing experimental data, let us outline an idea of a mechanism of the creation of localized magnetic moments under doping with nonmagnetic ions.

LOCALIZED MOMENTS FORMATION

Let the effective Hamiltonian of an electron system be

$$\mathcal{H}=t\sum c_{i\sigma}^{\dagger}c_{j\sigma}-2J\sum b_{ij}^{\dagger}b_{ij}, \quad b_{ij}=\frac{1}{\sqrt{2}}\sum_{\sigma}\sigma c_{i\sigma}c_{j-\sigma}, \tag{1}$$

with the constraint $\sum_{\sigma} n_{i\sigma}<2$. Then introduce slave bosons (holons) and fermions (spinons) [6]

$$c_{i\sigma}=e_i^{\dagger}s_{i\sigma}, \quad \sum s_{i\sigma}^{\dagger}s_{i\sigma}+e_i^{\dagger}e_i=1, \tag{2}$$

Here the operator c_i creates a hole and eliminates the spin from the site i. Having (2) in mind, rewrite the exchange term via spinon operators s and decouple it by introducing auxiliary fields Δ_{ij} and Δ_{ij}^{*}. Then

$$L=\sum\left(s_{i\sigma}^{\dagger}\frac{\partial}{\partial\tau}s_{i\sigma}+e_i\frac{\partial}{\partial\tau}e_i\right)+t\sum e_i e_j^{\dagger}s_{i\sigma}^{\dagger}s_{j\sigma}+\sum J^{-1}|\Delta_{ij}|^2+\mu\sum_{dopants}{}'e_l^{\dagger}e_l$$

$$+\sum[\Delta_{ij}^{*}(s_{j\downarrow}s_{i\uparrow}-s_{j\uparrow}s_{i\downarrow})+\text{c.c}]+i\sum_i\lambda_i(s_{i\sigma}^{\dagger}s_{i\sigma}+e_i^{\dagger}e_i-1). \tag{3}$$

The last term has been introduced to localize holons on substituted sites (strictly speaking one must also change the respective t and J). The Lagrangian (3) has the following local U(1) gauge symmetry

$$s_{i\sigma}\longrightarrow e^{i\theta_i}s_{i\sigma}, \quad e_i\longrightarrow e_i e^{i\theta_i}, \quad \Delta_{ij}\longrightarrow \Delta_{ij}e^{i\theta_i+i\theta_j},$$

$$\Delta_{ij}=|\Delta_{ij}|e^{i\phi_{ij}}, \quad \phi_{ij}\longrightarrow \phi_{ij}+\theta_i+\theta_j. \tag{4}$$

This means that the state resulting from action of the operator $e_i^\dagger$ on the ground state is not invariant. The gauge-invariant object is a string [7]

$$e_i^\dagger \Delta_{ij} \Delta_{jk}^* \Delta_{kl} \ldots$$

This string must terminate either on another holon (it is now implied that these holons are localized on dopant ions) or on a spinon. The energy of such objects increases with distance either logarithmically or linearly; the latter case corresponds to confinement [8]. In both cases, if the spacing between dopants is sufficiently large, strings connecting them must be torn via creation of pairs of spinons. The pieces of strings with the spinons on the free ends will contract. As a result, near each of the substituted ions there must be formed a spinon-holon pair which will reveal the properties of localized magnetic moments. From a rigorous point of view these speculations are rather simplified, but in our opinion they adequately illustrate the essence of the phenomenon.

EPR MEASUREMENTS

Following the general speculations presented above, in the fall of 1988 we started the experiments on EPR measurements in Zn-doped lanthanium superconductors. We aimed at searching for localized magnetic moments and finding out their nature, or ruling out the version c). The measurements have revealed that doping with Zn induces formation of localized magnetic moments sitting on Cu^{2+} ions. The results of our EPR measurements on the whole agree with the results of the recently published paper [9], where $La_{1.85}Sr_{0.15}(Cu_{1-x}Ga_x)O_4$ samples have been thoroughly studied in superconducting and normal phases. (The results of some analogous measurements performed on Zn-doped samples [10] are similar). In [9] it has been shown that the low-temperature magnetic susceptibility can fit the Curie-Weiss law. Besides, the normal state resistivity exhibits a minimum as the temperatures is lowered. After the appropriate data analysis, the authors of [9] have come to the conclusion that among other possible mechanisms, the Kondo effect can provide a better description of the resistivity behavior. However the interpretation suggested in [9] appears to be not quite correct. The authors have assumed that Ga doping induces localization of Cu 3d-holes. Assuming that magnetic moments reside on the Cu sites uniformally, they have extracted the value of the Cu effective moment which proved to be small. The EPR data which have obtained, favor another picture. Treating the integral intensity of the resonance line as a magnetic susceptibility of the individually localized spins and assuming their amount to be equal to the amount of Zn atoms, we get for the Zn-concentration of 1 at.% the magnitude of the effective magnetic moment of these spins to be very close to $3^{1/2}\mu_B$ the nominal value of the Cu-ion magnetic moment. Therefore, we conclude that each Zn-ion induces the formation of the localized magnetic moment sitting on a Cu-ion. At Zn-concentration of 3 at.% the integral intensity of the signal per Zn ion decreases somewhat, which points to the fact that in this region of impurity concentration the dopants can not be regarded as acting independently.

The EPR measurements have been conducted by us in the following way. Since we could not grow monocrystal samples, we applied a strong magnetic field to the small-grained powder. Owing to the anisotropy of the Lande g-factor, the magnetic

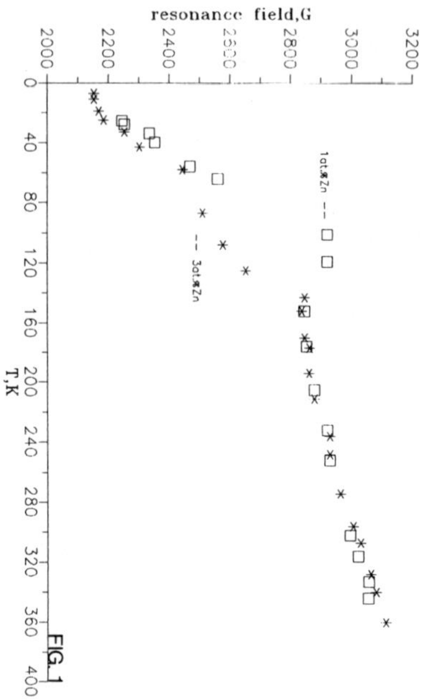

FIG. 1

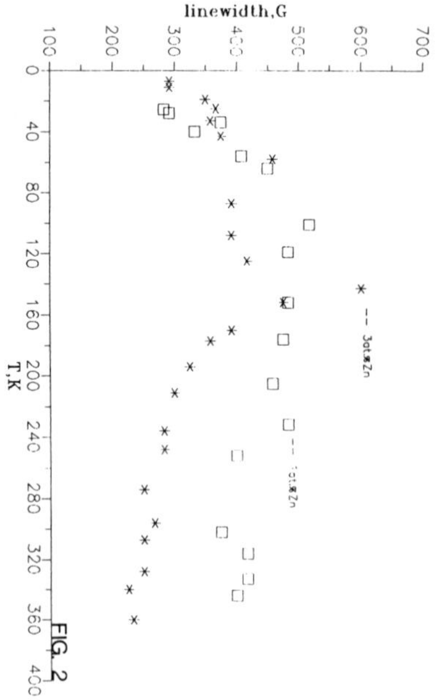

FIG. 2

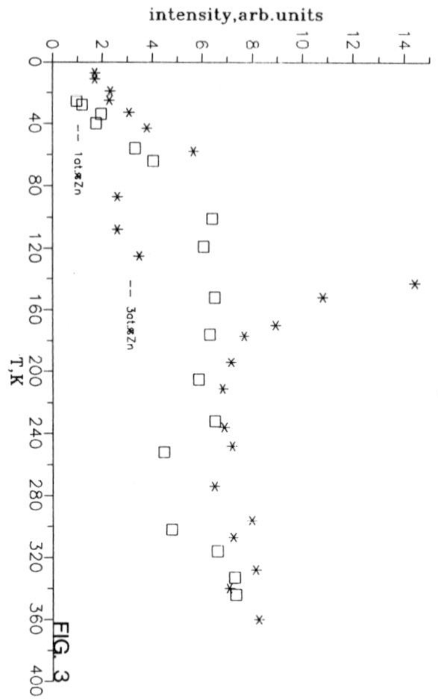

FIG. 3

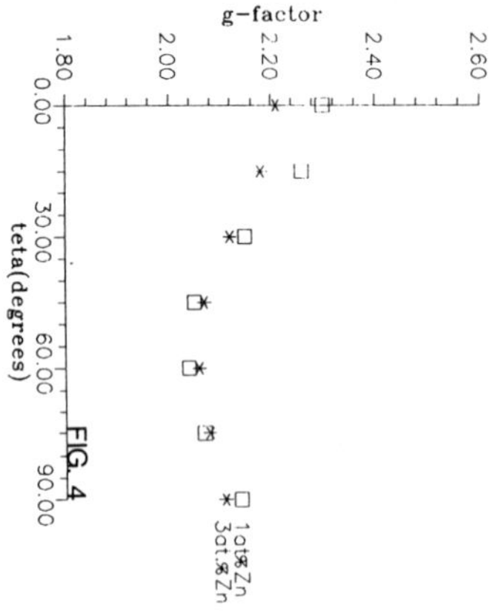

FIG. 4

field orients the particles in such a manner that the CuO_2 planes prove to be orthogonal to the magnetic field. Next, the thus-oriented particles were fixed by paraffin. The X-ray measurements made, comparing when possible, monocrystals with the thus-prepared oriented powders, testify that the accuracy of average orientation of the CuO_2 planes can be estimated to be not worse than 90 %. This technology is necessary since the resonance is not observed in non-oriented pellets owing to the g-factor anisotropy.

In Figures 1-4 results of the EPR measurements for Zn concentrations equal to 1 at.% and 3 at.% are displayed. The accuracy is not sufficient for analyzing the data for Zn concentration smaller than 1 at.%. If Zn dopants are absent, no traces of the resonance can be found.

Fig. 1 gives the magnitude of resonance field depending on temperature. The anomalies for 3 at.% concentration at $T \approx 150$K and respectively for 1 at.% concentration at $T \approx 80$K are alike those which are usually observed at spin freezing. Figs. 2 and 3 give the width of the resonance line and the intensity of the resonance signal, which also exhibit anomalies at the same temperatures.

The most interesting data are given in Fig. 4 for angular dependence of the g-factor. The existence of the anisotropy of this order of magnitude is rather natural for Cu ions. The corrections to the g-factor of the magnetic moment emerge because of the combined action of two terms, i.e. Zeeman splitting and spin-orbit interaction

$$\mathcal{H}_Z = \mu_B(\mathbf{L}+2\mathbf{S})\,H\,, \qquad \mathcal{H}_{sp-orb} = \lambda \mathbf{L}\cdot\mathbf{S}\,, \tag{5}$$

For a magnetic ion with orbitally nondegenerate ground state one gets in the second order of perturbation theory

$$g_{\mu\nu} = 2(\delta_{\mu\nu} + \lambda\Lambda_{\mu\nu}), \quad \Lambda_{\mu\nu} = \sum_{\Gamma'} \frac{<\Gamma|L_\mu|\Gamma'><\Gamma'|L_\nu|\Gamma>}{E_\Gamma - E_{\Gamma'}}\,. \tag{6}$$

In our case the Cu-ion is effected by a crystalline field of an elongated oxygen octahedron. Therefore the 3-d levels of the Cu^{2+}-ion are split and the unoccupied orbital has the symmetry x^2-y^2. Calculation of $\Lambda_{\mu\nu}$ within limits of 3-d orbitals yields $\Delta g(\theta) = \Delta g_\perp(1+3\cos\theta)$, where θ is the tilt angle of the magnetic field. For the Cu^{2+} ion $\lambda \sim 10^3$K, and the magnitude of the 3-d level splitting is ~ 1eV. Thus the observed anisotropy of the g-factor has the expected order of magnitude: $\Delta g \sim 0.1$, although this approach can hardly be applicable literally to the real system under consideration.

Finally note that in the presence of excess oxygen in the Zn-doped lanthanium the resonance was not observed, although the suppression of T_c was not influenced. We have not yet come to a clear understanding of this fact. It is quite probable that the excess of oxygen induces some sources of effective spin relaxation.

DISCUSSION

Let us now discuss how the localized moments whose existence was observed by us in EPR-experiments agrees with some current models employed in high-T_c superconductivity theory.

One of the popular models is a short range resonating valence bond theory (SRRVB) [11-15]. It is assumed in SRRVB that electrons on neighboring sites are paired in bonds, i.e. form dimers. This theory implies that the spin excitation energy E_{sp} is large (we mean the energy of the dimer dissociation, accompanied by creation of two spinons), and that spin excitations can be ignored. Quantum fluctuations generate the resonance between valence bonds and allow holes (i.e. sites nonoccupied by dimers) to move by tunneling. The important feature of this theory is the fact that on bipartite lattices, such as a square lattice, there are two topologically distinct types of holes: those that live on the "red" and the "black" sublattices, respectively [11]. The theory describing these quantum dimers appeared [13,15] to be related to the compact electrodynamics in 2+1 dimensions [8], and the two different types of holes have opposite gauge charge. There are reasons to believe [15] that this theory is confining, i.e. the interaction between holes depends on their distance linearly. In the opposite case, this interaction depends on distance logarithmically.

Let the system have only static holes, resulting from substitution for Cu-ions by Zn. Then the situation is appropriate for formation of localized magnetic moments. The necessary condition is that the interaction energy of two doped ions, at an average distance, should be sufficient to destroy a dimer and in result to form two spinon-hole pairs. If conduction electrons live mainly on the oxygen sublattice [16], then the notion that localized magnetic moments appear because of hole-spinon pair formation is consistent. However, the aim of this theory was to demonstrate a gauge interaction between free holes as a source of superconductivity [12,14,15]. Besides, after the discovery of a high-T_c superconductor with *n*-type carriers, the idea of the existence of a two-well-separated-electron subsystem was criticized, and the one-band model [17] got promulgated. Therefore we must discuss the behavior of dynamic holons on the background of static defects. Since the gas of holons is rather dense, there no longer exists a general reason for the ion to creat spinons and the formation of the hole-spinon pairs; perhaps we should go beyond the frame of the model and search for another mechanism for the appearance of localized moments. The mechanism under discussion can survive if the spinon mass considerably exceeds the mass of the holon (although in our opinion these masses are of the same order of magnitude). Suppression of holon superconductivity in the presence of the spinon+defect formations occurs because a holon from a superconducting pair, with a color opposite to that of the spinon, can exchange colors with it. This is the same mechanism as the suppression of the ordinary superconductivity by magnetic impurities, the only difference being that the role of spin is now assumed by color.

Another widely promulgated model [18-20] implies that in the spin system there is a locally antiferromagnetic environment, although the Neel ordering is absent; the corresponding correlation radius will be denoted as r_c. The physical picture emerging in the framework of this model is fairly similar to one emerging in the dimer model we have just discussed. This is largely accounted for by the fact that in the both cases the square lattice falls into two sublattices and there occur two types of spinless fermions, each living on one of these sublattices. The fermions are described by the field χ and the spin degrees of freedom are governed by Z-quanta. To excite Z-quanta, finite energy is required. Paramagnetic dielectrics of this kind exhibit a local symmetry, which corresponds to a certain transformation of fields χ and Z conserving the direction of local staggered magnetization [18]. Therefore in the physical spectrum there can be only those objects which are neutral in terms of the charge, corresponding to this symmetry. Among these neutral objects there is a pair $\chi - Z$. The article [19] describes the formation of two pairs of this kind, when two doped electrons are removed. This is in essence the same mechanism we are using (unfortunately, we learned about this interesting work only in the course of the preparation of this manuscript).

However, we are going to state now that all versions of the theory implying the existence of moderately short-range antiferromagnetism can hardly be in agreement with the results of EPR experiments discussed above. The thing is, that the spin 1/2 of Z-quantum can participate in the spin resonance only if the wave-length of the Z-quantum considerably exceeds the correlation radius r_c. Then the action of various regions of antiferromagnetically correlated spins on the spin of Z-quantum effectively vanishes. Therefore for the localized $X-Z$ pair to reveal itself in the EPR measurements, it is necessary that the radius R of such pairs should considerably exceed r_c. Since r_c should be at least several lattice constants we think that the dopant concentration $\sim 1\,at.\ \%$ is too high for the resonance on the individual localized magnetic moments to be observed. For this reason we think that versions of the theory of high-T_c superconductivity based on paramagnetic dielectrics [18-20] can scarcely be consistent with the results of our EPR experiments.

Thus, we have argued, that it is rather difficult to reconcile the observation of EPR in the Zn-doped lanthanium compounds with the two well-known theoretical models [12,14,15,18-20]. The analysis of flux phases [21,22] in view of this problem seems to be useful, and this is now being studied. Finally it worth noting that the theoretical constructions in which spins are united into quadruples with a zero total spin [23] make the formation of localized moments after doping with nonmagnetic ions possible.

We would like to conclude this paper with the appeal:
"The theory of the high-T_c superconductivity should be Zn - compatible!"

REFERENCES

1. J.G.Bednorz and K.A.Muller, Z.Phys. B 64, 188 (1986).
2. P.W.Anderson, Science 235, 1196 (1987).
3. P.W.Anderson, G.Baskaran, Z.Zou, T.Hsu, Phys.Rev.Lett. 58, 2790 (1987).
4. J.M.Tarascon, et al., Phys.Rev. B 36, 8393 (1987).
5. Gang Xiao, et al., Phys.Rev.Lett. 60, 1446 (1988), see notes added in proof.
6. S.E.Barnes, J.Phys. F6, 1375 (1976). P.Coleman, Phys.Rev. B29, 3035 (1984).
7. G.Baskaran, P.W.Anderson, Phys.Rev. B 37, 580 (1988); G.Baskaran, Physica Scripta T27, 53. (1989); J.F.Annet, N.Goldenfeld, S.R.Renn, Phys.Rev. B 39, 708 (1989).
8. A.M.Polyakov, Nucl.Phys. B 120, 429 (1977).
9. M.Z.Cieplak, Gang Xiao, A.Bakhshai and C.L.Chien, Phys.Rev. B39, 4222 (1989).
10. Gang Xiao, A. Bakhshai, M.Z. Cieplak, Z.Tesanovic and C.L. Chien, Phys. Rev. B39, 315, (1989).
11. S.Kivelson, D.Rokhsar and J.Sethna, Phys.Rev. B (R.C.) 35, 8865 (1987).
12. D.S.Rokhsar, S.Kivelson, Phys.Rev.Lett. 61, 2376 (1988).
13. N.Read, S.Sachdev, Phys.Rev.Lett. 62, 1694 (1989).
14. L.B.Ioffe, A.I.Larkin, Preprint, NSF-ITP-89-59.
15. E.Fradkin, S.Kivelson, Preprint, 1989.
16. V.J.Emery, Phys.Rev.Lett. 58, 2794 (1987).
17. F.C.Zhang, T.M.Rice, Phys.Rev. B37, 3759 (1988).
18. P.B.Wiegmann, Phys.Rev.Lett. 60, 821 (1988); Physica C 153-155, 103 (1988).
19. X.G.Wen, Phys.Rev. B 39, 7223 (1989).

20. P.A.Lee, Phys.Rev.Lett. 63, 680 (1989).
21. G.Kotliar, Phys.Rev. B37, 3664 (1988).
22. I.Affleck and J.B.Marston, Phys.Rev. B37, 3774 (1988).
23. X.G. Wen, F. Wilchek, and Z. Zee, Phys. Rev. B 33, 11413 (1989).

MARGINAL FERMI-LIQUID THEORY OF THE NORMAL AND SUPERCONDUCTIVE STATE OF Cu - O COMPOUNDS

C. M. Varma
AT & T Bell Laboratories, Murray Hill, New Jersey 07974

INTRODUCTION

The anomalous normal state properties of the superconducting Cu-O compounds raise fundamental questions about the limits of validity of the Landau theory of Fermi-liquids. This issue has been raised by Anderson * in connection with his ideas of Resonant Valence Bonds (RVB). The phenomenological resolution of the issue, that my colleagues and I have arrived at, however, seems not to bear relationship with RVB ideas.

We [1,2] have taken a careful look at the wide variety of universal anomalous experimental results in the Cu - O compounds, formulated a hypothesis about the excitation spectrum, and calculated properties in agreement with experiments. This work therefore uses a combination of western empiricism, Indian mysticism and Soviet mathematical tools.

THE MARGINAL FERMI-LIQUID HYPOTHESIS

The hypothesis [1] is that over a wide range of momentum **q**, there exist both charge and spin-density excitation with a polarizability which has a contribution of the form

$$\mathrm{Im}\tilde{P}(\mathbf{q},\omega) \sim -N(0)\,\omega/T\,, \quad \text{for } |\omega| < T\,,$$

$$\sim -N(0)\,\mathrm{sign}\,\omega\,, \quad \text{for} |\omega_c| > |\omega| > T\,. \tag{1}$$

Here $N(0)$ is the unrenormalized one particle density of states and ω_c is a cut-off. No qualitative difference in any calculated property arises if Eq. (1) is multiplied by a smooth function of **q**

The physical meaning of Eq. (1) is that there is no scale for low energy excitations other than the temperature itself. This is to be contrasted with a Fermi-liquid where (for most of the range of **q** space) the scale is E_f or holes in a Heisenberg antiferromagnet where the scale appears to be J.

Given (1), the leading contribution to the one-particle self-energy due to exchange of these charge and spin fluctuations can be calculated to be

$$\sum(\mathbf{k},\omega) \sim g^2N^2(0)(\omega\ln(x/\omega_c) - i\pi x/2)\,. \tag{2}$$

Here $x = \max(|\omega|, T)$ and g is a coupling constant. This is to be contrasted with a Landau Fermi-liquid, where $\mathrm{Re}\sum \sim \omega$, $\mathrm{Im}\sum \sim -\omega^2$. Then a quasi-particle representation for the one-particle Green's function is valid:

$$G(\mathbf{k}, \omega) = \frac{1}{\omega - \varepsilon_{\mathbf{k}} - \sum(\mathbf{k}, \omega)} = \frac{Z_{\mathbf{k}}}{\omega - E_{\mathbf{k}} + i\Gamma_{\mathbf{k}}} + G_{\text{incoh.}} \tag{3}$$

In the present case, we find

$$Z_{\mathbf{k}}^{-1} = \left[1 - \frac{\partial \mathrm{Re} \sum}{\partial \omega}\right]_{\omega = E_{\mathbf{k}}} \sim \ln |\omega_c / E_{\mathbf{k}}| , \tag{4}$$

near the "Fermi-surface" $E_{\mathbf{k}} = 0$. Therefore according to (4), the quasi-particle weight vanishes logarithmically as $E_{\mathbf{k}} \to 0$, and G is entirely incoherent. We refer to this as representing a *marginal* Fermi-liquid. Note that the Fermi-surface is still given by band-structure calculations.

The prediction from Eq. (2) that the width of $\mathrm{Im}\, G(\mathbf{k}, \omega)$ increases linearly with the deviation from the Fermi-energy has recently been verified [3] by angle resolved photoemission measurements. Using Eqs. (1) and (2) we have calculated [1] the resistivity, thermal conductivity, tunneling conductance as a function of bias voltage, optical conductivity as a function of frequency, Raman scattering intensity as a function of frequency and the Cu - nuclear relaxation rate, in agreement with the experimental results. (The oxygen nuclear relaxation rate differs qualitatively from the copper rate. From a general hypothesis like (1) we can not obtain such differences.) Three parameters are involved in such a fit: ω_c, $\lambda_\rho = g_\rho N(0)$ - the charge-coupling constant and, $\lambda_\sigma = g_\sigma N(0)$ - the spin-coupling constant: $\omega_c \approx 2$–3×10^3K, $\lambda_\rho \approx 3$ and $\lambda_\sigma \approx 1$ are deduced.

A few things about the phenomenology ought to be stressed. The central hypothesis is Eq. (1) about $\mathrm{Im}\,\tilde{P}$, not Eq. (2) for self-energy $\sum$. The former implies the latter but not vice-versa. It is necessary that $\tilde{P}$ has the assumed form over most of $\mathbf{q}$ space. The experiments test both $\tilde{P}$ and $\sum$. The assumption about $\mathrm{Im}\,\tilde{P}$ is not an assumption that there is a constant density of states of low energy excitations. This would be enough to give $\mathrm{Im}\sum \sim \max(\omega, T)$, but not enough to understand all the experiments. The important thing is that there are low energy excitations whose characteristic scale is the temperature itself.

INSTABILITIES OF THE MARGINAL FERMI-LIQUID

It is possible that Eq. (1) and (2) are only the leading terms in a (singular) expansion about the Fermi-gas. We have found two instabilities following from the hypothesis, Eq. (1): s-wave superconductivity and localization below a temperature determined by the *bare* impurity scattering rate. Note that Eq. (1) implies a logarithmically diverging contribution to (any finite $\mathbf{q}$) compressibility. Both s-wave superconductivity and localization are related to this property.

Superconductive Pairing

The particle-particle scattering amplitude, which is the pairing kernel in a theory of superconductivity, is given approximately by $V(\mathbf{q})\,\mathrm{Re}\,\varepsilon^{-1}(\mathrm{q},\omega)$ where $V(\mathbf{q})$ is the bare interaction. As we discussed, the measured $\mathrm{Im}\,\varepsilon^{-1}(0,\omega) \sim \mathrm{Im}\,\tilde{P}(0,\omega)$ in Raman scattering has precisely the form of Eq. (1) up to a (soft) cutoff, say ω_c of order few tenths of eV. The agreement with experiment which we have found supports our initial hypothesis of the smooth dependence of $\tilde{P}$ on $\mathbf{q}$ over a substantial range of the Brillouin zone. The sign and behavior of $\mathrm{Im}\,\varepsilon^{-1}$ which follows from these considerations leads, by Kramers-Kronig, to a negative contribution to $\mathrm{Re}[\varepsilon^{-1}(\mathrm{q},\omega)-1]$ over a sizable region of $\mathbf{q}$ and $\omega < \omega_c$. It thus appears likely that the excitations represented by Eq. (1) are responsible both for s-wave s-wave pairing consistent with observations, and for the anomalous normal state properties. Contributions from other processes, in general ensure that $\mathrm{Re}\,\varepsilon^{-1}(\mathrm{q},0) > 0$, as required for stability. These contributions presumably extend on a scale of the bandwidth $\gg \omega_c$, so that the attractive contribution for $\omega < \omega_c$ can win out for pairing.

If this is true, the spin fluctuation channel of $\tilde{P}$ is deleterious for pairing. This is consistent with the observation that superconductivity is found for compositions suitably distant from the antiferromagnetic part of the phase diagram.

While the normal state is full of surprises, the superconducting state of the Cu-O metals presents, as far as I know, only one qualitatively new feature. This is the absence of the BCS-coherence peak in the nuclear relaxation rate T_1^{-1} just below T_c. No coherence peak arises in p-wave or d-wave superconductors with a finite density of low energy excitations. There is convincing evidence, however, that in good samples of the 123-O_7 compound, there are no low energy excitations at $T \ll T_c$. The superconducting state is quite consistent with s-wave pairing. A quantitative surprise is the high ratio of the zero temperature gap to T_c, $2\Delta(0)/T_c \approx 8$.

How is one to reconcile the absence of the coherence peak in T_1^{-1} with s-wave superconductivity? Kuroda and I [4] have calculated the properties of the superconducting state which follow an s-wave attraction between quasiparticles arising from exchanging the excitation spectrum, Eq. (1). As already discussed, this excitation spectrum, quite obviously, encourages one to look for pairing in the s-wave channel. This spectrum leads also to strong inelastic scattering both in the charge and spin-fluctuation channels which, in turn, implies strong pair-breaking and gapless behavior near T_c. Below T_c a superconducting gap and a gap in the excitation spectrum of Eq. (1) must be looked for self-consistently. The latter then leads to much reduced pair-breaking. Kuroda and I have found that with parameters which are consistent with the normal state properties, one can calculate, besides the right order of magnitude of T_c, $2\Delta(0)/T_c \approx 8$ as well as $(d \ln T_1^{-1} / d \ln T)_{T_c} \approx 5$, consistent with observations in 123–O_7. We also predict that the mean-field specific heat jump at T_c is smaller than the BCS value and the slope of the specific heat at T_c^- much larger. This sort of behavior is unlike the customary phonon-induced superconductors and comes about solely from the excitation spectrum, Eq. (1).

Localization

Kotliar and I [5] have found that, if Eq. (1) is true, the electron-impurity scattering vertex is divergent. The elastic mean free path is $l(T) \approx l_{\mathrm{born}}/\ln^2(\omega_c/T)$, where l_{born} is the mean free path for non-interacting electrons calculated in the Born approximation. Using $k_F l(T) \approx 1$ as the criteria for localization, the states near the Fermi-energy are localized below a temperature $T_l \approx \omega_c \exp(-k_F l_{born})^{1/2}$, unless superconductivity intervenes at a higher temperature. One of the remarkable properties of the high T_c materials is that a significant suppression of T_c by impurities is always accompanied by upturn of resistivity at low temperature. When $T_c \to 0$, $\rho \to \infty$, at $T \to 0$. These results are consistent with our findings.

PHYSICS OF THE ANOMALOUS POLARIZABILITY

The basic hypothesis, Eq. (1), leads to such a rich havest of results in agreement with experiments that one must regard it as the answer which a correct microscopic theory must give. I think it also constrains microscopic models. Eq. (1) and the fit to experiments, say, optical conductivity, resistivity, and nuclear relaxation rate demand that charge and spin-excitations have a similar energy scale. There is no separation of charge and spin degrees of freedom as is envisaged in RVB ideas.

$\tilde{P}(q, \omega)$ of Eq. (1) implies an attractive contribution to particle-particle scattering in the *s*-wave channel. It is hard to imagine that a Hubbard model or its progenies like the t-J model can lead to this behavior.

The idea of regarding the Cu - O metals as doped antiferromagnets implicit in the t-J model is discredited directly by angle resolved photoemission experiments, which give a Fermi-surface consistent with Luttinger's theorem. Such ideas appear ill motivated anyway, since they seek to understand the metallic state through expansion (in density of holes) about a long-range ordered state not analytically connected to the metallic state.

At this point it is worth recalling the properties of $BaPb_xBi_{1-x}O_3$ and $Ba_xK_{1-x}BiO_3$. In the normal metallic phase, they have the same anomalies in tunneling characteristics, Raman scattering and optical properties as their Cu - O counterparts. They must also be classified as *marginal Fermi–liquids.* They are neither two-dimensional nor do they have local magnetic moments in any phase. The insulating nature of these compounds near 1/2 filling and the superconductiivty away from 1/2 filling arise in a model [8] with effective attractive electron-electron interactions at low energies on an energy scale of tenths of eV.

A microscopic model [6] for Cu - O compounds has been proposed with both Cu - Cu and Cu - O repulsions. This model has an antiferromagnetic insulator phase at 1/2 filling and a metallic phase with effective particle-particle attraction at low energies. This model uses the special chemistry of the Cu-Oxides (and of the Bi-Oxides) that the ionization level of oxygen and the affinity level of copper, at the self-consistent charge distribution achieved in the solid, are nearly coincident. In this situation the Madelung energy terms, represented in the model by Cu - O repulsion, have an important dynamical role to play in the metallic state. They mix particle-hole excitation near the Fermi-surface with inter-band charge transfer resonances leading to an effective particle-particle attraction in the *s*-wave channel.

Given such an attraction, how is Eq. (1) to be obtained? Quite generally, the polarizability can be divided into two parts represented graphically in Fig. 1. The first part is due to non-interacting but fully renormalized particle-hole pairs while the second contains all the vertex corrections. The hypothesis, Eq. (1), represent the leading (lowest) order frequency (temperature) contribution from the second part of Fig. 1. Given that, the leading contribution to the self-energy, Eq. (2) is obtained simply by perturbation theory, Fig. 2. The usual features of the dielectric properties of a metal, a plasma frequency, screening etc., still follow from the first part of Fig. 1 with only logarithmic corrections.

As already noted, $\tilde{P}$ leads to a marginal departure from Fermi-liquid. This implies that the correction about the Fermi-gas are non-perturbative or singular already in the normal metallic state. The obvious singularity to look for in Fig. 1 is in the vertex Γ. This singularity should be distinct from the superconducting singularity and be present at a very high temperature.

Fig. 1. Division of the irreducible particle-hole polarization into two parts.

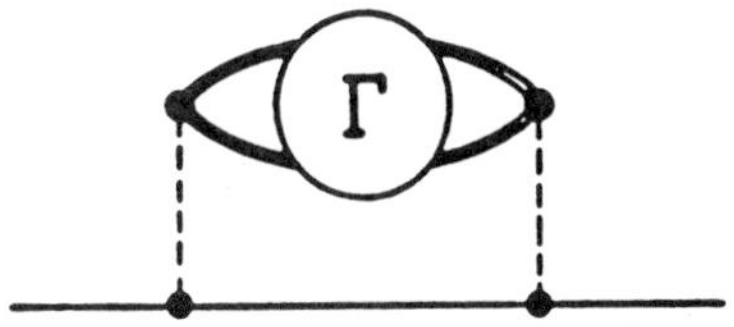

Fig. 2. Leading one particle self-energy graph.

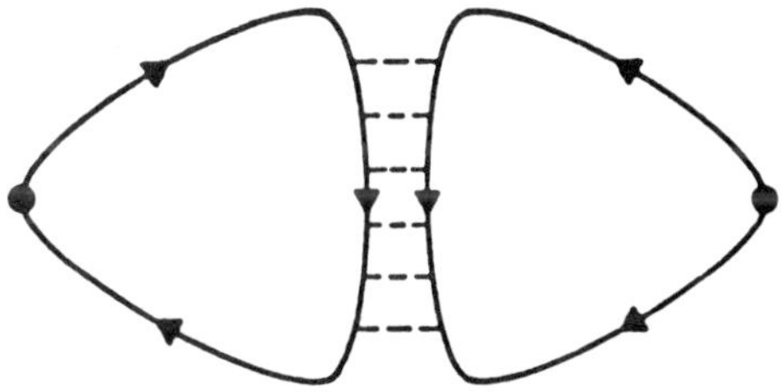

Fig. 3. Speculation on the source of the magic polarizability, Eq. (1) and second graph of Fig. 1. If Γ in Fig. 1 has bound states in the particle-particle channel as shown in this figure, the assumed polarizability may be obtained.

The singularities in Γ may be looked at in the horizontal channel or in the vertical channel in Fig. 1. In the vertical channel, a singularity in the particle-particle channel has interesting properties. It is drawn in Fig. 3. If the particle-particle channel has bound states over substantial part of (twice) the Brillouin zone, then the contribution of Fig. 3 appears to have the form of Eq. (1) multiplied by $n_B(T)$, where $n_B(T)$ is the number of pair-bound states thermally occupied. If this quantity is roughly temperature independent,. all requirements seem to be met. (The corresponding Aslamazov-Larkin diagram would have the same property as Fig. 3.) The physics then would be that, if the normal state is a mixture of bound pairs of electrons and "free" electrons, then their mutual scattering leads to a polarizability of the right form. Obviously, such bound states may only be found in attractive interaction models. My guess would be that attractive interaction models, in the normal state, probably have such a mixed phase solution. This may be so above a critical value of coupling constants in three dimensions, (Bi-O family in mind). But in lower dimensions perhaps any finite coupling may do, as bound states in the Fermi-sea form there for arbitrary coupling constants [9].

These are very speculative suggestions. The next important thing to do is to find reasonable solutions for an attractive interacting electron gas model.

REFERENCES

*. See the paper by P. W. Anderson in the proceedings of this symposium.
1. C. M. Varma, P. B. Littlewood, S. Schmitt-Rink, A. Puckenstein, and E. Abrahams, Phys. Rev. Letts., **63**, (1989).
2. C. M. Varma, Int. J. Mod. Phys. (to be published).
3. C. Olson *et al.*, Phys. Rev. (to be published).
4. Y. Kuroda and C. M. Varma, Phys. Rev. Letts., (submitted).
5. G. Kotliar and C. M. Varma, (unpublished).
6. C. M. Varma, S. Schmitt-Rink, and E. Abrahams, Phys. Rev. B **62**, 681 (1987), and in *Novel Mechanism of Superconducting*, V. Kresin and S. Woef, Editors (Plenum, New York, 1987).
7. P. B. Littlewood, C. M. Varma, and E. Abrahams, Phys. Rev. Letts., **63**, 2602 (1989).
8. C. M. Varma, Phys. Rev. Letts., **61**, 2173 (1988).
9. S. Schmitt-Rink, C. M. Varma, and A. Puckenstein, Phys. Rev. Letts., **63**, 445 (1989).

THE ELECTRONIC STRUCTURE AND PROPERTIES OF HIGH-TEMPERATURE SUPERCONDUCTIVITY AT HIGH DOPING LEVELS

D. I. Khomskii
P. N. Lebedev Physical Institute, Moscow, USSR

The electronic structure of copper metallo-oxides of the type $La_{2-x}Sr_xCuO_4$ and $Nd_{2-x}Ce_xCuO_4$ at high doping levels is discussed. Specifically, different possible explanations of the behavior of kinetic properties (notably the Hall effect) are suggested. It is concluded that the study of highly doped samples may provide useful information on the existence of electronic correlations in these systems and on their role in superconductivity.

INTRODUCTION

Two competing pictures discussing the properties of high-temperature superconductors are usually used. The first is based more or less on the standard band approach and ascribes the superconductivity, for example, to very strong electron-phonon coupling. The second approach starts from the picture of strong electron-electron correlations (Hubbard model and its generalizations).

This second situation is definitely realized in insulating compounds (undoped La_2CuO_4 and Nd_2CuO_4, $YBa_2Cu_3O_6$ etc.) However, as to the situation in the metallic and superconducting region the question of the importance of electron-electron correlations remains open. In this respect, the study of systems like $La_{2-x}Sr_xCuO_4$ and $Nd_{2-x}Ce_xCuO_4$ at high doping levels may be very informative. There now exist very interesting experimental data about the behavior of these systems in this region (although, unfortunately, many experimental findings are still controversial).

In my paper I address the question of the behavior of copper metallo-oxides in the high doping region. This study is inspired by the recent observation of the behavior of the Hall constant in $La_{2-x}Sr_xCuO_4$ and $Nd_{2-x}Ce_xCuO_4$ at large x -notably, the rapid drop and change of sign of R_H at $x>0.3$ in "La" and $x>0.2$ in "Nd" compounds [1-3]. I want to discuss possible explanations of such behavior and possible implications of these findings. In particular I want to analyze whether this behavior is a signature of the transition from strongly correlated to band-like electron states.

THE CHANGE OF ELECTRONIC STRUCTURE AND HALL EFFECT: "MOTT TRANSITION" UNDER DOPING ?

As is well known, undoped La_2CuO_4 and Nd_2CuO_4 are antiferromagnetic insulators; the accepted explanation of this fact is that d-electrons (d-holes) on Cu are localized so that these systems may be treated as Mott insulators.

There is still some controversy about whether one can describe these systems with the one-band Hubbard model. This model should definitely apply in the "electronic" superconductor $Nd_{2-x}Ce_xCuO_4$, where extra electrons introduced by doping have no choice but to occupy empty d-levels, thus creating some Cu^+ (d^{10}) ions. As for the "hole" system $La_{2-x}Sr_xCuO_4$ it is now believed that the holes go mainly to oxygen p-orbitals (hybridized with Cu d-orbitals). So the complete model should include not only d-orbitals but also p-orbitals in an obvious way (so called $d-p$ or Emery model [4]):

$$H = \sum E_d d^\dagger_{i\sigma} d_{i\sigma} + E_p p^\dagger_{j\sigma} p_{j\sigma} + U_d n_{di+} n_{di-} + (t d^\dagger_{i\sigma} p_{j\sigma} + \text{h.c.})\,. \tag{1}$$

Here $d^\dagger_{i\sigma}$, $p^\dagger_{j\sigma}$ are d and p hole orbitals; we omitted for simplicity some terms like $p-p$ or $p-d$ Coulomb interaction. (For electron doped systems the corresponding Hamiltonian would have the same form but with $(E_p - E_d)_{\text{el}} = U - (E_p - E_d)_{\text{hole}}$.)

It is possible that even in the hole case the system may be described by the effective one-band Hubbard or t–J model [5-7], where the symmetric combination of p-orbitals of 4 oxygen ions around a given Cu ion would serve as an elementary entity [6]. The detailed numerical results show that such a one-band model gives a good description of both "electron" and "hole" systems and that the effective parameters of the corresponding one-band Hamiltonian are not very different:

$$H_{\text{eff}} = \sum t_{\text{eff}}\, c^\dagger_{i\sigma} c_{i\sigma} + U_{\text{eff}}\, c^\dagger_{i+} c_{i+} c^\dagger_{i-} c_{i-}\ , \tag{2}$$

and $t_{\text{eff}} = 0.41$ eV for "La" and 0.44 eV for "Nd" systems, $U_{\text{eff}}/t_{\text{eff}} \approx 6$ [7].

It is clear that when we start to dope these Mott insulators, the charge carriers in $La_{2-x}Sr_xCuO_4$ should be holes and in $Nd_{2-x}Ce_xCuO_4$ they should be electrons. This corresponds to the following experimental result: at small x the Hall constant $R_H > 0$ for the La compound and < 0 for Nd [2,3]. If, on the other hand, the standard band approach was applicable, the situation would be just the opposite: pure La_2CuO_4 and Nd_2CuO_4 would be metals with half-filled bands and doping would give electron-like carriers in the La case (less than half-filled band), $R_H < 0$, and hole-like $R_H > 0$ for the Nd system.

Now, very interesting recent studies show that at high enough doping levels ($x > 0.3$ for La and $x > 0.2$ for Nd case) when these systems are metallic but no longer superconducting the Hall constant decreases rapidly and changes sign [1-3]. There is still some uncertainty about the experimental situation: these systems are rather unstable at large x, so one may doubt the quality of the samples used. There are also reports [8] that good single crystals of $Nd_{2-x}Ce_xCuO_4$ have $R_H > 0$ and are already in the superconducting phase, $x = 0.15$. Since on general grounds one may argue that for small x, R_H should be negative, this would mean that it changes sign at smaller x. If these experimental observations turn out to be true, there may be important consequences.

How can one explain such behavior? There are two possibilities. The first is that with increasing x there occurs something like an insulator-metal transition (Mott-Hubbard transition) accompanied by the disappearance of the energy gap, so that at $x > x_c$ these systems behave more or less like normal metals (by "insulator" we mean doped Mott insulators with partially filled Hubbard bands which may actually behave as metals and even superconductors, but in which the electron correlations are still very significant and produce these split Hubbard bands). As discussed above, in this case we can expect just such behavior of the Hall constant, in a band picture R_H would be negative for the La system and positive for Nd one. If so one would come to an important conclusion: our systems are high-T_c superconductors just when they are still "on the insulator" side of this transition; but when they are transformed to ordinary metals without significant electron correlations, they simultaneously lose their superconductivity. This would still not be proof that the superconductivity itself has a "magnetic" origin, but it would mean that electronic correla tions are somehow vitally important for it (although, may be, in directly - e.g. by influencing some charge fluctuations or even electron-phonon interaction).

However, it is rather difficult to see what could be the origin of such an insulator-metal transition. In the simple Hubbard model such a transition is driven by the change of the ratio U/t. If for $La_{2-x}Sr_xCuO_4$ one can imagine that $U_{\rm eff}$ decreases with increasing doping (Here $U_{\rm eff}$ is the effective Coulomb interaction on an O_4 cluster around Cu. It can be partially screened by doping) it is more difficult to visualize how one can screen by doping the real d–d Coulomb interaction for $Nd_{2-x}CeCuO_4$.

There exists in principle another possibility for inducing such a transition - the increase of $t_{\rm eff}$ with x. As is known from the study of the Hubbard model (see Eq. [9]) the effective bandwidth W depends on the background magnetic structure, or magnetic correlations, one may crudely model this dependence as [10]

$$W = W_0 \frac{1/4 + \langle S_0 S_1 \rangle}{(1/4 - S^2)^{1/2}} , \tag{3}$$

where $\langle S_0 S_1 \rangle$ is the nearest neighbor spin correlation function and S is the antiferromagnetic sublattice magnetization. In this case, as long as there is antiferromagnetic correlation the effective bandwidth is diminished, and the energy gap $E_g \approx U - W$ is large. With doping, however, the antiferromagnetic correlations are suppressed and consequently W increases. If $U < W_0 = 2zt$, one may get a insulator-metal transition (gap shrinking) at a certain x_c (cf. the description of the insulator-metal transition in Mott-Hubbard antiferromagnets in Refs. [10,11]).

The experimental situation in this respect is also rather uncertain. The first possibility (the decrease of $U_{\rm eff}$ with x) seems to contradict optical data [3] which shows that charge transfer transitions remain at the same energy for different x and become only broader and weaker. If $U_{\rm eff}$ changes with x, one would expect a change of the position of the charge transfer peak.

Progressive broadening of the bands (or Hubbard subbands) may actually occur; there are some experimental indications that effective mass of the carriers in $La_{2-x}Sr_xCuO_4$ decreases with x [12]. As $U_{\rm eff}$, E_p, and E_d in (1) remain in this case unchanged, there would be no contradiction with optical experiments. The real reason for a change in $W_{\rm eff}$ (if there is such an effect), a change in magnetic correlations or some other factor, remains to be seen.

It seems also difficult to reconcile with the picture of "Mott transition" the observations that at $x > x_c$ $La_{2-x}Sr_xCuO_4$ seems to remain magnetic (and in a sense looks "more magnetic" than the metallic and superconducting phase at $x < x_c$ [2]). Thus, we should think of another possibility to explain electronic and magnetic properties of these systems at high doping levels. Namely, another plausible scenario is that the system remains in a relatively strongly correlated state even at $x > x_c$. But then how can one explain the change of sign of Hall constant at $x > x_c \approx 0.2$~0.3? To understand this we should look more carefully at the band structure of the Hubbard model. At

$U=0$ we have the standard noninteracting electron situation; in our case of 2-d square lattice the dispersion law is

$$E(k) = 2t(\cos k_x + \cos k_y) , \tag{4}$$

where the effective mass at the Fermi-level is $m^* > 0$ for less than a half-filled band ($n < 1$, where n is the electron density) and $m^* < 0$ for $n > 1$.

In the opposite case of strong interaction we have two Hubbard subbands, and the dispersion in each of them looks qualitatively similar to (4). But each subband has a place for only one electron per site and not two as in the free electron case (4).

One can get these results more formally making a simple Hubbard I type decoupling [13] (this approximation ignores subtle details of the dependence of the energy spectrum of holes on the background magnetic structure and takes it into account only on the average. However it is presumably qualitatively correct at not too small x - i.e. not too close to the half filling of the band). We obtain in this way [13]

$$E_{\pm}(k) = \frac{E(k)+U}{2} \pm \frac{1}{2}[(E(k)+U)^2 - 4E(k)U(1-n_{-\sigma}]^{1/2} . \tag{5}$$

The condition that there is at most 1 electron per site in a lower subband is taken care of by the residues of the Green's function [13]:

$$G(E, k) = \frac{1}{2\pi}\left[\frac{A_+(k)}{E-E_+(k)} + \frac{A_-(k)}{E-E_-(k)}\right], \tag{6}$$

where

$$A_{\pm} = \pm \frac{E_{\pm} - U(1-n_{-\sigma})}{E_+ - E_-} . \tag{7}$$

One can see that at $U \gg t$ the dispersion law for each subband is similar to the free electron dispersion (4), so when one changes the occupation of these subbands one can also have a change of sign of m^* and R_H. (The question of the Hall effect in a Hubbard model is not a simple one [14-16], but qualitatively one finds that the sign of R_H is also determined here by the sign of m^* (the curvature of the dispersion law) at E_F as in the ordinary metals or semiconductors). For $U \gg t$ this change of sign would occur close to half filling of each subband i.e. for $x \approx 0.5$ or 1.5 (doping level $x_c \approx 0.5$), but not at $x_c \approx 0.2$~0.3. However, it is the case only for very strong interaction. As we have seen, for $U=0$ this point moves to half filling, $n \to 1$ (i.e. $x_c = 0$). Thus it is clear that in the intermediate case one can easily get $x_c \approx 0.2$~0.3 even in the case of relatively strong correlations.

One can check this by using the results of (5)-(7). The critical concentration x_c is given by the expression

$$x_c = \frac{1}{2\pi^2} \int_{E_-(k_x,k_y)<E_{FC}} dk_x\, dk_y\, A_-(k_x, k_y) , \tag{8}$$

where E_{FC} is defined so that the effective mass changes sign at this energy. (Crudely m^* changes sign close to half-filled subbands $\pm k_x \pm k_y = \pi$ for all U/t although not exactly (see (5)). We find in this way that the critical value x_c for which $m^*(E_F)$ changes sign does indeed depend on U/t: e.g. $x_c = 0.1$ for $U/t = 0.5$, $x_c = 0.16$ for $U/t = 1$, $x_c = 0.24$ for $U/t = 2$, $x_c = 0.35$ for $U/t = 5$, $x_c = 0.48$ for $U/t = 50$. Thus we see that x_c is equal ~0.3 (the value probably applicable for $La_{2-x}Sr_xCuO_4$) at relatively strong interaction $U/t \approx 4$–5. This is just the range of the parameters used most often for this system [5-7].

For $Nd_{2-x}Ce_xCuO_4$ presumably U/t is some what smaller. (One can deduce from rather large correlation length, $\xi_0 \sim 70$–120Å in the Nd system [17] that m^* here is probably smaller and t and bandwidth are larger). This may explain the smaller values of x_c (~0.2 [3] or even smaller [8]) in $Nd_{2-x}Ce_xCuO_4$.

If this picture is correct it would mean that we can get the change of sign of m^* and R_H without completely losing electron correlation. In this case the system should exhibit some non-trivial magnetic properties even for $x > x_c$. Probably, this case could explain the observation of Curie-like behavior of magnetic susceptibility at $x > x_c$ in $La_{2-x}Sr_xCuO_4$ [2].

The situation for the hole doping would be somewhat more delicate if we use the full model (1) and do not make the mapping into 1-band model. The energy spectrum for the holes with spin σ in this case would be given in the same Hubbar I decoupling scheme by the equation

$$(E - E_d)(E - E_p) - t^2(k)\frac{E - E_d - U(1 - n_{-\sigma})}{E - E_d - U} = 0\ ,$$

where $t(k) = 2t_{pd}(\cos k_x + \cos k_y)$. For $t_{pd} \ll \varepsilon = E_p - E_d$, $U - \varepsilon$ it would consist of three well separated bands at $E \approx E_d$, $E_d + U$ and E_p. The upper and lower bands correspond to the Hubbard subbands and the intermediate band with the dispersion

$$E_p(k) = E_p + t^2(k)\frac{\varepsilon - U(1 - n_{-\sigma})}{\varepsilon \cdot (\varepsilon - U)} \tag{10}$$

is the p-band. It is interesting to note that the sign of the dispersion of $E_p(k)$ depends on the ratio of ε and $U - \varepsilon$: for $\varepsilon < U - \varepsilon$ $E_p(k) \geq E_p$ and for $\varepsilon > U - \varepsilon$ p-band lies below E_p, $E_p(k) \leq E_p$ (in half-filled paramagnetic case, $n_\sigma = n_{-\sigma} = 1/2$).

The holes introduced by doping occupy in this case this "p-band". As we treat p-electrons as uncorrelated, R_H would be positive up to very high doping, formally up to $x \approx 1$. However in the real situation the p–d hopping matrix element t_{pd} is not much smaller than $E_p - E_d$, so that there is rather strong admixture r_d of d-states in the p-band. Correspondingly, these composite band states will acquire some effective repulsion $U_d r_d^2$. There exist also direct p–p repulsion although its strength is rather uncertain (different authors give for U_p values from < 1 eV up to 4 eV). As a result this effective p–d band may also have significant correlation so that one can presumably apply to it the same arguments as above.

Another possibility is that with increasing x, ε decreases or t_{pd} increases so that chargetransfer gap closes and we go to the metal of mixedvalence type [19]. But the evolution of the electronic spectrum in this case and the behavior of Hall effect is unclear at present and deserves a special study.

CONCLUSION

Summarizing, one can say that the investigation of $La_{2-x}Sr_xCuO_4$ and $Nd_{2-x}Ce_xCuO_4$ at high doping levels may prove very useful for understanding the electronic structure and properties of high-T_c superconductors and for elucidating the conditions for obtaining high values of T_c. There are in principle two possibilities. One is that at high doping levels the system to a large extent loses electronic correlations and becomes more or less standard band metal. But it is also possible to explain the main features observed, including the behavior of the Hall effect, with the picture of reasonably strong (but not too strong) correlations which are preserved even at large x. To discriminate between these two possibilities, some extra experiments would be very helpful e.g. direct measurements of spin excitations, or the study of copper NMR (Korringa law?) for $x>x_c$. But in any case, if it turns out that the Hall constant really changes sign at large x and the disappearance of superconductivity correlates with this change of normal state properties, the finding would be very significant. It would mean that the electronic correlations are indeed vital for high-T_c superconductivity.

This work was partially done during the author's stay at the Institute for Scientific Interchanges in Torino, Italy. I am grateful to the staff of the ISI for hospitality and to the participants of the Program on High-T_c superconductivity at ISI, and also to Professor S. Uchida for useful discussion.

REFERENCES

1. N.P.Ong, Z.Z.Wang, J.Clayhold, J.M.Tarascon, L.H.Green and W.R.Mackinon, Phys.Rev. B35, 8807 (1987).
2. H.Takagi, Y.Tokura and S.Uchida, Mechanisms of High Temperature Superconductivity, Ed. H.Kamimura and A.Ochiyama, Springer Series in Material Sciense, v.11, p.238, SpringerVerlag, Berlin, Heidelberg, 1989.
3. S.Uchida, The report at the International Seminar on High Temperature Superconductivity, Dubna, USSR, 28June1July 1989, World Scientific, Singapore, in press.
4. V.Emery, Phys.Rev.Lett. 58, 2794 (1987).
5. F.C.Zhang and T.M.Rice, Phys.Rev. B37, 3759 (1988).
6. A.Ramsak and P.Prelovsek, Phys.Rev. B40, 2239 (1989).
7. M.S.Hybertsen, E.B.Stechel, M.Schluter and D.R.Jennison, preprint (1989).
8. V.Emery, private communication (citing the results due to J.M.Tarascon)
9. D.I.Khomskii, Fiz. Metallov i Metalloved. (Sov.Phys.Phys. of Metals and Metallogr.) 29, 31 (1970).
10. L.N.Bulaevskii and D.I.Khomskii, ZhETF 52, 1603 (1970) (Sov.Phys.JETP 25, 1067 (1967)).
11. L.N.Bulaevskii and D.I.Khomskii, Fiz.Tverd.Tela 9, 3070 (1968) (Sov.Phys.Solid State Phys. 9, 2422 (1968)).
12. V.Z.Kresin, private communication.
13. J.Hubbard, Proc.Roy.Soc. A276, 238 (1963).
14. D.C.Langreth, Phys.Rev. 148, 707 (1966).
15. W.F.Brinkman and T.M.Rice, Phys.Rev. B4, 1566 (1971).
16. I.P.Kogutyuk, V.M.Nitsovich and F.V.Skripnik, Ukrainsk.Fiz.Zh. (Ukr.Phys.Jorn.) 25, 1629 (1980).
17. Y.Hidaka and M.Suzuki, Nature 338, 635 (1989).
18. J.Zaanen, J.W.Allen and G.A.Sawatzky, Phys.Rev.Lett. 54, 418 (1985).

UNDERSTANDING THE QUANTUM HALL EFFECT USING THE S-MATRIX APPROACH

A. Douglas Stone, A. Szafer, P.L. McEuen and J.K. Jain[1]
Department of Applied Physics, Yale University, New Haven, CT 06520-2157

The origin of the quantum Hall effect is discussed in terms of the transmission probabilities of states at the fermi-level. A new argument for the existence of extended states near the center of the Landau level is given. Disequilibrium between edge and bulk currents in the extended state region is shown to give rise to non-local and geometry-dependent behavior of the Hall and longitudinal resistance. A model is discussed in which a new intensive "resistivity" is defined and used to generate quantitative predictions for the resistance of GaAs Hall conductors.

INTRODUCTION

In this paper we discuss the conceptual outlines of an approach to understanding the integer quantum Hall effect and particularly the dissipative regime, based on a formulation in which the Hall resistance is determined by the transmission properties of the states at the fermi surface. In a high magnetic field it may happen that all or some of those states are edge-states, i.e. states localized within a few magnetic lengths $l = \sqrt{\hbar c/eB}$ of the edges of the sample and propagating in the direction determined by the magnetic field; the first analysis of the quantum Hall effect in terms of these states was given by Halperin[1]. More recently several authors have discussed the Hall effect in terms of edge states and transmission probabilities[2-4] using the approach developed by Landauer[5] and extended by Büttiker[6]. In addition a number of experiments in high-mobility GaAs using either small or inhomogeneous samples, or point-contact current and voltage probes, have corroborated some implications of this new viewpoint[7-11]. However, except in the original work of Halperin, little effort has been made to reconcile the new point of view with the older theory used to describe the Hall resistivity of macroscopic samples, based on the theory of two-dimensional localization in a high magnetic field[12]. Indeed it has been unclear from both an experimental and theoretical point of view how to define a resistivity which is truly intensive and geometry-independent within the "edge-state" picture. Moreover, neither approach has given a full quantitative description of the "extended-state" regime and our physical understanding of the origin of these extended states remains incomplete. In this paper we present in brief outline an extension of the "fermi-level picture" to address these issues; in particular we present some results from a new model[13] of the extended-state region in high-mobility Hall conductors which quantitatively predicts the behavior in the

[1]Department of Physics, SUNY at Stony Brook, Stony Brook, NY 11794

dissipative regime, and allows the definition of a new "resistivity" which *is* intensive. We purposely avoid the term "edge-state picture" since a major point of our analysis is that in macroscopic disordered samples it is rarely the case that all the states at the fermi-level are localized within a few magnetic lengths of the edge, and it is the interplay of edge and bulk currents which explains the recent experimental results.

THE S-MATRIX APPROACH

Rolf Landauer[5] pioneered the approach in which the transport properties of a sample are derived by treating it as a single composite scattering center between measuring reservoirs which inject current incoherently into the sample. The basic idea is that such a reservoir at electrochemical potential μ_1 injects a current $N(e/h)\mu_1$ into the sample, where N is the number of states at the fermi-level; each channel gives the same contribution to the current because each corresponds to a different quasi-one-dimensional sub-band, and there is a well-known cancellation of velocity and density of states factors in this case. We note that this exact cancellation persists in the presence of arbitrary magnetic field. If an ideal *two-probe* measurement is made in which the sample is attached between two perfect reservoirs with electrochemical potentials, μ_1 and $\mu_2 = \mu_1 + eV$ respectively, then the net current between the two reservoirs is just $I = \mathcal{T}[N(e/h)(\mu_1 - \mu_2)]$, where $\mathcal{T}$ is the transmission *probability* averaged over all channels at the fermi surface, $\mathcal{T} = (1/N)\sum_{i,j}^{N} T_{ij}$, and T_{ij} is the transmission *coefficient* from channel j to i at the fermi surface. This leads to a two-probe conductance of the form

$$G \equiv \frac{e^2}{h} g = I/V = \frac{e^2}{h} \sum_{i,j}^{N} T_{ij} \tag{1}$$

In the limit in which there is no scattering at all in the sample, $T_{ij} = \delta_{ij}$ and $g = N$, i.e. the conductance is quantized in units of e^2/h per channel. This effect has recently been clearly observed in high mobility GaAs samples separated into two regions by a constriction of variable width[14]. This "ideal contact resistance"[15] arises becauses the sample can only carry a finite current per channel proportional to $\Delta\mu = eV$, so if the electrochemical potential difference is measured between the source and sink then of necessity the ratio V/I is non-zero.

Since we know that it *is* possible to measure arbitrarily small resistances experimentally, even in small samples, clearly it is necessary to extend this approach. This important extension, made by Büttiker[6], turned out to be particularly useful for discussions of magnetotransport. The key observation was that in typical experiments (and in almost all Hall effect measurements) current is run from a source to a sink and the induced electrochemical potential is measured by additional voltage probes which one can think of as measuring reservoirs[6,16]. If the voltages on all the probes are fixed, then in general current will flow into or out of all the probes and a straightforward extension of the two-probe argument given above yields the relation

$$I_m = e^2/h \sum_{n}^{N_L} g_{mn} V_n \tag{2}$$

where I_m is the total current into lead m, V_n is the voltage applied at lead n, and $g_{mn} = T_{mn}$ is the total transmission coefficient (summed over all channels i, j at the fermi surface) for an electron injected at lead n to be collected at lead m, and $g_{nn} = T_{nn} - N$. One obtains the result of any resistance measurement from this formula by assuming that for the current source $I_m = I$, for the sink $I_m = -I$, and for all voltage probes $I_m = 0$, and then inverting this linear relation.

It is very important to bear in mind the physical interpretation of these boundary conditions on the currents: when a current is injected initially from the source, part of that current flows transiently into the the voltage probes, whose electrochemical potential are then adjusted until they draw no net current, i.e. the outgoing currents exactly balance the ingoing currents. Different probes will receive different amounts of the injected current and in general will acquire different voltages once they have adjusted to null out the transient current. In general the probes can be *invasive* in this approach, i.e the measured resistance can depend on transmission probabilities between regions of the sample normally considered as passive measuring leads. Such behavior is indeed observed e.g. in resistance fluctuations of mesoscopic devices[17], however it has typically been assumed that such effects are only important in systems small enough that transmission between leads is phase-coherent; we shall see below that this is not the case.

RELATION TO LINEAR RESPONSE THEORY AND EDGE-STATES

This formulation provides a simple and intuitively appealing fermi-level expression for the Hall resistance which apparently should be valid for an arbitrarily strong magnetic field. Its plausibility is supported by the fact that it naturally generates the reciprocity symmetries for resistance measurements in a magnetic field[6], which are experimentally observed even in for the complex magnetoresistance patterns observed in mesoscopic conductors[17]. Nonetheless the apparent generality of Eq. (2) was somewhat puzzling as conventional expressions for the Hall resistivity from quantum-mechanical linear response theory involved sums over *all* states up to the fermi-level, and in the theory of the quantized Hall effect it is often stated that in the Hall plateaux the current is carried by extended states below the fermi-level[12]. Recently Baranger and Stone[18] re-examined this question using an exact eigenstate approach and were able to demonstrate from standard linear response theory that for an infinite system with edges the conventional approach and the S-matrix approach are equivalent. In particular, the following statements were proved: 1) The coefficients $g_{mn} = \int dS_m \cdot \sigma(x_m, x_n) \cdot dS_n$ where $\sigma(x, x')$ is the local conductivity tensor related to the local current density by $< J(x) >= \int dx' \sigma(x, x') \cdot E(x')$ and the integrals dS_m, dS_n are over the cross-sections of leads m,n. 2) $\sigma(x, x')$ which is the familiar current-current correlation function is *not* a fermi-level quantity, since it describes local circulating currents as well as transport currents. 3) These circulating currents give no contribution when integrated through any cross-section of the system, hence they do not contribute to g_{mn} or the total Hall resistance. 4) When the expression for g_{mn} in terms of $\sigma(x, x')$ is transformed by appropriate applications of the integral equations of scattering theory, exactly Büttiker's result, Eq. (2) is obtained.

It must be noted that in this case the channels at the fermi-level are labelled by the asymptotic solutions in the infinite straight leads *with* a magnetic field. With an appropriate choice of gauge, these solutions are of the form $\phi_a^{\pm}(x,y) = e^{\pm k_a x}\chi_a^{\pm}(y)$, where x is the direction along the wire and y is the transverse direction, and we consider only two dimensions for simplicity. The transverse wavefunctions satisfy the reduced Schrödinger equation

$$[(\frac{-\hbar^2}{2m})\frac{\partial^2}{\partial y^2} + \frac{1}{2}m\omega_c^2(k_a l^2 \pm y)^2 + U(y)]\chi_a^{\pm}(y) = \epsilon(k_a)\chi_a^{\pm}(y) \tag{3}$$

where $U(y)$ is the confining potential in the transverse direction. When $kl^2 \ll W$ where W is the width of the lead, then $U(y)$ is negligible, and we can redefine our origin of coordinates and obtain the familiar Landau levels with energy $(n+1/2)\hbar\omega_c$ independent of k. However when $kl^2 \geq W$ then the energy of the state is pushed up by the confining potential, the energy depends on k, and the state acquires a substantial longitudinal velocity determined by the sign of k and the direction of the magnetic field. The states in this energy range are the so-called edge states and they have large probability density only within a few magnetic lengths of the edges. As shown in Fig. 1, *if* the fermi level is between the energies $(n+1/2)\hbar\omega_c$ of the Landau levels, then the *only* states at the fermi level *of the ordered leads* will be edge states, and the transmission coefficients in Eq. (2) must be understood as connecting these edge states.

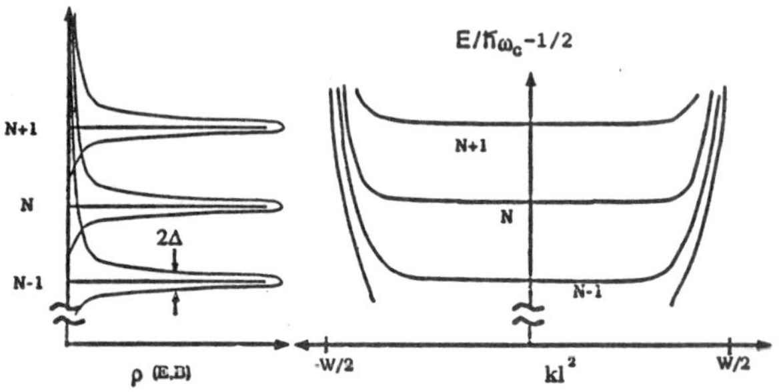

Fig. 1: Schematic of the dispersion relation and density of states for the ordered system with edges. Disorder-broadening of DOS is indicated.

ORIGIN OF QUANTIZED HALL EFFECT IN THIS APPROACH

It is easy using this approach to specify sufficient conditions on the transmission coefficients leading to the integer quantized Hall effect(the specification of necessary conditions is slightly more complicated, and will not be discussed here). Suppose that we have a standard Hall geometry as shown in Fig. 2a and a current $Ne/h\mu_1$ is being fed into the sample from reservoir 1 along the upper edge. In the sample there is disorder which in general can cause scattering and also pin the fermi level(we shall discuss this complication shortly); however *if* all the current continues to flow along that edge and into voltage probe $2(T_{21} = N)$, then in order for the total current into probe 2 to vanish we must have $N(e/h)(\mu_1 - \mu_2) = 0$, i.e. $\mu_1 = \mu_2 = \mu_{source}$. By repeating this argument for μ_3 one immediately sees that all voltage probes on this side of the current path will adjust their chemical potentials to be equal to that of the source. The current sink on the other hand is being maintained at a different chemical potential μ_{sink}, and by the same argument all probes on the lower side of the current path will be at the potential μ_{sink}. Thus we see that a net current $I = N(e/h)(\mu_{source} - \mu_{sink})$ can flow from source to sink with no voltage appearing between any two probes on the same side of the current path, i.e. the longitudinal resistance R_L vanishes. The Hall resistance is just the ratio of the voltage induced between any two probes separated by the current path, which is $(\mu_{source} - \mu_{sink})/e$, divided by the net current, giving $R_H = (h/Ne^2)$. Since N the number of states at the fermi-level is equal to the number of Landau levels below the fermi surface(Fig. 1), this formula yields the familiar quantized Hall resistance.

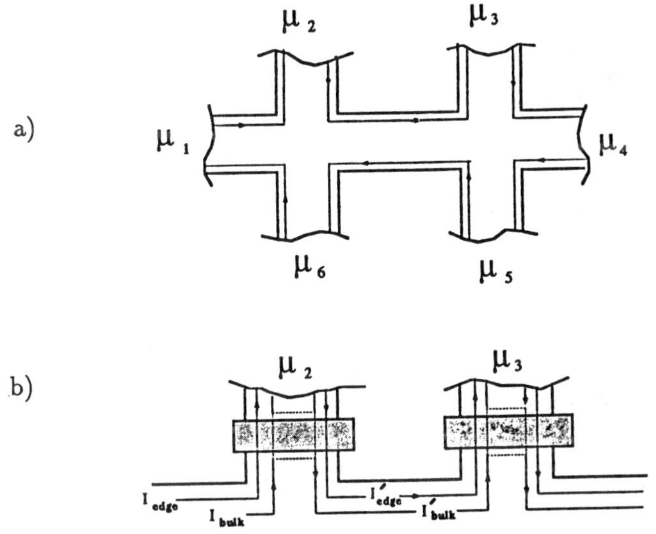

Fig. 2:a) Schematic of the ideal flow pattern of currents leading to the quantized Hall effect. b) Schematic showing that voltage probes alter the current distribution if the initial distribution is not in equilibrium.

This argument establishes that if the transmission coefficient $T_{n,n-1}$ are quantized to N, and all other transmission coefficients are zero, then one obtains the usual quantized Hall effect. It is easy to see how this can occur when all the states at the fermi-level are true edge-states. Electrons are injected near one edge, and in order to backscatter they must be removed at the other edge; as long as the cyclotron radius is short compared to the width of the sample, and the the disordered potential is weak, it will be unable to scatter them across the sample[2-4]. An arbitrary amount of forward-scattering between edge-states will not break the quantization, since by unitarity the transmission coefficient must still sum to N. This type of argument has several virtues. First, it shows that the precise quantization of the Hall resistance arises from precise quantization of transmission probabilities, and the exact cancellation of velocity and density of states factors in Eq. (2) which occurs for any fermi liquid system. Second it explains the coincidence of quantized R_H and $R_L = 0$. Third it clarifies the origin of dissipative of quantization, which occurs when the transmission coefficients depart from the quantized values they will have if the electrons just follow the Lorentz force with no backscattering[2-4].

EFFECT OF DISORDER AND FERMI-LEVEL PINNING

The weakness of the argument is that it is only intuitively obvious that the transmission coefficients *will* take on the appropriate quantized values if all the states at the fermi-level are true edge-states, so that backscattering would involve transmission against the Lorentz force across the entire sample. However for a disordered macroscopic 2d sample(i.e. a sample with width much greater than l), in which the energy-broadening of the Landau level, $2\Delta \ll \hbar\omega_c$, the ratio of the density of edge-states to that of bulk states is of order l/L where L is the sample size. Since the electron density, n, is approximately fixed, then as the magnetic field varies, the chemical potential is determined by the condition $\int d\epsilon\rho(\epsilon, B)[e^{\beta(\epsilon-\mu)}+1]^{-1} = n$, where $\rho(\epsilon, B)$ is the density of states which of course depends strongly on the field; hence the fermi level will be pinned to the regions of high density of states near the topmost (Nth) Landau level. At such an energy there will be N-1 true edge states and one "bulk" state, i.e. one state which in the absence of disorder would propagate in a strip of width l through the bulk. In the presence of disorder, unlike the edge-states, this bulk state mixes strongly with the large number of states exponentially close in energy, leading to a complicated bulk state which need not propagate within a few magnetic length of the physical edge of the sample. When this state is extended, it is the quantum analog of the classical percolation cluster[19]. This state may be perfectly transmitting, perfectly reflecting or partially transmitting; *only in the latter case will the presence of such a state break quantization of* R_H. Let us analyze the qualitative behavior as the Nth Hall plateau is traversed in the direction of decreasing B. For roughly the first half of the plateau, the fermi-level will remain pinned within Δ above $(N + 1/2)\hbar\omega_c$ as the magnetic field is decreased; in this region the Nth state must be perfectly transmitting, even though it is not a true edge state, in order to account for the quantization of R_H. Only as the midpoint of the plateau is approached does the Nth state become a true edge-state, at which time the fermi-level changes rapidly to $\varepsilon_f \geq (N+1+1/2)\hbar\omega_c - \Delta$. Now the $(N+1)$st state becomes the "bulk" channel, which initially is completely reflected[20], so that the resistance is still quantized to $1/N$ for the lower half of the plateau. Finally, when the center of the $(N+1)$st Landau level is approached the transition to plateau $(N+1)$ occurs, and in this interval the $(N+1)$st state is partially transmitted.

Hence the existence of plateaux in R_H in two dimensional systems can only be explained by considering the transmission properties of this topmost or "bulk" state, which then requires treating the 2d localization problem in high magnetic field, in which one has a highly degenerate manifold of states mixed by the disorder potential. The actual nature of this "bulk" state, and whether it mixes strongly with other states at the fermi-level to eliminate all true edge states, depends on the nature of the microscopic disorder in the system. In our opinion, further work is required to understand this behavior fully. However an explanation of the quantum Hall effect based on the fact that true edge-states are insensitive to disorder appears insufficient to explain the plateaux, because as a plateau is crossed, the fermi-level is rarely in a region in which all the states at the fermi surface are true edge-states(for which the disorder potential is a weak perturbation).

ORIGIN OF EXTENDED STATES IN THIS APPROACH

Although the occurence of plateaux is not trivial in the fermi-level picture, one can use this approach to understand the existence of extended states whose energy does not go to infinity as the system size goes to infinity for a two-dimensional system in a high magnetic field, whereas no such states are believed to exist at $B = 0$. Previously Halperin[1] has adapted Laughlin's gauge arguments[12] to this purpose, but no such argument has been given using the S-matrix approach. Briefly, we have argued above that the transition from $R_H = 1/N \rightarrow 1/(N+1)$ occurs when the (N+1)st state goes from being perfectly reflected to perfectly transmitted, as a function of B. Thus in an operational sense the topmost state becomes "extended" when its transmission probability differs substantially from zero *or* unity. Physically, the transition occurs in the following manner. When ε_f is below $(N + 1 + 1/2)\hbar\omega_c - \Delta$, then the only transmission is by tunneling through the rare sites in the sample with binding energy greater than Δ, and the reflection coefficient is exponentially close to unity. When ε_f is above $(N + 1 + 1/2)\hbar\omega_c + \Delta$, then the (N+1)st state is a true disorder-insensitive edge-state, and the *transmission* of the topmost state is exponentially close to unity. For any finite-size disordered sample, somewhere in between, roughly at the center of the Landau level, this transmission probability must pass through all values between zero and unity by the continuity of the S-matrix with energy; hence there must be a set of energies in the interval $[(N + 1 + 1/2)\hbar\omega_c - \Delta, (N + 1 + 1/2)\hbar\omega_c + \Delta]$, in which the states are extended. Since the energies of the Landau level, and of the edge-states is essentially independent of the system size, this energy interval must remain bounded as the system size goes to infinity, and there must exist some interval of extended states per Landau level even in the infinite system. Presumably, the width of this interval vanishes as the system size goes to infinity, and it would be interesting to attempt to prove this within the S-matrix approach. Note that the crucial point in the argument, which is obviously invalid in weak or zero magnetic field, is the existence of perfectly transmitting disorder-independent edge-states within a finite energy interval above the center of the Landau level. Intuitively, one can argue that the edge-states, even though they are effectively one-dimensional, are not subject to localization effects because there are simply no backwards traveling states in their vicinity, and thus the destructive interference which leads to localization does not occur. We believe that these observations provide the basis for a rigorous argument for the existence of extended states in the usual sense.

CONSEQUENCES OF CURRENT DISEQUILIBRIUM

We have argued that the non-quantized regions of the Hall resistance correspond to the intervals in which the topmost (Nth) state goes from being perfectly transmitted to perfectly reflected. In these intervals one may consider two limiting behaviors for the interchannel scattering: 1) The $N-1$ edge-states are strongly mixed with the bulk state so that electrons injected into any state at the fermi-level have equal reflection probability. 2) The $N-1$ edge-states are *decoupled* from the bulk state, so that only electrons in the bulk state have non-zero reflection probability, and in steady-state the current is unequally distributed between the bulk channel and the edge channels. Recent experiments[10,11] which use point-contacts to inject current selectively into different channels have conclusively shown that in high mobility GaAs the second limit is approached for *macroscopic* ($\sim 0.5mm$) samples. Theoretically, it appears that the occurence of this decoupling depends crucially on the shape of the confining potential at the edge. It will be shown elsewhere that under realistic assumptions this potential has a long flat tail extending into the bulk[21]. If this is true, and the disordered potential fluctuations are much less than $\hbar\omega_c$ and smooth on the scale of the magnetic length, then it can be shown that there will be a *parametrically* larger spatial separation of the bulk and edge channels, and an exponential suppression of scattering between the $N-1$ edge channels and the bulk channel[21]. Here, taking this decoupling as an experimental fact, we wish to describe briefly some physical consequences of the disequilibrium between edge and bulk currents which apparently occurs in GaAs Hall conductors. In particular, we show how current disequilibration in the extended state region causes the measured resistance to violate ohmic scaling(R_{xx} not proportional to length over width), depend on the shape of the voltage probes, and become non-local over macroscopic distances.

Consider the longitudinal resistance measurement illustrated in Fig. 2b. A current I is flowing, driven by the magnetic field, in the clockwise direction into voltage probe 1. Assume that this current is *not* equally shared by the $N-1$ edge channels and the one bulk channel, because at some earlier point a portion of the bulk current has been backscattered leaving the edge current unaffected. We represent this by writing $I = I_{edge} + I_{bulk} \equiv (e/h)[(N-1)\mu_E + \mu_B]; \mu_E \neq \mu_B$ (where μ_E and μ_B are fictitious "chemical potentials" for the edge and bulk channels). Assume that the transmission probability of edge channels into probe 1 is unity, but for the bulk channel it is T_1. If probe 1 is at potential μ_1, then the current out of probe 1 is $I' = (e/h)\{N\mu_1 + [T_1\mu_1 + (1-T_1)\mu_B]\} \equiv I'_{edge} + I'_{bulk}$. Since the current into probe 1 is known and it draws no net current ($I' = I$), μ_1 can be determined and eliminated to obtain the change in the edge and bulk currents across probe 1. Neglecting small terms of order $(T_1/N)^2$ yields

$$I_{bulk} - I'_{bulk} = -(I_{edge} - I'_{edge}) = (e/h)[T_1(\mu_E - \mu_B)]. \tag{4}$$

Thus we see that although the presence of the voltage probe does not change the total right-moving current, it alters the *distribution of the current betweeen edge and bulk channels* unless $\mu_E = \mu_B$, i.e. unless the incident current is in equilibrium. This new current distribution will then couple differently to probe 2 than the original distribution did to probe

1, leading to a longitudinal resistance. This is why the geometry of the probes strongly influences the measured resistance as long as one has unequal edge and bulk currents (per channel).

To be more explicit, Eq. (4) tells us the new current distribution in terms of the old one, which we can then use to calculate μ_2, the potential of probe 2, if the transmission probability T_2 of the bulk channel into probe 2 is known. One finds

$$\mu_1 - \mu_2 = \frac{T_2(1-T_1)}{N}(\mu_E - \mu_B). \tag{5}$$

We conclude that ***even without any backscattering in the sample***, a longitudinal resistance will appear as long as the incident current is disequilibrated and $T_2 \neq 0, T_1 \neq 1$. Clearly the geometry of the probes is crucial in determining the size of this resistance and will not cancel out under typical experimental conditions even for normal probes of macroscopic width.

Note that this argument is completely independent of whether there is a *net* current flowing in the segment to which probes 1 and 2 are attached(i.e. there could have been an equal left-moving current and still a voltage difference would appear between probes 1 and 2). Hence even if the voltage probes are removed from the current path, there will be disequilibrated circulating currents of the type shown in sufficiently high field. Thus such a decoupling of edge and bulk currents implies that as the field increases there will appear *non-local longitudinal resistances* throughout the Hall resistor! This is dramatically confirmed by the data in Fig 3, reprinted from McEuen et al.[22]. This non-locality has nothing to do with phase-coherence as should be clear from the above discussion, and is on a length-scale three orders of magnitude longer than the non-local effects observed previously in mesoscopic systems.

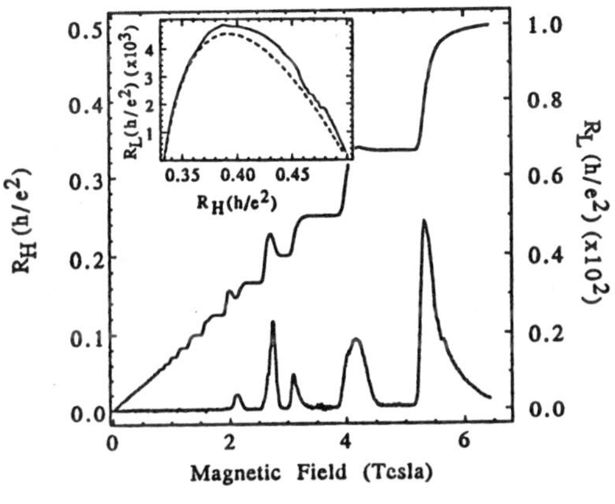

Fig. 3: R_L and R_H measured in a GaAs Hall conductor in a probe configuration equivalent to choosing probes 2 and 6 in Fig. 2a as current probes and probes 3 and 5 and 1 and 4 respectively as "longitudinal" and Hall probes. In the experimental device probes 3 and 5 are $\approx 0.6mm$ removed from the current path. Inset: Comparison of experimental and theoretical curves for R_H vs. R_L in this non-local measurement.

DECOUPLED NETWORK MODEL OF NON-QUANTIZED REGION

It is obvious from the above example that in these systems measured resistances will depend on scattering properties of both the "sample" (i.e. the current-carrying segments between the voltage probes) and the "leads", the contribution from each region depending on both its local transmission properties and on the imbalance of the incident current distribution. Thus it is not surprising that many experiments[23,24] have shown that in high mobility Hall conductors longitudinal resistance measurements do not scale linearly with the ratio of the length to width of the sample; this behavior has been previously interpreted as arising from the existence of edge currents in the system[24]. One is tempted to conclude that there is no measurable intensive parameter which plays the role of a "resistivity" in such systems. However, recently Szafer et al.[13] proposed a model of the completely decoupled limit which contains just such a resistivity parameter, and very recent experiments by McEuen et al.[22] convincingly corroborate the hypothesis that such a parameter is indeed geometry-independent.

In this *decoupled network model* each segment j of the conductor (except for the junction of two segments) is modeled by a barrier which completely transmits the N-1 edge channels, but transmits only a fraction T_j of the topmost(bulk) channel. The transmission probability T_j can be related to a "longitudinal resistivity" *for the Nth channel only* by the single-channel Landauer formula: $\rho_N(L_j/W_j) = (h/e^2)(1-T_j)/T_j$, where L_j, W_j are the length and width of the jth segment. When $\rho_N = 0, (T_j = 1)$, the topmost channel behaves as an edge channel and $R_H = 1/N$; as backscattering increases ($\rho_N \to \infty$), T_j decreases at a rate that depends only on ρ_N and the aspect ratio of the segment, and $R_H \to 1/(N-1)$ with a concurrent peak in R_L measured across that segment. If we neglect phase-coherence in the scattering between different segments, then for fixed ρ_N and fixed geometry there will be a linear relation between bulk "chemical potentials" on opposite sides of each segment; if we add the conditions that $I_{source} = -I_{sink}$, and voltage probes draw no current, we have a closed set of equations for any resistance measurement as a function only of the intensive parameter ρ_N. Sincc as the $N \to N-1$ Hall step is traversed the parameter ρ_N varies from zero to infinity, for any given geometry the parameter ρ_N can be eliminated from the system of equations by plotting any one resistance measurement vs. any other, e.g. $R_L(B)$ vs.$R_H(B)$. Thus the model generates zero-parameter prediction for such curves. In Fig. 3 the comparison of the model-generated curve and experimental results are given for the non-local resistance measurement, and remarkably good agreement is found. Conversely, given any resistance measurement the parameter $\rho_N(B,T)$ can be inferred, and should be independent of which set of probes is used. This prediction of the model was also found to be strikingly confirmed by the experiments of McEuen et al[22]. The model also accounts simply for the puzzling result[25] that the height of the σ_{xx} peaks in GaAs are approximately independent of the Landau level index[13]. These results indicate that in high-mobility quantum Hall conductors it is the parameter $\rho_N(B,T)$ that appears in the framework of the decoupled network model which is the appropriate parameter for experimental and theoretical study.

An important question raised by this work is whether an intensive "resistivity" parameter can be defined other than in the completely decoupled or completely equilibrated limit, and more generally whether it is possible to construct a model with predictive power which interpolates between the decoupled and equilibrated limit. It does appear however that very different conduction mechanisms may be involved in the quantum Hall steps depending on the mobility of the system, and this calls into question the existence of a single "universal" description of the localization-delocalization transition in these systems.

ACKNOWLEDGEMENTS

We acknowledge very useful discussions with H. Baranger, L. Glazman and P. Streda. This work was partially supported by NSF grant DMR-8658135 and ECS-8509135. P.L.M. is supported by an AT&T Bell Laboratories Fellowship.

REFERENCES

1. B.I. Halperin, Phys. Rev. B **25** , 2185 (1982).
2. P. Streda, J. Kucera, and A.H. MacDonald, Phys. Rev. Lett. **59**,1973 (1987).
3. J.K. Jain and S.A. Kivelson, Phys. Rev. Lett. **60**, 1542 (1988); J.K. Jain and S.A. Kivelson, Phys. Rev. B **37**, 4276 (1988).
4. M. Büttiker, Phys. Rev. B **38**, 9375 (1988).
5. R. Landauer, Philos. Mag. **21**, 863 (1970).
6. M. Büttiker, Phys. Rev. Lett. **57**, 1761 (1986).
7. R.J. Haug, A.H. MacDonald, P. Streda, and K. von Klitzing, Phys. Rev. Lett. **61**, 2797 (1988).
8. S. Washburn, A.B. Fowler, H. Schmid, and D. Kern, Phys. Rev. Lett. **61**, 2801 (1988).
9. B.J. van Wees, E.M.M. Willems, C.J.P.M. Harmans, C.W.T Beenakker, H.van Houten, J.G. Williamson, C.T. Foxon and J.J. Harris, Phys. Rev. Lett. **62**, 1181 (1989).
10. B.J. van Wees, E.M.M. Willems, L.P. Kouwenhoven, C.J.P.M. Harmans, J.G. Williamson, C.T. Foxon and J.J. Harris, Phys. Rev. B **39**, 8066 (1989).
11. B.W. Alphenaar, P.L. McEuen, R.G. Wheeler, and R.N. Sacks, Phys. Rev. Lett. **64**, 677 (1990).
12. "The Quantum Hall Effect", Ed. by R.E. Prange and S.M. Girvin, Springer Verlag, New York, 1987.
13. A. Szafer, P.L. McEuen, J.K. Jain and A.D. Stone, in preparation.
14. B.J. van Wees, H. van Houten, C.W.J. Beenakker, J.G.Williamson, L.P. Kouwenhoven, D. van der Marel, and C.T. Foxon, Phys. Rev. Lett. **60**, 848 (1988); D.A. Wharam, T.J. Thornton, R. Newbury, M Pepper, H. Ajmed, J.E.F. Frost, D.G. Hasko, D.C. Peacock, D.A. Ritchie, and G.A.C. Jones, J. Phys. C.**21**,L209 (1988).

15. Y. Imry, in "Directions in Condensed Matter Physics", Ed. by G. Grinstein and G Mazenko (World Scientific, Singapore, 1986), p. 101.

16. H.L. Engquist and P.W.Anderson, Phys. Rev. B **24**, 1151 (1981).

17. A.D. Benoit, S. Washburn, P. Umbach, R.B. Laibowitz, and R.A. Webb, Phys. Rev. Lett. **57**, 1765 (1986).

18. H.U. Baranger and A.D. Stone, Phys. Rev. B **40**, 8169 (1989).

19. S.A. Trugman, Phys. Rev. B **27**, 7539 (1983).

20. If the zero of the potential in the ordered leads is the same as in the "sample", then strictly speaking there is no propagating state with Landau index $N+1$ in the leads. However this corresponds to the assumption of no disorder-broadening of the Landau levels in the reservoirs, which is unphysical. This problem can be removed in practice in several ways, e.g. by putting a small potential difference between the leads and the sample.

21. L.I. Glazman and A.D. Stone, in preparation.

22. P.L. McEuen, A. Szafer, C.A. Richter, B.W. Alphenaar, J.K. Jain, A.D. Stone, R.G. Wheeler, and R.N. Sachs, preprint.

23. H.Z. Zheng, K.K. Choi, D.C. Tsui, and G. Weimann, Phys. Rev. Lett. **55**, 1144 (1985).

24. B.E. Kane, D.C. Tsui, and G. Weimann, Phys. Rev. Lett. **59**, 1353 (1987); R.J. Haug and K. von Klitzing, Europhys. Lett. **10**, 489 (1989).

25. J.K. Luo, H. Ohno, K. Matsuzaki, T. Umeda, J. Nakahara, and H. Hasegawa, Phys. Rev. B **40**, 3461 (1989).

SCALING OF THE TRANSMISSION AND REFLECTION COEFFICIENTS NEAR THE MOBILITY EDGE

Michael J. Stephen
Rutgers University, Piscataway, NJ 08855

The behavior of the optical transmission and reflection coefficients near the mobility edge of a disordered medium which elastically scatters light are discussed using a scaling theory. The transmission coefficient shows an interesting dependence on slab thickness. The coherent back scattering peak has a width determined by the correlation length and a height which also depends on this length. Both become small near the mobility edge. The effects of absorption are found to be important.

INTRODUCTION

Recently there has been considerable experimental and theoretical interest in using light scattering as a method of observing localization effects of waves in disordered media. Most of this work has been in the regime of weak localization where coherent back scattering (cbs) is now well established [1]. It would be of considerable interest to carry out light scattering experiments on a system near the mobility edge and there have been several sugestions as to how this may be achieved [2]. The purpose of this article is to discuss the behavior of the transmission and reflection coefficients in light scattering from a medium near its mobility edge. Here we present simple scaling arguments. These arguments have been confirmed by microscopic calculations near two dimensions [3].

SCALING DESCRIPTION

Light of unit intensity and wavelength $\lambda = 2\pi/k$ is incident normally (in the z direction) on a slab of the disordered medium of cross sectional area A and thickness L (assumed much greater than the mean free path ℓ). In the region of weak localization $\lambda < \ell$ the transmission coefficient $T \sim \ell/L$. Microscopic calculations [3] indicate that this is more correctly written $T \sim D/Lc$ where c is the velocity of light. This result also follows from the analogy between the transmission in the optical case and the conductance of an electronic system. The scaling properties of T near the mobility edge then follow from the scaling properties of the diffusion constant or the conductance [4]. Then

$$T \sim \frac{D^*}{Lc} \left(\frac{\ell}{\xi}\right)^{d-2} f_1(L/\xi) \qquad (2.1)$$

when ξ is the correlation length, $D^* = c\ell/d$ is the microscopic diffusion constant and d is the number of spatial dimensions and f_1 is a scaling function. We will express all results in terms of the correlation length but if t-t* is a measure in parameter space of the distance from the mobility edge then $\xi \sim (t-t^*/t^*)^{-\gamma}$ where the exponent $\gamma = 1/d-2$ close to two dimensions.

The dependence of the transmission coefficient on L then follows from (2.1)

$$T \sim \frac{\ell}{L}\left(\frac{\ell}{\xi}\right)^{d-2} \qquad L > \xi \qquad (2.2a)$$

$$\sim \left(\frac{\ell}{L}\right)^{d-1} \qquad L < \xi \qquad (2.2b)$$

These results agree with those of Anderson [5] for d=3 and apply in the region of extended states. In the region of localized states if $L > \xi$ we would expect $T \sim e^{-L/\xi}$ while (2.2b) which is independent of ξ should also apply in the region of localized states.

We now consider the light diffusely reflected from the slab at a small angle θ to the backward direction (coherent back scattering (cbs)). Let $q=k \sin \theta$ denote the transverse wave vector of the reflected light at angle θ and R_q its intensity. In the region of weak localization [6]

$$R_q \sim R_o - f_2(q\ell) \qquad (2.3)$$

where R_o is the background intensity. It is convenient to normalize the background so that $f_2(o) = 0$ i.e. R_o is the intensity in the backward direction (excluding any specular reflection). The form of f_2 depends on the boundary conditions at the surface of the slab. The important feature is that for a semi-infinite slab $f_2(q\ell) \sim q\ell$ for $q\ell < 1$ leading to the characteristic triangular shape for the cbs peak. Microscopic calculations [3] indicate that this is more correctly written $f_2(q\ell) \sim qD/c$ where again D is the diffusion coefficient. The scaling properties of f_2 near the mobility edge then follow again from those of D i.e.

$$R_q = R_o - \left(\frac{\ell}{\xi}\right)^{d-1} f_3(q\xi) \qquad (2.4)$$

where f_3 is a scaling function.

From this equation we obtain the following results for cbs near the mobility edge. At small angles $f_3(q\xi) \sim q\xi$ $(q\xi < 1)$ and the cbs peak has the characteristic triangular shape and its angular width $\theta \sim \lambda/\xi$ gets small. The height of the cbs peak relative to the background is of order $\left(\frac{\ell}{\xi}\right)^{d-1} f_3(1)$ and is also quite small (in the region of weak localization $\xi \sim \ell$ it is of order unity). At larger angles ($q\xi > 1$), ξ will cancel in (2.4) and $R_q = R_o - (q\ell)^{d-1}$ which gives the shape of the cbs peak at larger angles.

The triangular shaped cbs peak only occurs for a semi-infinite slab. Any perturbation that limits the length of the multiple scattering paths will have the effect of rounding this peak. As an example we consider the effects of finite thickness L of the slab. In this case $f_3(q\xi)$ in (2.4) is replaced by $f_3(q\xi,qL)$ and for small angles

$$\begin{aligned} f_3(q\xi,qL) &\sim q^2L\xi && L > \xi \\ &\sim (qL)^2 && L < \xi \end{aligned} \tag{2.5}$$

and thus the cbs peak is parabolic. These results apply in the region of extended states. In the region of localized states the cbs peak will be rounded as in (2.5) with $L \sim \xi$.

The effects of absorption can be important. If ℓ_a is the absorption length in the medium as we are dealing with a diffusion process the characteristic length determining the effects of absorption is

$$L_a = \left(\frac{D\ell_a}{c}\right)^{1/2} = \left(\frac{D^*\ell_a}{c}\right)^{1/2} \left(\frac{\ell}{\xi}\right)^{d-2/2} \tag{2.6}$$

Thus near the mobility edge L_a gets small and can become the most important length in the problem. Then if $L_a < L$ the transmission coefficient will no longer be of the form (2.2) but decrease exponentially $\sim e^{-L/L_a}$. Cbs will only be observable if $L_a > \xi$ which can be an important limiting factor. The rounding of the cbs peak due to absorption is also described by (2.5) with L replaced by L_a.

CONCLUSIONS

We have discussed the behavior of the transmission and reflection coefficients for light scattered from a disordered medium near the mobility edge. Both the transmission and coherent back scattering show interesting scaling behavior. Observation of these effects require that the absorption in the sample be small.

ACKNOWLEDGEMENT

This work was supported in part by the National Science Foundation under Grant No. NSF 4-20508. I am grateful to I. Edrei for useful discussions.

REFERENCES

1. M. P. van Albada and A. Lagendijk, Phys. Rev. Lett. 55, 2692 (1985); P. E. Wolf and G. Maret, Phys. Rev. Lett. 55, 2696 (1985).
2. S. John, Phys. Rev. Lett. 58, 2486 (1987); V. M. Agranovich, V. E. Kravtsov and I. V. Lerner, Phys. Rev. A125, 435 (1987); U. Sivan and A. Sa'ar, Europhys. Lett. 5, 139 (1988).
3. I. Edrei and M. J. Stephen, to be published.
4. E. Abrahams, P. W. Anderson, D. C. Licciardello and T. V. Ramakrishnan, Phys. Rev. Lett. 42, 673 (1979).
5. P. W. Anderson, Phil. Mag. B52. 505 (1985).
6. A. A. Golubentsev, Zh. Eksp. Teor. Fiz. 86, 47 (1984); E. Akkermans, P. E. Wolf and R. Maynard, Phys. Rev. Lett. 56, 1471 (1986); M. J. Stephen and G. Cwilich, Phys. Rev. 34, 7564 (1986).

ELECTRON LIGHT SCATTERING IN SUPERCONDUCTING SINGLE CRYSTALS OF $Tl_2Ba_2CaCu_2O_8$

A. A. Maksimov, I. I. Tartakovskii, L. A. Fal'kovskii*, V. B. Timofeev
Institute of Solid State Physics, The USSR Academy of Science,
142432 Chernogolovka, Moscow dist.
* L. D. Landau Institute for Theoretical Physics,
117334 Moscow GSP-1, USSR.

The observed temperatural singularities of electron light scattering (ELS) spectra for $Tl_2Ba_2CaCu_2O_8$, measured in different crystallographic directions and for different light polarizations, are interpreted as manifestations of strong anisotropy of the superconducting gap ($2\Delta_{min} < 50\,cm^{-1}$, $2\Delta_{max} \approx 300\,cm^{-1}$).

INTRODUCTION

Interest in electron light scattering (ELS) in superconductors was stimulated by the well-known work of Abrikosov and Fal'kovskii [1] (see also [2-3]). In theory, an ELS spectrum of metal in the normal state must have the form of a bell-shaped continuum and begin with zero recoil frequencies. At $T \ll T_c$ this spectrum must demonstrate the occurrence of a dip related to the superconducting gap Δ.

This work is devoted to a detailed study of ELS in high T_c crystals of $Tl_2Ba_2CaCu_2O_8$ with the superconducting transition temperature $T_c = 110K$. Investigation of ELS in Tl-based superconductors has a number of specific features as compared with analogous investigations performed in $YBa_2Cu_3O_7$ and $Bi_2Sr_2CaCu_2O_8$ crystals [4-12]. Firstly, in the actual frequency range $130 < \omega < 500\,cm^{-1}$ $Tl_2Ba_2CaCu_2O_8$ single crystals exhibit no intense phonon lines in Raman Scattering (RS) spectra which are attributed to small distortions of cuprate planes [13]. Secondly, the shape of phonon components of RS and their spectral position are weakly dependent on temperature; that is very important on measurement of temperature variations in an extended continuum in the superconducting gap region. A stronger dimpling of CuO_2-planes in $YBa_2Cu_3O_7$ crystals gives rise to an intensive RS line of B_{1g} symmetry with frequency $\sim 340\,cm^{-1}$, meeting antiphase vibrations of oxygen atoms in cuprate planes along the c-axis of the crystal and having a specific temperature dependence [7]. Finally, in Tl-2212 crystals of the tetragonal symmetry $D_{4h}^{17}(I4/mmm)$ twinning is absent, which simplifies the performance of polarization measurements.

EXPERIMENTAL RESULT

This ELS study was performed in Tl-2212 single crystals grown from melt of stoichiometric composition with slowly cooling in an oxygen atmosphere. The samples were thin plates with a smooth developed basel **ab**-plane measuring $3{\times}3{\times}0.2\,mm^2$. The crystallographic structure and the unit cell parameters (**a** = 3.856Å, **c** = 29.34Å) were established by X-rays. The superconducting transition temperature was $T_c \approx 110K$. The data on the variation of the magnetic susceptibility χ, for one of the samples in question, in the temperature range $T=50$–130K, are shown in the insert of Fig. 1, wherefrom one can easily estimate T_c and the uniformity of the superconducting transition.

RS spectra were taken from the same site on the crystal surface with a guaranteed accuracy $\sim 2\,\mu$m, controlled by means of a microscope and were registered in backscattering geometry by a spectrometer Dilor XY, provided with an optical multichannel detector and specially devised microscopic attachment for low temperature measurements. The laboratory axes x, y, and z coincide with the direction of the crystallographic axes **a**, **b** and **c**, respectively. Fig. 1 demonstrates the RS spectra at the excitation of the basal **ab**-plane for two value of T_0 in a thermostat, $T_0 > T_c$ and $T_0 \ll T_c$. A temperature decrease is seen to lead to a noticeable drop of the intensity I_s of the extended background in the region of frequencies $\omega < 400\,cm^{-1}$, the dip value at $T_0 \approx 5$K and $\omega \approx 50\,cm^{-1}$ being 30% ($\omega = \omega_i - \omega_s$, the spectral shift of ELS, counted from the Rayleigh line and characterizing the recoil energy in one-particle act of electron scattering). For different samples this value may vary within a rather wide range: from $\sim 10\%$ to $\sim 80\%$. However, even for minimal T_0 values in the most perfect samples we did not observe any drop of I_s down to zero in the studied frequency range. It follows that the continuous background represents a broad structureless line in the frequency range from $30\,cm^{-1}$ (lower limit of the registration range) to $\sim 2000\,cm^{-1}$. In the range to $\sim 400\,cm^{-1}$ this background exhibits the temperature dependence.

Characteristic features of the intensive extended background, observable in RS spectra, enable one to interpret it as intraband one-particle electron light scattering by the Fermi-liquid. This interpretation may be supported by the following arguments.

1) The background length is more than $2000\,cm^{-1}$; that, as in $YBa_2Cu_3O_7$ crystals [6], noticeably increases the actual frequency region wherein contributions from one- and two-phonon transitions of light RS may be manifested.

2) Phonon lines in RS spectra whose symmetry is coincident with ELS continuum have strongly asymmetric contours (Fano resonances) that indicate a sufficiently strong electron-phonon interaction [13].

3) The observed peculiarities in the range $\omega < 400\,cm^{-1}$, arising with decreasing temperature $T_0 < T_c$ in the ELS spectrum are naturally to be connected with the presence of the superconducting gap, since a decrease of the scattering intensity in this frequency range unambiguously correlates with the superconducting transition.

Polarization measurement (Fig. 2) of ELS spectra in the normal and superconducting states suggest that for all polarizations, except (zz) polarization, a decrease of the ELS intensity I at $T_0 < T_c$ is observed, with small variations, starting, approximately, from frequencies $\omega < 400\,cm^{-1}$. For (zz) polarization the peculiarity in the spectrum manifests itself at a smaller frequency $\omega < 160\,cm^{-1}$.

The temperature dependence of ELS intensity I alteration is illustrated in Fig. 3 wherein are demonstrated the intensity ratios $I(T)/I(180K)$ obtained by directly dividing the ELS experimental spectra, measured for several $T < T_c$ values. It is seen that with $\omega > 400\,cm^{-1}$ this ratio is practically temperature independent and close to 1 (with experimental accuracy $\pm 10\%$). At lower frequencies ω one can observe a dip, that disappears at $T \approx T_c$, and a further increase in temperature does not lead to noticeable spectral changes. An analogous behavior is observed in the (zz) polarization, which is indicative of the similar nature of this phenomenon. We emphasize that in the dip region the spectra intensity ratio as a function of T does not have an activation

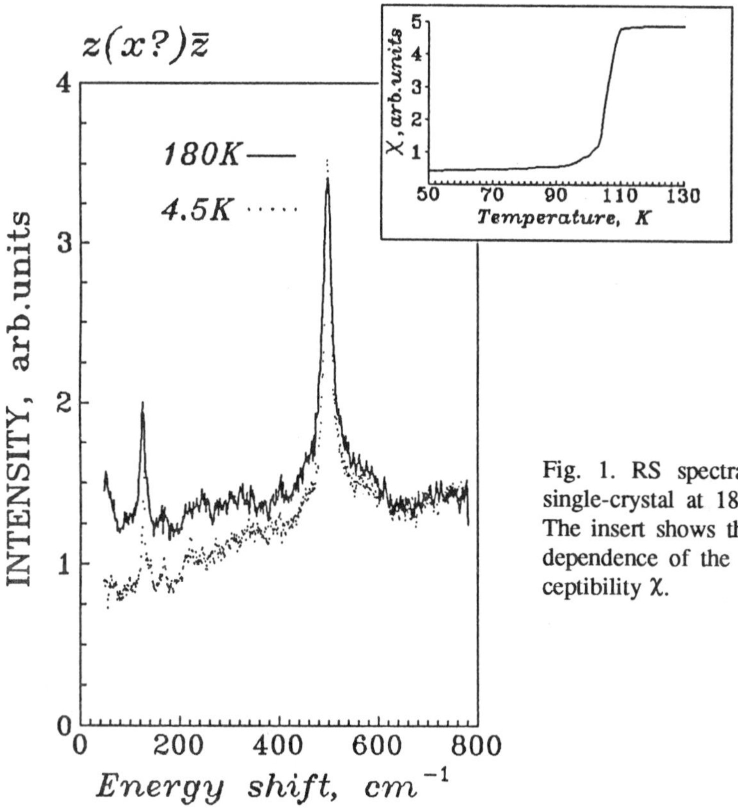

Fig. 1. RS spectra for Tl-2212 single-crystal at 180K and 4.5K. The insert shows the temperature dependence of the magnetic susceptibility χ.

character. Note, also, that along with a decrease of the dip value with increasing temperature $T \to T_c$ the dip region tends to shift towards lower frequencies.

THE THEORY

We consider now ELS in a superconductor and restrict ourselves to the "pure" case. Here one has to distinguish two qualitatively different situation for traditional superconductors: the limit of large and small recoil vectors where $V/\delta \gg \Delta$ or $V/\delta \ll \Delta$. Here V is the Fermi velocity, δ is the penetration depth (so that the recoil vector is of order $1/\delta$). The equations of Refs. [1,2] are then applicable. Note, that we have to take for δ the light penetration depth into the metal at a frequency $\omega_i \approx \omega_s$, which is large compared to Δ in the optical range. Naturally, the δ value is not affected by superconductivity in this case. According to [1,2], there is a threshold $\omega_{th} = 2\Delta_{min}$ in the spectrum, determined by a minimal gap value on the Fermi-surface belt, where the electron velocity is parallel to the sample surface. We emphasize that close to the threshold the cross section does not change in a jump-like manner. This cross section approaches zero continuously in both isotropic and anisotropic cases. (In the isotropic case it is necessary to sum electron interaction diagrams to all orders).

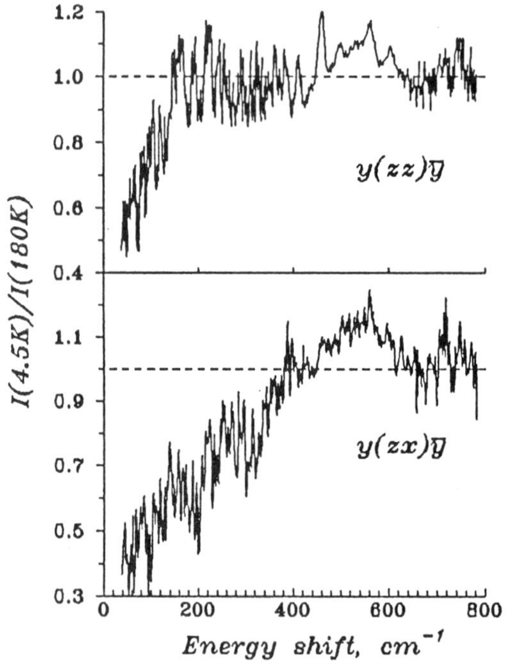

Fig. 2. Intensity ratio in ELS spectra at temperature $T \ll T_c$ and $T > T_c$ under two polarizations of the exciting and scattering light.

Just now we are interested in the high-T_c superconductor, when the limit of small recoil vector $V/\delta \ll \Delta$ is realized. If a minimal gap value on the Fermi-surface Δ_{min} is other than zero, then in this case, too, at zero temperature there is a threshold $\omega_{th} = 2\Delta_{min}$ in the scattering spectrum [3]. Note that here the threshold depends neither on the direction of the normal to sample surface nor on the polarization vectors.

The frequency dependence of the cross section is determined to a large extent by the angular dependence of the energy gap. We shall consider there three different cases.

1) The gap anisotropy is small. Then there exists a threshold value of the recoil frequency $\omega_{th} = 2\Delta$, below which scattering is absent. Above the threshold the following asymptotic formulae are applicable

$$d\sigma = d\sigma_0 \begin{cases} \dfrac{4}{\pi}\dfrac{\xi^3}{\delta^3}\dfrac{\ln \delta/\xi}{\ln(\xi_1/\delta)\ln(\xi_1/\xi)}, & [\omega^2 - 4\Delta^2]^{1/2} \ll V/\delta \qquad (1) \\ \dfrac{4}{\pi}\dfrac{\xi^3}{\delta^3}, & V/\delta \ll [\omega^2 - 4\Delta^2]^{1/2} \qquad (2) \\ 4\left[\dfrac{V}{\omega\delta}\right]^3, & \omega \gg 2\Delta \qquad (3) \end{cases}$$

where $\xi = V/\Delta$, $\xi_1 = V/[\omega^2 - 4\Delta^2]^{1/2}$. We emphasize that the Eqs. (1-3) describe a wide maximum. Its width is of order of Δ, i.e. the same as the distance to the unshifted line.

2) The superconductor has a significantly anisotropic gap, not becoming zero on the Fermi-surface. In this case the threshold is determined by a gap value minimal on the Fermi-surface. It is independent of the polarization of the incident and scattered light. In the proximity of the threshold the cross section behaves differently depending on whether a minimum Δ is attained in the point on the Fermi-surface or, else, on the line (the latter is possible, in particular, for quasi-2D superconductors).

$$d\sigma \sim d\sigma_0 \begin{cases} [(\omega - 2\Delta_{min})/\Delta_{min}]^{1/2}, & \text{minimum in the point} \\ \left[\ln \dfrac{\Delta_{min}}{\omega - 2\Delta_{min}}\right]^{-2}, & \text{minimum on the line} \end{cases} \qquad (4)$$

Away from threshold, the cross section diminishes when the recoil frequency ω is increased by a value, equal in the order of magnitude to a maximal (on the Fermi-surface) gap value of $2\Delta_{max}$.

3) The gap become zero in certain points or lines belonging to the Fermi-surface. The cross-section has no threshold. The cross section exhibits a maximum at the recoil frequency of the order of a maximal gap value. With decreasing frequency it decreases by linear law if the gap becomes zero on the line, and quadratically if it occurs in isolated points.

DISCUSSION

A marked feature is the behavior of the scattering cross section at $\omega \rightarrow 0$. From our observations it does not tend to zero, whereas according to theory the electron scattering has to drop to zero irrespective of whether the metal is in the normal or superconducting state. However, for the normal metal the drop can not start from $\omega > V/\delta$. For a "dirty" normal metal a decrease starts from even smaller ω. The available data permits only a rough estimate $V/\delta \approx 20\,\text{cm}^{-1}$; that indicates that our observations pertain to the region wherein the cross section does not have to diminish yet towards smaller ω. If we take into account this circumstance and also the fact that the dip value in the cross section at low temperatures for $\omega \approx 50\,\text{cm}^{-1}$ varies in different samples from ~10% to ~80%, then it follows that our samples comprised extraneous nonsuperconducting phases (a local content of extraneous phases may noticeably affect the light scattering despite the fact that an X-ray analysis guarantees the volumetric content of the phase Tl-2212 with an accuracy of ~2%).

Nevertheless, for some samples the cross section dropped to nearly zero at small frequencies. In principle, this might be attributed to the existence of excitations below the gap. However, if we are dealing here with electronic excitations, then their number decreases with the temperature in proportion with $\exp(-\Delta/k_B T)$, and this factor is very small at low temperatures. It is to be assumed that the superconducting gap becomes zero somewhere on the Fermi-surface or, else, its minimal value, as suggested by our data, does not exceed $2\Delta_{min} < 50\,\text{cm}^{-1} \approx 0.6\,k_B T_c$. In order to explain the linear dependence of the cross section in the region of small recoil frequencies, one should accept that the gap disappears (or assumes a sufficiently small value) along the lines, lying on the Fermi-surface.

We shall consider the possibility of determination of a maximal gap value from

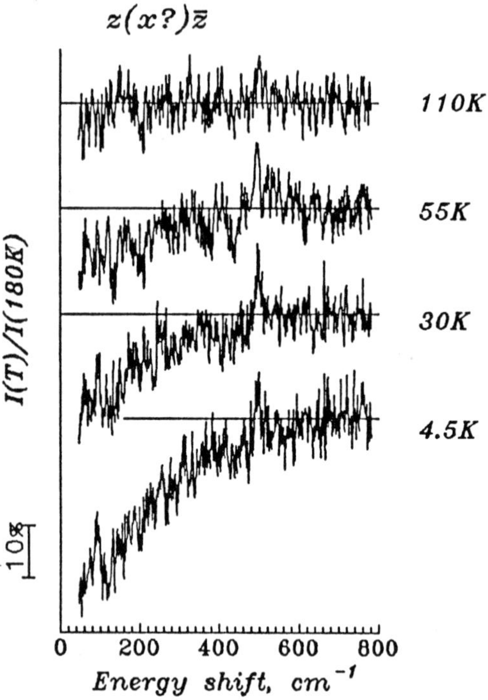

Fig. 3. Temperature dependence of the intensity ratio I/I(180K) in ELS spectra; the polarization of the exciting light $E_i \parallel x$, the scattering light polarization has not been analyzed.

the electron scattering data. Scattering in the superconductor is a wide continuum with a maximum approximately at $\omega \approx 2\Delta_{max}$. The maximum position, as contrasted from the threshold value of $2\Delta_{min}$, depends on integral characteristics of the electronic spectrum, for example, on the factor, that is determined by the tensor of inverse effective masses and by the direction of polarization vectors [2,14].

The position of the maximum in the ELS spectra (or the threshold where the difference in scattering is observed in the case of superconductor or normal metal) does not suggest (as is sometimes believed) that the gap depends on the polarization direction. One can, however, estimate a maximal gap value. As seen from Figs. 2, 3 this estimation yields $2\Delta_{max} \approx 300\,cm^{-1} \approx 4k_BT_c$.

In conclusion, the authors wish to express their deep gratitude for helpful discussion to A. A. Abrikosov, G. M. Eliashberg and D. E. Khmelnitskii for useful discussion.

REFERENCES

1. A. A. Abrikosov, L. A. Fal'kovskii, Sov. Phys.-JETF **13**, 179 (1961).
2. A. A. Abrikosov, V. M. Genkin, Sov. Phys.-JETF **38**, 417 (1974).
3. A. A. Abrikosov, L. A. Fal'kovskii, Physica C, **156**, 1 (1988).
4. A. V. Bazhenov *et al.*, JETP Lett. **46**, 32 (1987).
5. K. B. Lyons *et al.*, Phys. Rev. B, **36**, 5592 (1987).
6. S. L. Cooper *et al.*, Phys. Rev. B, **37**, 5920 (1988).
7. C. Tomsen and M. Cardona, in "Physical Properties of High Temperature

Superconductors 1'', edited by D. M. Ginsberg (World Scientific Press, Singapore, 1989) Ch. 8.
8. Yu. A. Ossipyan *et al.*, Physica C, **153-154**, 1133 (1988).
9. R. Hackl *et al.*, Phys. Rev. B, **38**, 7133 (1988).
10. S. L. Cooper *et al.*, J. Opt. Soc. Am. B, **6**, 436 (1989).
11. S. L. Cooper *et al.*, Phys. Rev. B, **38**, 11934 (1988).
12. D. Kirillov *et al.*, Phys. Rev. B, **38**, 11955 (1988).
13. L. V. Gasparov *et al.*, Physica C, **160**, 147 (1989).
14. L. A. Fal'kovskii, Zh. Eksp. Teor. Fis., **95**, 1146 (1989).

THE NORMAL STATE OF HIGH T_c SUPERCONDUCTORS: A NEW QUANTUM LIQUID

P.W. Anderson and Y. Ren
Joseph Henry Laboratories of Physics, Jadwin Hall
Princeton University, Princeton, NJ 08544

A renormalization group analysis of interacting Fermion systems shows that there are two fundamentally different fixed points, Landau Fermi liquid theory and what has been called "Luttinger Liquid Theory" by Haldane, a type of state in which fluctuations of different quantum numbers—*e.g.*, charge and spin—acquire distinct spectra, and Fermi surface correlations have unusual exponents. The Luttinger liquids comprise most interacting 1-dimensional systems, and some higher-dimensional systems in which, as in the Hubbard model, the band spectrum is bounded above. All higher-dimensional systems with true Mott-Hubbard gaps and an upper Hubbard band, including most 2d Hubbard models, are Luttinger liquids. We summarize the experimental evidence in favor of the 2d Hubbard model for high T_c superconductors and the constraints pointing towards its non-Fermi liquid nature. We describe progress towards explaining normal state and superconducting properties and elucidating the high T_c mechanism.

1. INTRODUCTION

We find that the "normal" metal from which high T_c arises in the cuprate larger compounds is in a new metallic state for which we borrow Haldane's phase "Luttinger Liquid", which is appropriate both because it has a pseudo-Fermi surface satisfying Luttinger's theorem—*i.e.*, it is a "$Z = 0$" generalization of a Landau Fermi liquid—and because it is a higher-dimensional generalization of the soluble one-dimensional interacting Fermi systems described in this way by Haldane. It is also a generalization of the Bethe-ansatz soluble systems.

We will first describe the experimental facts which require us to consider the two-dimensional Hubbard model as basic to the physics, and show that the state of this system is not a Fermi liquid. Next we show that the 2D Hubbard Model, as well as any higher-dimensional system which has an "upper Hubbard band", can not be a normal Fermi liquid, because the wave-function renormalization parameter $Z = 0$. This central proof forces us to examine the only previously known case of $Z = 0$, the 1D "Luttinger Liquid" and in particular, the 1D Hubbard model, its exact solution and the general Haldane formalism. We then show how to generalize this formalism to higher dimensions, retaining the separation of charge and spin and the Luttinger liquid formalism. Finally, we relate this state to the experimental response functions of the normal cuprate systems, in particular, proposing a superconductivity mechanism. A final section lists experimental phenomena, and assesses the state of agreement with the theory, in quantitative or qualitative terms.

2. EXPERIMENTAL CONSTRAINTS ON THE NORMAL STATE OF THE CUPRATES

Two core questions must be decided before a serious approach to the high T_c problem can be mounted: "Is the normal state a Fermi liquid?", and "What is the appropriate model?". Although there is no consensus in the community on either, the evidence on both is quite overwhelming if one examines the situation as a whole and does not narrow one's focus to one individual measurement to the exclusion of others. Any single fact can always be explained away, but the overall picture has only one plausible meaning.

I shall only mention the broad range of computational, structural, and chemical evidence which force us to the one-band Hubbard model, with dominantly repulsive interactions, as an underlying model; much more detailed discussions are in my three sets of lectures on this subject[1-3]. Most conclusive are the cluster calculations of Stechel and Schluter[4], which parametrize perfectly to such a model, but the overall structural evidence—such as the structural integrity of the CuO_2 planes, the magnetic nature of all undoped phases, the cleavage of the BISCO compounds, the wide optical and photoelectric gaps even after doping, and many other indications, leave one with no alternative to the following conclusions: that the strongest bonds are the $d_{x^2-y^2} - Op_\sigma$ hybrids which hold the planes together, and push the corresponding antibonding state up to the neighborhood of the chemical potential. The only other bond which plays an electronic role is the d_z^2 antibonding hybrid, which must be beginning to empty in overdoped $(La - Sr)_2CuO_4$. (And which must also be split off from the $d_{x^2-y^2}$ hybrid which it would touch in the absence of second-neighbor $O-O$ hopping integrals). Of course, by dint of hybridizing into the active bands, many other electronic states play a role, but one must reject the many band theorists' suggestions of "pockets" of carriers in Bi or other orbitals, in that these are totally incompatible with the very high c-axis resistivity, far above the Mott limit, which is shown by all compounds, and with the excellent infrared data which are compatible with only one set of carriers (and confirm zero c-axis metallic conduction).

The chains in YBCO are a special problem. Clearly in the lower O members of the series their oxidation state moves from Cu^+ to Cu^{++}, indicating that at $\sim O_{6.5}$ Cu^+ has just moved above μ. It is implausible therefore, since on the planes Cu^{+++} is now at the chemical potential, that many carriers ever flow on the chains, which, I would propose, are simple Heisenberg chains throughout the superconducting region (with, of course, 1d spinon excitations mimicking the magnetic properties of carriers, as well as, very likely, chain-end magnetic centers confusing the magnetic properties except for $O_{6.9-7}$). Many lines of evidence, not least the NMR and Raman data described at Los Alamos by Millis, make it clear that electronic parameters do not shift with doping, which makes almost inescapable the dominance of repulsive Hubbard interactions in the planes. (These "Hubbard" interactions are, of course, not to be taken literally as "U" 's but result predominantly from $Cu - O$ bonding integrals and Hilbert space restrictions)

Next let us take up the other question. "Is it a Fermi liquid?" Many of the relevant facts were adduced in a recent publication[5] but a more complete summary is in my own lecture notes of June this year. I will list a sequence of types of measurement with "pros" and "cons" on the Fermi liquid question. Let me make it clear that the question is about the normal state; superconductivity I only touch on.

a. Tunneling density of states.

Here the relatively few measurements on the normal state are almost the whole of the worthwhile story, since they are immune to the ubiquitous surface problems (themselves a strong piece of indirect evidence that superconductivity is an interlayer phenomenon and not intrinsic to the individual layers). Tunneling into the normal state invariably produces the very reproducible "V" shaped curves, with $\sigma \simeq A+B|V|$. (The few studies which conflict were point contact measurements primarily concerned with superconducting results). Pro: there is a finite zero-bias conductance, indicating a finite angle-averaged density of states at the Fermi level. Con: but the $|V|$ behavior is unique (except for the equally mysterious case of $BaBiO_3$-based materials.) Really the only reasonable explanation is that of Zou-Anderson, that electrons are composite, not elementary, in the normal layers; and no other reasonable suggestion has been made. **First constraint:** *particles are composite.*

b. Resistivity and Transport

Pro: the resistivity in the *ab* plane is of metallic magnitude, coming down to $\sim 50\mu\Omega$ cm at $\sim 100°$K. Con: almost all other facts about transport are unusual. Strangest of all is that the non-stoichiometric cases are as good conductors as pure ones, and that no residual resistance occurs except in the cases where doping is in the planes. The potential scattering resistance calculated from a reasonable estimate for most of these substances should be of least 10 times larger and should appear as a residual resistance. I calculated $\hbar/\tau \sim 1/4ev$ for BISCO, about equal to the mean free time extrapolated to 2000° K. Even stranger is that ρ_c is 300-10^4 times larger (well above the Mott limit) and usually has inverse T-dependence. Localization theory for a Fermi liquid does not permit anisotropic localization. The only viable suggestion is that of Zou and Anderson. **Second constraint:** *the real excitations are confined* to 2d planes.

The sign and behavior of the Hall resistance are anomalous. It is nearly certain that the Fermi surface is a continuous ring about the origin, *i.e.*, electron-like; yet R_H becomes increasingly positive at low T. This is among the key facts which should essentially be explained by any theory. Clearly, however, we have a **third constraint**. The carriers obey lower (or possibly upper) Hubbard band counting, not conventional. What little is known about the thermoelectric power concurs. An interesting fact about thermoelectric power is that, in general, the entropy carried per carrier does not extrapolate to zero at $T = 0$.

Heat conductivity seems to obey a Wiedemann-Franz law with the constant > 1, above T_c; there is an anomalous absence of phonon transport for the *ab* plane which suddenly reappears at T_c. In the c-direction, confirming ρ_c, there appears to be no electronic heat conduction.

c. Spectroscopy

In the infrared for the normal state, this mimics conductivity if we allow $\omega \leftrightarrow T$—e.g., σ_{ir} extrapolates to σ_{dc} at $\omega = 0$. This means that the resistivity is a good measure of the inelastic electron scattering, and hence that $\frac{\hbar}{\tau} \simeq E_k$, or that $Im\sum(\omega) \simeq \omega$. This, in turn, implies that $Re\sum \sim \omega \ln \omega$, $Z \equiv 0$, *i.e.*, that the renormalization constant of any possible Fermi liquid theory is zero and no quasiparticles exist in the usual sense **(4th constraint)**. Incidentally, the superconducting state is revealed by the IR to be conventional in the sense that the sum-rule agrees with the penetration depth λ. (Also, IR reveals a reasonably good gap, of order $2\Delta \stackrel{\sim}{>} 6kT_c$. It also reveals no sign of an infrared excitation strong enough to cause high T_c.)[6] A strong Raman background, very nearly frequency-independent and obviously a bulk electronic property, is observed. This is a second strong evidence for $Im\sum \sim \omega$, which, as far as we can see, implies composite quasiparticles.

d. Angle-resolved Photoemission

Here we have one pro: a sharp Fermi surface is revealed. But again, it is far too sharp: its breadth Δk seems to be less than .1 of the Brillouin zone, yet the trivial calculations on inelastic scattering carried out above show that $l \sim 1/k_F$ so that $\Delta k \sim k_F$. *The entities which form the Fermi surface cannot be scattered by charged impurities!* Hence, as above, $Z \equiv 0$ and $F_0^s \rightarrow \infty$: the Fermi surface must be formed by particles with spin only. Important **constraint 5:** *There is a Luttinger-Theorem Fermi surface*, formed by spin-only entities.

When we look at the detailed spectra (see e.g., Fig. 1), we see a remarkably sharp cusp dispersing away from the Fermi surface at a reasonable clip, a broad background extending to high energies both above the cusp and even above the Fermi surface, and a linear decrease to a sharp second cusp at the Fermi energy. We will see that this spectrum is easily explained by a composite picture, while I do not feel that the conventional interpretation attempted by the experimentalists can be made consistent with all the facts. Photoemission in the superconducting state roughly confirms the large IR gap.

Angle-averaged photoemission gives us no new information on density of states which is not available from tunneling into the normal state. It confirms sensible estimates of the hybridization of the carrier bands, and that the carriers are in a smaller band above the *Op* non-bonding mass of bands.

e. NMR

NMR relaxation and Knight shifts in the normal state reveal the **sixth constraint:** *there are two different suceptibilities.* A conventional spin susceptibility seems to control the relaxations on O^{17} in the plane and on Y, while a second, very unconventional and large, relaxation process occurs in addition (not alone) on the *Cu* II sites in the planes. This term has no large Knight Shift or susceptibility

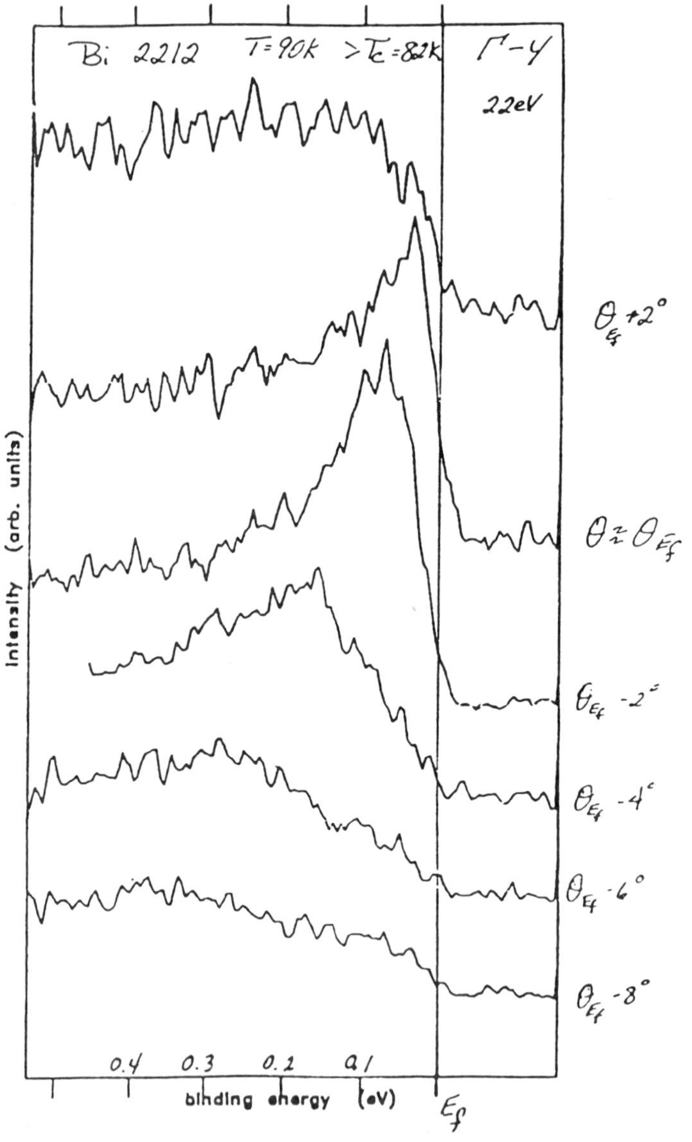

FIGURE 1. Angle-resolved photoemission data for BISCO at $T > T_c$.

associated: we guess it is an orbital effect, since χ is remarkably constant in the best superconducting materials, yet this term must correspond to something which fluctuates very slowly. Antiferromagnetic spin fluctuations seem utterly implausible to me: there is no independent evidence for them and they could not behave this way.

The 60° YBCO is a special case which has so many anomalies (*e.g.*, IR "knees", residual resistance, strange knight shift, etc.) that it clearly is not generic.

f. Heuristics of T_c.

Here we make our one foray into superconducting properties with a **seventh constraint:** superconductivity occurs as a result of interlayer contact. That is, the isolated layer has no mechanism to cause superconductivity. Some hints of evidence: One-layer BISCO material is metallic but has T_c 6° or (for many crystals) 0°. BISCO T_c's *vs.* layer number can be fit well empirically by the layer scheme.

$$\Delta_i(T) = \chi^i_{pair} \sum_{\theta} \frac{t^2_{ij}}{J} \Delta_j(T) \tag{1}$$

where $\chi_{pair} \propto \frac{1}{T}$ and i, j are layer numbers. That is, χ_{pair} is only moderately large but rapidly T-dependent relative to BCS; and it does not diverge at $T > 0$. (1) also fits all known pressure-dependences if we add the premise that close layers with strong coupling are not much affected by pressure, only distant couplings. *Tl* systems fit as well, but we must assume that the *Tl* layer has a pretty big $T_\perp$.

Finally, epitaxial experiments and the simple observation that $\Delta \to 0$ at surfaces are incompatible with an intralayer mechanism.

These seven constraints alone practically direct us unequivocally to the theory we shall now discuss.

3. "LUTTINGER LIQUID" BEHAVIOR OF THE NORMAL METALLIC STATE OF THE 2D HUBBARD MODEL

Haldane[7] has characterized the behavior of a large variety of one-dimensional quantum fluids by the term "Luttinger liquid", showing that they can all be solved by common techniques based on transforming to phase and phase-shift variables for the Fermi surface excitations (a procedure often called "bosonization" even though some of the Luttinger liquids start out as bose systems). These systems are characterized by fractionation of quantum numbers — *e.g.*, in the Heisenberg spin chain the excitations are spin 1/2 Fermion-like, while in the Hubbard model they are spin 1/2 chargless spinons and $\pm e$ spinless holons with Fermion-like properties — and, often, a Fermi surface with nonclassical exponents, and unusual exponents for correlation functions.

I will here restrict the term, for my purposes, to systems based on Fermions — preferably ordinary electrons — and argue that the "Luttinger liquid" is a fixed point, or a manifold of fixed points, of the same renormalization group which, "usually", leads to the Landau-Fermi liquid as a unique fixed point. (The interaction parameters of Landau Fermi liquid theory are well-known[8] to be marginal operators around a single fixed point, the effectively free Fermi liquid.)

Some years before, Luther[9] showed that the bosonization techniques used to solve these one-dimensional models are equally applicable to d-dimensional Fermi gases, and they describe certain facts slightly more accurately than Fermi liquid theory — the existence of $2k_F$ singularities in correlation functions, for instance, for the free-particle systems. But Luther did not consider the possibility that the interacting d-dimensional problem could lead to new physics.

The first new point I want to make is that two of the reasons usually given for the unique nature of one-dimensional Fermi systems are untenable. The first is that in $1d$ one has only forward scattering, or backward scattering where the momentum of one particle is maintained, if not its spin. This is indeed the correct reason for viability of the Bethe Ansatz. But after renormalization the Landau theory has only forward or exchange scattering, and the renormalized particles indeed obey a Bethe ansatz of the simplest form. This is the essence of Luther's argument, that the excitations can be bosonized in each direction around the Fermi surface.

Second, it is argued that particles cannot be interchanged in $1d$ without encountering phase-changing interactions, hence statistics are meaningless in $1d$: but none of Haldane's arguments seem to fail in the slightest if we introduce weak longer-range hopping integrals in any of the examples, and such hopping integrals can allow a Berry process. No one argues, in fact, that real electrons living in $3d$ space in the presence of a chain of ions, which know perfectly well that they are Fermions, will not obey the models and show the fractionization effects.

The unique effect in $1d$ is one which is also present in a class of higher-d models, specifically $2d$ repulsive Hubbard models, and in some strong coupling higher d cases. This is the presence of an unrenormalizable Fermi surface phase shift. Such a phase shift signals that the addition of a particle changes the Hilbert space for the entire system of particles — it requires a net motion of field amplitude through the distant boundary of the system, or a net change of wavelengths. The effects of such phase-shifts were explored thoroughly in connection with the "x-ray edge problem"[10] and are summarized in the "infrared catastrophe" theorem[11].

$$\langle VAC\,(V) \mid VAC\,(0)\rangle \propto e^{-\frac{1}{2}\left(\frac{\delta}{\pi}\right)^2 \ln N} \tag{2}$$

where $|VAC\,(V)\rangle$ is the non-interacting Fermi sea in the presence of a potential V which causes a phase shift δ. The singularity is the result of the shifting of the entire spectrum of k values (in the presence of fixed boundary conditions) or of the displacement of wave-function nodes (for scattering boundary conditions), and is independent of the finite contribution which may ensue from local modifications of the wave-functions.

In the conventional higher d, free electron gas cases to which Landau liquid theory applies, it is implicitly assumed—and indeed, self-consistently so—that the phase-shifts caused by adding or removing a single particle can be made to vanish in favor of a renormalization of all the quasiparticle mean field energies. It is assumed that there is an effective mean field energy whose eigenstates are the precise k-states of the appropriate free particle system[12]. The formal result of this process is that the wave-function renormalization constant Z is finite: that is, the overlap integral

$$\sqrt{Z} = \langle c^{+}_{k\sigma} \Psi_G(N) \mid \Psi_{k\sigma}(N+1)\rangle > 0 \tag{3}$$

where Ψ_k is the exact wave function of the $N+1$ particle state with one quasiparticle added, and in particular

$$\langle c^{+}_{k_F\sigma} \Psi_G(N) \mid \Psi_0(N+1)\rangle > 0 \tag{4}$$

(where in a Fermi system, the $N+1$ particle system necessarily has one particle added near the Fermi surface, hence its ground state is quasi degenerate). Eq. (4) cannot be true if there is a phase-shift due to the addition of $c^{+}_{k\sigma}$.

In one dimension, for interacting particles, such a phase shift is unavoidable, since the effective range of interactions is necessarily (for real interactions) of the order of the wavelength ($\delta = k_F a$). Thus in all the realistic one-dimensional systems, $Z = 0$, the Fermi liquid fixed point is excluded, and the phase-shifts due to interactions must be taken into account as relevant variables—in fact, in many cases renormalization invariants. $Z \equiv 0$ implies that the Fermi sea excitations—which may still exist—do not carry charge (but may carry spin and be spinons). I will summarize Haldane's analysis of the spectrum of the $1d$ Hubbard model shortly.

In $2d$, the scattering length for free particles and repulsive interactions diverges only as $\frac{1}{k \ln k}$ as $k \to 0$, and for higher dimensions it is $\sim \frac{1}{n} \sim k_F^{-\frac{1}{d}}$; in both cases no serious problems need ensue for shorter wavelengths. But there is one type of problem where finite phase-shifts are inevitable, namely systems with a single-particle spectrum bounded above and below in energy. In this case the introduction of an extra particle may cause a bound state to split off from the top of the spectrum (an "anti-bound state")[13]. By Levinson's theorem[14], the presence of a bound state, either above or below the band, is signalled by a difference π in phase shift in the appropriate channel between top and bottom of the band. This corresponds to the fact that one state must be removed from the band to make up the bound state, and to Friedel's identity[15]

$$\Delta n(k) = \sum_{\ell} (2\ell + 1)\, \delta_{\ell}(k) \tag{5}$$

for the change in number of states to be found below a wave vector k due to a phase shift $\delta(k)$. Continuity—or the fact that any bound state must be a superposition of all states in the band—tells us that some δ must remain finite at all energies in the band.

For any dimension, a repulsive interaction U sufficiently strong to split off an upper Hubbard band adds one state to that band for each added electron; the upper Hubbard band is the manifold of antibound states and where it is present we must have $Z \equiv 0$ in the occupied lower Hubbard band, since the Hilbert space changes when we add a carrier. In the $2d$ Hubbard model (with one band—the generalized Hubbard models recently introduced are an irrelevancy) any potential whatever will split off bound states of holes (for which the interaction potential is attractive) above the band, because of the well-known fact that in two dimensions all potentials bind. Thus, although for very low or very high occupancies the relevant singularities come in with small coefficients which are non-analytic in interaction ($\sim e^{-t/u}$) or density ($e^{-n^2 \ln n}$), $Z = 0$ in all cases. These terms will not be picked up in series expansions.

We can identify the relevant interactions by thinking of the upper Hubbard band as a kind of "ghost" condensate in a channel of $2k$ total momentum and zero total spin reflecting the fact that each particle of down spin prevents some state of the same momentum and up spin from being occupied, so that the "condensate" represents both states being occupied. The interactions which maintain the non-occupancy are attractive in the particle-hole channel, and are of the form, in k-space

$$V_{2k_F+q}\ \rho_{2k_F+q\uparrow}\ \rho_{-2k_F-q\downarrow} \tag{6}$$

which contains terms like $c^+_{k\uparrow}\ c_{k\downarrow}\ c^+_{-k+q\downarrow}\ c_{-k+q\uparrow}$ attractive to electron-hole pairs of approximately zero momentum, *i.e.*, spin fluctuations.

The best representation is to pick an arbitrary origin and use angular momentum eigenfunctions, $\varphi_{k,m,\sigma}(r)$. Consider scatterings of electrons in k, m and k', m'. $m + m'$ is the total angular momentum and is a function of choice of origin; the phase-shifts at a given energy are independent of it. $m - m'$ is the relative angular momentum and the strong Hubbard interaction acts primarily in the $m = m'$ channel, $m - m' = 0$. The antibound state is the top state in this channel and the relevant phase shift occurs between ingoing and outgoing waves in this channel at the Fermi energy. (Corrections due to the lattice are minor and easy to add)

It is worth discussing the resulting state in terms of a "renormalized Bethe ansatz" picture. If we, following Gallivotti[7], use a "poor man's renormalization group" procedure to eliminate k states far from the Fermi surface, we will end up with a shell of low-energy excitations with momenta near the Fermi surface. Even if $Z = 0$ and even in the presence of our new interactions, for the thin shell of states near k_F every real scattering is non-diffractive in that the two k-vectors never change: charge is always scattered forward. When $Z = 0$, however, the original k-states of the Fermi liquid are not adequate to contain all the particles, and the Bethe ansatz wave-function contains a continuous spectrum of k's through the Fermi surface, exactly as in one dimension, where the Hubbard model solution for large U may be written $\sum_Q (-1)^Q \det \|e^{k_i x_{Q_j}}\| \times$ spin function, and the spectrum of k_i's extends continuously to $|Q|$, $Q > k_F$. We presume that the same form is valid for renormalized particles in 2d, near the Fermi surface.

The interaction terms (6), just as in one dimension, can be rewritten as a coupling between ingoing and outgoing spin fluctuations

$$V \sum_{\substack{k\sim k_F \\ k'\sim k_F \\ q}} c^+_{k\uparrow}\ c_{k+q\downarrow}\ c^+_{-k'\downarrow}\ c_{-(k'+q)\uparrow} \tag{7}$$

and under bozonization this turns into a term proportional to $\nabla\theta_\uparrow\ \nabla\theta_\downarrow$ in the Hamiltonian for the phase-shift variables θ_σ, defined by $\nabla\theta_\sigma = 2\pi\rho_\sigma$. This term in the effective Boson Hamiltonian must be transformed away by a Bogoliubov transformation. But as in one dimension, when transformed back into Fermion variables the new dynamical variables, even though their equations of motion have linear dispersion relations, do not correspond to simple Fermion or Boson excitations, and have Green's functions with non-classical exponents. Charge and spin separate, the low-energy spin excitations being Fermions at the original k_F, the charge excitations centering around the spanning vectors $2k_F$.

The actual correlation functions and Green's functions in the two-dimensional case have not yet been calculated: fortunately, many experimental data can be calculated by using the photoemission data[16] to describe a semiempirical fit, and by using the one-dimensional Hubbard model as an appropriate guide to understanding[17]. The actual calculation of physical properties will be described separately.

At present, all experimental observations seem compatible with this point of view, and many puzzling ones receive almost unique explanations.

4. SOME NEW AND OLD RESULTS ON THE 1-DIMENSIONAL HUBBARD MODEL

All of our formal structure will depend on understanding the 1-dimensional repulsive Hubbard model, since this is the model for our "Luttinger Liquid" state and since we believe the fixed point for the 2-dimensional problem maps onto this model in each separate angular momentum channel.

The repulsive Hubbard Model in one space dimension was solved, in principle, many years ago by Lieb and Wu, who calculated the ground state and the spectrum of elementary excitations using Bethe Ansatz methods. Ogata and Shiba have written the Lieb-Wu solution for large, but not infinite, U in a suggestive form:

$$\Psi_{LW} = \sum_{Q} (-1)^Q \ \text{Det} \ ||e^{ik_i x_{Qj}}|| \times \Psi(y_1 \ldots y_\alpha \ldots y_{n/2}) \tag{8}$$

Here Q is a permutation of the N occupied sites $x_1 \ldots x_N$ in a chain of N_a atoms, and the y's are a labeling of the positions $x_1 \ldots x_{N/2}$ of the up spins (we assume $N/2$ up spins, $N/2$ down spins) by simple ordering. Ψ is a solution of the "Squeezed" Heisenberg model in which all hole sites are omitted, which depends on $N/2$ momentum-like variables Λ_α, which are labelled by a set of integers or half-odd integers J_α, $\alpha = 1 \ldots N/2$. The k_i's in turn, are related to a set of integers or half-odd integers I_j. As $U \to \infty$ the k's become evenly spaced. The energy is $\sum \cos k_i$ and the momentum is $\sum k_j = \frac{2\pi}{N_a} \sum (I_j + J_\alpha)$. The excitation spectrum for, say, holes is obtained by dropping out one I and one J, and the energy is the sum of non-interacting spinon (J) and holon (I) contributions. Both spectra are linear for low energies:

$$E_h(k) = v_h \, | \, k_h - 2k_F \, | \quad E_s(k) = v_s \, | \, k_s + k_F \, | \tag{9}$$

Here $v_s \sim 2J = 2t^2/u$ and $v_h \simeq 2t >> v_s$. The rules for shifting the I and J spectra from integer to half-integer assure that the holon and spinon have opposite velocities near k_F (There is a second zero of energy at $3k_F$, where they are parallel). Electron-like spectra are similar with "antiholon" and "antispinon" behaving as such. The spectrum is shown in Fig. 2, and a schematic of the "thermal" density of states —the number of actual one-electron-like or one-hole-like eigenstates—is shown in Fig. 3.

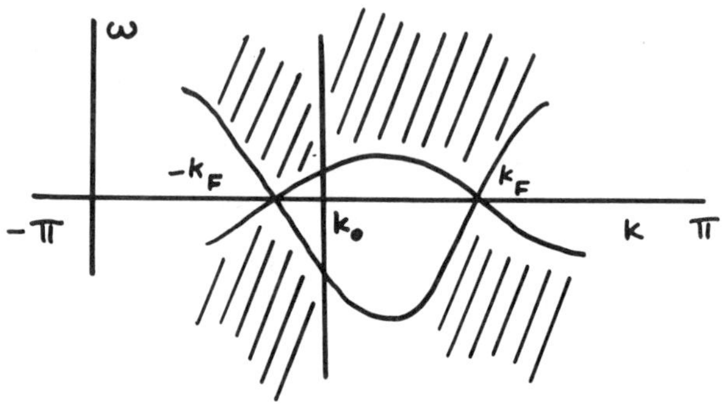

FIGURE 2. F vs. k for holons and spinons in one dimension.

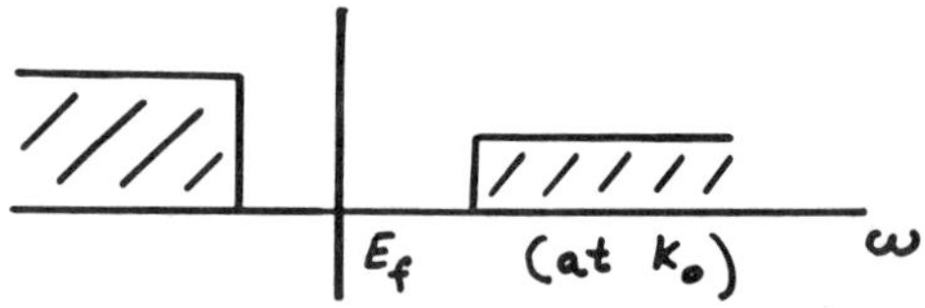

FIGURE 3. Thermal density of states for a given k in the 1d Hubbard Model

Clearly, an angle-averaged thermal density of states would rise linearly with ω, as we postulated as an explanation for the linear tunneling DOS (which, oddly, is probably not caused by this—see Section 5.) One may assure oneself by analytical arguments that the corrresponding $n(k)$ would have only a simple linear cusp at $k = k_F$.

The actual one-particle density of states which would be measured by PES or tunneling cannot be calculated exactly from Lieb and Wu's solutions, and has been the subject of a number of numerical calculations and formal approaches, the latter using one version or another of the "bosonization" methods used to solve the Tomonaga-Luttinger models. Calculations by Ogata and Shiba, and by Parinello et al assure us that the actual n_k behaves like

$$n_k \simeq 1/2 - c|k - k_F|^\alpha \text{sgn}\,|k - k_F|$$

where $\alpha \simeq .125$ for $U = \infty$, less for finite U. This requires, essentially, that there be a $k^{-1+\alpha}$ singularity in $Im\, G_1$ at k_F, since

$$n_k = \frac{1}{\pi} \int_{-\infty}^{0} Im\ G_1(k, \omega) d\omega$$

where

$$\frac{1}{\pi} Im\, G_1\,(k, \omega) = \rho(k, \omega) = \sum_{E_n} |\,(n\,|\,c_{k\sigma}\,|\,0)\,|^2 \delta\,(\omega - E_n) \tag{10}$$

must vanish for $\omega < |E_k^h|$ or $|E_k^s|$ on the respective sides of k_F. Parinello et al, and Haldane claim from "Luttinger liquid" theory that $\alpha = 1/8$ for all n and $U = \infty$. We have calculated this exponent, on the basis of the wave-function (8) and the ΔI shift rules, also, to be 1/8: see below. In any case, it is clear that the same exponent holds for the tunneling density of states

$$\int dk\, \rho\,(k, \omega) \propto |\omega|^\alpha \quad \alpha \sim 1/8 \tag{11}$$

This tunneling DOS has been observed by Dynes and Gurvitch for oxide junctions with YBCO; we interpret these observations as coupling to normal planes. The "Luttinger liquid" (Haldane) or "Tonanaga-Luttinger scaling" theory (Parinello et al) can only be sketched here. One starts by introducing two conjugate variables "ρ" and "φ" to describe a boson field

$$\Psi_{b\sigma}^{+}(x) = \sqrt{\rho_\sigma\,(x)} e^{i\varphi_\sigma(x)}$$

and a Fermion $\psi_\sigma^+(x)$ may be created from this by Jordan-Wigner transformation. ρ_σ, in turn, may be written as the gradient of a second phase variable, $\theta_\sigma(x)$, $\rho_\sigma = 2\pi\nabla\theta_\sigma$, and θ_σ is now to be thought of as a "phase-shift" variable—i.e., we work in terms of a kind of dynamical Friedel theorem, where the number of particles to, say, the right of a given point is given by the local Fermi surface phase shift. φ and ρ are conjugate dynamical variables, so $[\varphi(x), \theta(x)] = i\theta(x - x')$. We focus on excitations which can be described in terms of density waves of constant velocity, in which the local phase and phase-shift are the position and momentum analogs. This representation is well-suited to the strong, unrenormalizable interactions in which we are interested, which can be expressed in terms of finite Fermi surface phase shifts—in fact, one can see that the Hubbard interaction behaves like $U\nabla\theta_\uparrow\nabla\theta_\downarrow$.

Details of a transcription of the large-U Hubbard model into this language, using Lieb and Wu to aid in the determination of the relevant parameters, are given in a later paper. The singular behavior is, it seems to us, unequivocally determined, but, of course, we cannot get any exact idea of multiplicative constants or behavior far from the pseudo-Fermi-surface.

The outcome is that the Green's function near k_F has the form

$$\frac{1}{\pi} Im\, G_1(k,\omega) = \frac{(\omega - qv_s)^\alpha}{(\omega + qv_h)} \tag{12}$$

where $q = k - k_F$. This has singularities at the two cuts in the spectrum at $-kv_h$ and $+kv_s$ and nowhere else, and must be unique except for the apportionment of the powers α and 1 to the two cuts.

Eq. (12) may be normalized by recognizing that $n = 1/2$ at k_F and introducing an upper cutoff, Λ, which must be of order several $\times J$. Then

$$\frac{1}{2} = N \int_0^\Lambda \frac{d\omega}{\omega^{1-\alpha}} = N \frac{\Lambda^\alpha}{\alpha}$$

$$N = \frac{\alpha}{2\Lambda^\alpha}$$

so that

$$\rho(k,\omega) = \frac{\alpha(\omega - qv_s)^\alpha \Delta^{-\alpha}}{2(\omega + qv_h)} \tag{13}$$

We will use this result repeatedly in what follows.

It is perhaps instructive to derive the exponent $\alpha = 1/8$ in the $U = \infty$ case directly from the Lieb-Wu wave function. First let us try to calculate the overlap between a state from which we have removed an electron from site i with spin σ, and the ground state with $N-1$ particles. There are two pieces to this overlap. First, we have removed a spin from the spinon wave function $\psi(y, \ldots y_{N/2})$, which is very close, as shown by Shastry, to a Gutzwiller-projected free electron gas. Removal of an electron and Gutzwiller projection seem likely to commute so that the new state has an overlap proportional to $N^{-1/2}$ with the ground state, simply the overlap of φ_i with a φ_k. This is a simple way of seeing that the spin function renormalizes into a simple but uncharged free Fermion liquid with a conventional Fermi level.

The two ground states, however, differ not only in the removal of one plane wave, but in the fact that because of the Lieb-Wu shift rules every I_j has shifted to $I_j + 1/2$. The reason for this is that even if the extra plane wave is removed at k_F, all of the plane waves have to shift phase by $\pi/2$ to return to a uniform distribution—as we see in Figure 4

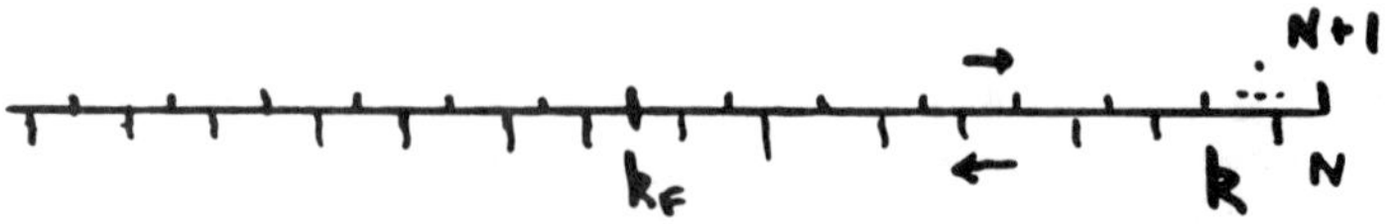

FIGURE 4. Interleaving of k values between N and $N+1$ ground states.

the new distribution interleaves the old one. But this encounters our old friend the orthogonality catastrophe, which leads to an additional overlap reduction by

$$(\, 0N \,|\, c_k^+ \,|\, 0N-1\,) = e^{-(1/2)^2 \cdot 1/2 \ln N} = N^{-1/8}\ . \tag{14}$$

Now a thought-experiment familiar in the x-ray edge problem[18] is carried out: we realize that N and $v_F t$ are the same variable in effect if all waves move at velocity v_F, thus we find that $\langle 0|\psi^+ \; 0,t)\,\psi(0)\,|\,0\rangle \propto t^{-9/8}$ (a second $N^{-1/2}$ comes from the second $\psi^+(0)$. The velocity of the 1/8 power part is the charge velocity. Recognizing that t must enter either as $x - v_n t$ or $x + v_s t$, we recover (12).

Responses to charge perturbations or, for instance, to a pair potential must be dealt with carefully—perhaps the best way being to use variational derivatives of G_1, i.e., Ward identities, to get the correct responses; or one may calculate them directly from Luttinger Liquid Theory. Such calculations are obviously called for.

5. PHYSICS OF THE CUPRATES IN THE LUTTINGER LIQUID MODEL

Attempts to calculate experimental results using the above outline scheme can be based on two ideas which, one assumes, are in the end equivalent. The first and preferable one is to start from first principles using the partial wave free particle representation k, m in which the Green's function is diagonal in the low-energy limit—*i.e.*, the singular terms in the Hamiltonian are correctly handled. In this representation the correlation functions are effectively one-dimensional, in that only ingoing and outgoing waves are mixed. Many quantities—*e.g.*, pairing, NMR relaxation, tunneling, angle-averaged photoemission densities of states, and Raman background intensity may be calculated directly in this representation since they are summed over all channels. All of these are essentially the same as the one-dimensional Hubbard model would give. Transport properties such as resistivity and Hall effect, however, must be calculated in linear momentum representation and involve an extra step of partial wave resolution which we have not yet carried out.

A second technique, much less rigorous, for these transport properties is to fit to the experimental angle-resolved photoemission results and use this to estimate the Green's functions in k representation. This was the method of our original papers[1,2]. Of course the angle-resolved photoemission is then an input as far as its detailed shape is concerned, but its singularities follow from the kinematic D.O.S. which we can estimate; and, in principle, resistivity, etc., can be calculated this way. In so far as quantities calculated by the two schemes agree—and they do—this verifies the Luttinger Liquid Hypothesis for the angle-resolved D.O.S., which is, of course, the most fundamental measurement of the electronic state.

First let us calculate the first category of effects. In the preceding section we have conjectured the one-dimensional one-particle D.O.S.

$$\rho(k,\omega) = \alpha \frac{(\frac{\omega - k v_s}{\Lambda})^{\alpha}}{(\omega + k v_h)} \tag{13}$$

The corresponding space-time correlation function is $(x + v_h t)^{-1-\alpha} \times \theta(x - v_s t)$. This is now to be assumed valid in each channel m, so that we assume $\rho_m(k,\omega)$ is of the same form, and the total D.O.S. is

$$\rho(\omega) = \sum_m \sum_k \rho_m(k,\omega)$$

Let us start with NMR relaxation. It is important to recognize that one of our predictions is that there will be two entirely separate processes, as is actually observed. The first is mediated by the conventional nuclear spin-electron spin coupling I.S., and involves the scattering of a spinon by the nuclear spin:

$$\hbar/T_1 \propto \sum_{kk'} | A_{kk'} |^2 \, f_k(1 - f_{k'}) \times \delta(\epsilon_k - \epsilon_{k'}) \ \propto K^2 T \tag{15}$$

This is a pure Korringa process and should be indistinguishable from the corresponding process due to ordinary electrons. The removal of a spinon from k, m leaves behind the opposite-spin partner and does not affect the charge k_i's.

This process, when drawn in diagram form, is rather strange: it represents the scattering and reabsorption of not an electron, but a spinon acting independently of its accompanying holon. We may, apparently, for incoherent processes like this, divide the Green's function into spinon and holon parts (of course, there can be no net flow of the two because of gauge couplings!) and either can be created so long as it is reabsorbed in the same channel—that is to say, there are preexisting spinon or holon-like fluctuations which violate conventional quantum number conservation notions.

There is also the other possibility, direct coupling of orbital current to nuclear spin. In this case we couple into the orbital current and we can think of it as scattering a holon from k, m to k', m' and back, leaving the spin in a singlet state. The diagram looks like the conventional diagram

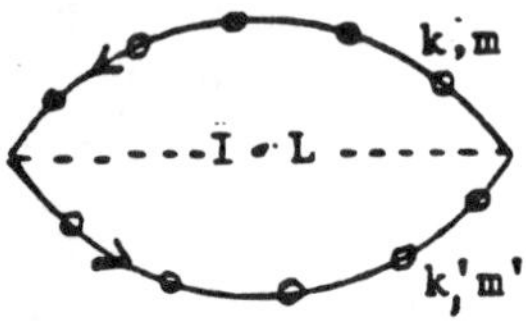

but we have to recognize that as for the spinon the scattering involves change of m as well as k, so that there can be no coherence effects between the two channels. The holon fluctuation is sharply localized and we get an enhanced relaxation.

Our best bet for the holon Green's function is

$$Im\, G_h = \frac{(\omega - k v_s)^\alpha}{\Lambda^\alpha \epsilon_k^h}$$

The interaction is a direct $I \cdot (L)$ with $L_{kk'}^{mm'}$ being an orbital moment matrix element (vanishing for $m = m'$). The Golden Rule give

$$\begin{aligned} \frac{\hbar}{\tau} \propto \int d\omega \sum_{\substack{kk' \\ mm'}} \ & |L_{kk'}^{mm'}\ |^2 \, Im\, G_h(k, \omega) \times \\ Im G_h(k', \omega)\, f(\omega)(1 - f(\omega)) & \\ = N(0)^2 \int d\omega \int d\epsilon_k \int d\epsilon_{k'} & \frac{(\omega - k v_s)^\alpha \ (\omega - k' v_s)^\alpha}{\epsilon_k \, \epsilon_k} \ f(1 - f) \\ \propto T \ln^2 \frac{J}{T} & \end{aligned} \tag{16}$$

neglecting the k-dependence of the wavevector. Fig. 5 gives a typical fit of the sum of (15) and (16) to NMR data in the normal state.

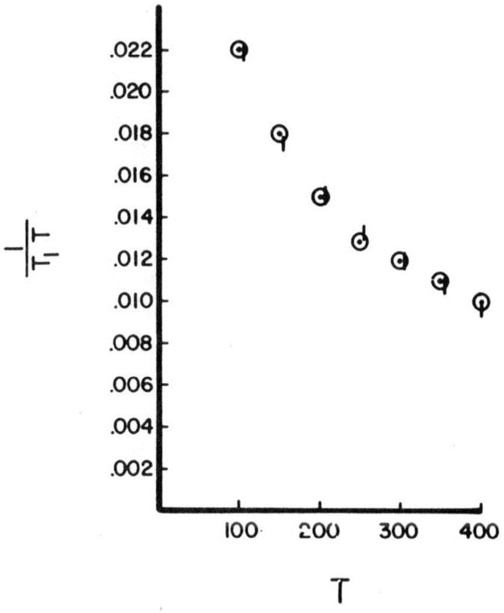

FIGURE 5. Fit to NMR relaxation data on YBCO.

Next we consider the pairing problem. Let us strictly confine ourselves to the interlayer tunneling mechanism of Hsu, Wheatley and Anderson, on the basis of the "experimental constraints" which seem to lead this way. The philosophy is as follows: the interlayer tunneling matrix elements $t^{ab}_{kk'}$ between layers a and b transport full electrons (both holon and spinon), and do not just scatter charge or spin without actual particle transport. We hypothecate a pairing self-energy $\Delta^a(k,\omega)$ in layer a, and ask whether the proximity effect induces a corresponding self-energy in layer b next door. $\Delta^a_k\tilde{\tau}$ is a self-energy insert between a k_σ electron line and a $-k_\sigma$ hole line. The corresponding anomalous part of the Green's function in layer a may be written to first order:

$$\tilde{F}_{k,-k} = \frac{\delta \tilde{G}^a_k}{\delta \Delta^a \tau}\, \Delta^a_k \tau \tag{17}$$

where $\tilde{G}^a$ is the matrix Green's function in layer a. Then the corresponding self-energy insert in layer b comes from the process of tunneling into layer a and back

$$\Delta_b\,\tilde{\tau}(k',\omega) = \sum_{k'} |t^{ab}_{kk'}|^2 \frac{\delta \tilde{G}^a(k')}{\delta \Delta^a} \Delta_a \tilde{\tau}$$

and, cancelling, T_c is determined by

$$\sum_{k'} |t_{kk'}|^2 \frac{\delta \tilde{G}^a}{\delta \Delta^a} = 1$$

Using (13), we find

$$1 = t^2 N(0) \int d\epsilon_k \frac{1}{(\omega^2 - \epsilon_k^2 \pm i\delta)^{1-\alpha}} \frac{\alpha}{J^\alpha}$$

$$1 = \frac{|t|^2 \; N(0)}{2T_c} \alpha \left(\frac{T_c}{J}\right)^\alpha \tag{18}$$

Thus T_c is roughly $(\frac{m_{||}}{m_\perp}) \cdot t_\perp \cdot \alpha$ If so, we are already close to the extreme limit of T_c. It is not at all clear that all normalization factors are correct in (18), but this seems surely the smallest possible value.

One clear implication is that the same mechanism can be operative in one- and two-dimensional organic chains, and should be tested accordingly. Is it significant that pressure brings out the superconducting state in that case?

The $\frac{1}{T}$ dependence of the effective χ_{pair} means that the steepness of the rise of $\Delta^2(T)$ and the decrease of C_{sp} below T_c will both be $\sim 3\times$ greater than in simple BCS. This prediction is in good agreement with experiment. The superconducting state below T_c is not available from this method.

Since there is a great deal of confusion about the meaning of the HWA superconductivity mechanism, we should say a word about dimensionality. Of course, two isolated layers as occurs in YBCO or BISCO 2-layer materials, have a Kosterlitz-Thouless superconducting transition, *as would a single layer if it had an appropriate attractive interaction.* What we postulate is that the coupling between 2 single layers provides that attraction: above T_c, the two exist as two independent, incoherent 2-dimensional Luttinger liquids, while below T_c they are a single coherent system with no low frequency Luttinger liquid behavior. Whether charge-spin separation persists is a separate issue. The issue is not primarily dimensional but of relative sizes of interactions: this regime requires $t_\perp < J << t$.

Angle-averaged density of states is easily calculated from $Im G_1$. This, rather disappointingly, is quite directly simply proportional to ω^α—and this behavior has been observed by Dynes and Gurvitch. We note, however, two possibilities for the "linear $|V|$" behavior. The most interesting is that, as we pointed out, the k_F points are not the only singularities in k-space. As observed by Ogata and Shiba, there also is a singularity at $3k_F$. This is one order of ω less singular than k_F, and gives an averaged density of states $\propto (\omega/J)^{1+\alpha}$, which is of a correct form and order of magnitude to fit the $|V|$ behavior. It will be interesting to study the slight curvature in the Dynes results to see if it is compatible with this expression.

A second mechanism is simply spin-orbit tunneling, which breaks holon-spinon coherence. This will tunnel into the thermal, $|\omega|$ DOS. It seems likely that this will not have the universal magnitude and occurrence seen in the normal state tunneling results.

Many calculations call out to be done on the one-dimensional system: for instance, Raman background scattering, behavior in the presence of a pair gap, tunneling conductance (to explain ρ_c), etc.

At least two experimental phenomena require the transformation back from k, m partial waves to plane waves $e^{i(k_x x + k_y y)}$. These are angle-resolved photoemission and resistivity: in the k, m representation of $G_1(r, r')$, spinon and holon represent circular waves emanating from the origin (but, oddly, one is incoming and the other outgoing!). In the idealized case of a circular Fermi surface they have a Fermi point at k_F and $2k_F$, respectively, for each m, and in the limit of infinite time ($\omega \to 0$) they have become effectively plane waves. This picture can clearly be modified to account for only approximately circular Fermi surfaces, but it is still true that the Fermi surfaces per se are well-defined in plane wave k-space, and occur at the "Luttinger" $k_F(\theta)$ for spinons and the spanning vectors $2k_F(\theta)$ for holons. For finite frequencies or shorter times, however, the curved nature of the waves comes out, and a given electron plane wave k may be composed of holon and spinon travelling in different directions, with

$$\vec{k} = \vec{k}_s - \vec{k}_h \tag{19}$$

the kinematics of the superposition of different plane-wave equivalent holons and spinons is shown in Fig. 6.

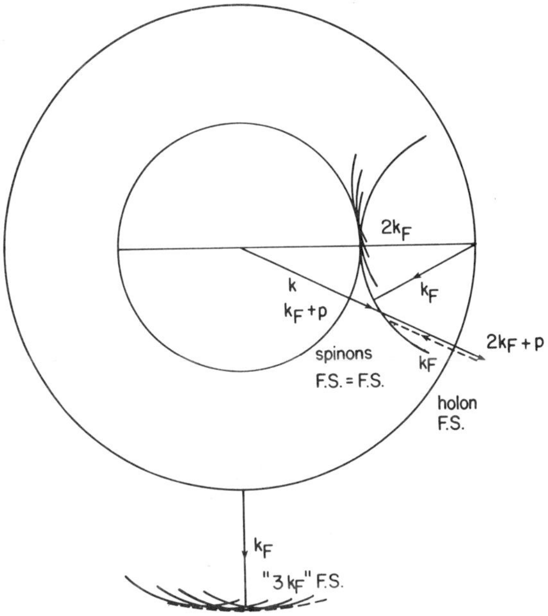

FIGURE 6. Kinematics of possible energy associated with $\vec{k} = k_F + \vec{q}$ in 2 dimensions.

It is clear that for a given k, the colinear holon and spinon represents an osculation singularity, and we expect a cusp singularity when the spinon is on the Fermi surface, at its minimum energy, and the holon has all the extra momentum $|k_F - k|$. We interpret this as the cusp singularity in Fig. 1. The kinematic density of states is inversely related to the area of the triangle formed by the three k's by a Bessel function integral formula. On the other hand, as shown, it is possible for every k to have either a hole-or an electron-like excitation at zero energy, with both holon and spinon momenta on the Fermi surface, up to $3k_F$. $3k_F$ represents a second osculation singularity which may show up in photoemission spectra (these singularities reflect in zone boundaries)

We can sketch the thermal DOS spectrum but we have not calculated the detailed form of Fig. 1 yet, dynamically. Instead, we have taken the schematic fit to Fig. 1 shown in Fig. 7 as a semi-empirical starting point for the calculations of real physical properties. Fig. 7 is constrained by the following requirements.

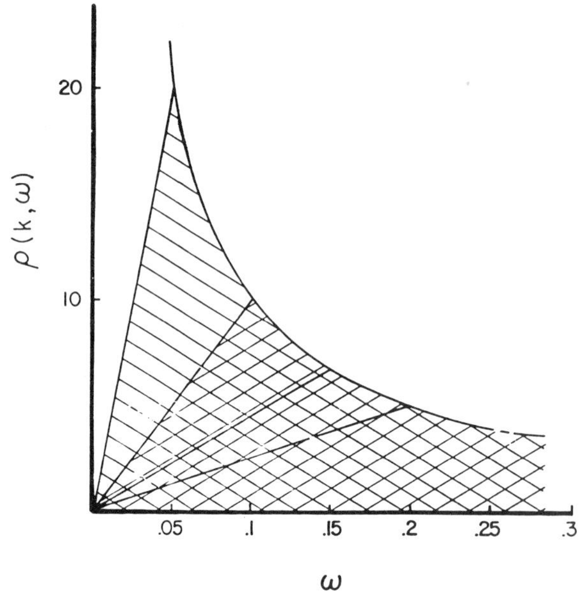

FIGURE 7. Semi-empirical fit to $\rho(k,\omega)$

(1) For $k = k_F$, we are probing the longest-range (in space) correlations, which must come from colinear holon and spinon; this we expect to have the same ω-dependence as the one dimensional case, $\omega^{-1+\alpha}$.

(2) $\omega = \epsilon_k^h$ must be some kind of cusp singularity, since below this point the spinon can no longer be colinear with the holon, and a decreasing volume of configuration space becomes available. In the thermal DOS, in fact, this is a vertical slope singularity.

(3) Finally, at $\omega = 0$ there is only one possible configuration of spinon and holon and the DOS vanishes: again, more sharply in the thermal DOS. We have no

explanation for the observed linear slope as yet, but wonder whether the perfect lattice assumption we have so far been using is really appropriate for the actual substances.

This point is worth emphasizing and discussing. The Fermi surface is established by $Z = 0$ spinons, behaving as a real spinless Fermion, and, since they refer only to the wave-function of the "squeezed Heisenberg" problem, relatively unaffected by potential scattering. The appropriate model wave function, however, is probably

$$\sum_Q (-1)^Q \det \| \varphi_n(x_{Qj}) \| \Psi_{SH}(y_1 \dots y_{N/2})$$

where the φ_n are not plane wave but scattered plane-wave wave functions. They are roughly complete in the neighborhood of k_F, but the mean free path must be assumed very short. Under these circumstances holons away from $2k_F$ (this momentum is sharp and established by the spinon Fermi surface) will be severely scattered and the concept of "colinearity" and of m-values may rapidly become meaningless. A charge density wave can have a sharp frequency—the linear spectrum is relatively unperturbed—but scattering will probably act to connect in non-colinear holons and spinons more effectively.

Thus we have an excellent qualitative, but little quantitative, understanding of the main features of the angle-resolved PES.

Conductivity is another linear-momentum dependent calculation. What is on the face of it clear is that the linear-T, linear ω conductivity is easily understood from the one-particle Green's function. The external vector potential excites only the charge current to be sure, but implicit in the Lieb-Wu theory (though hidden in our independent spinon and holon conceptual structure) is a gauge-like interaction traceable to the constraint structure, enforcing equal real currents of spinons and holons. Thus we must assume the current vertex excites both holon and spinon parts of the electron; then $G_1(r,t)$ measures the amplitude that they arrive together at point r and time t. Any amplitude which is not in this coherent piece is instantly scattered away by the strong disorder potential. For this purpose, then, the apparent breath of G_1, due to the electron's decay into holon and spinon, represents a true electronic mean free time satisfying $\hbar/\tau \simeq E^h_k \simeq$ the greater of T, ω.

In terms of standard diagnostics we arrive at the same conclusion. I am indebted to E. Abrahams for help with this derivation. We write

$$\sigma = 1/\omega\langle j,j\rangle = 1/\omega \sum_{\omega_1,k} G(k,\omega_1+\omega)G(k,\omega_1)$$

$$\simeq \frac{k_F^2}{\omega}\sum_k \int d\nu\, d\nu' \frac{\rho(k\,\nu)\,\rho(k,\nu')}{[\nu-(\omega_1+\omega)]\,(\nu'-\omega_1)}$$

$$\frac{k_F^2}{\omega}\int d\nu \sum_k \int d\nu' \frac{\rho(k\,\nu)\,\rho(k\,\nu')(th\beta(\nu-\omega)-th\beta\nu')}{\nu-\nu'-\omega}$$

and inserting the experimental fit for $\rho(k,\omega)$:

$$\begin{aligned}\frac{\rho(k,\omega)}{\alpha} &\simeq \frac{\omega}{(\epsilon_k^h)^2} \quad \text{for} \quad \omega < E_k^h \\ &\simeq 1/\omega \qquad \omega > E_k^h\end{aligned} \tag{20}$$

we indeed find $\sigma \propto 1/T$. (Not $(\ln J/T)^3$ as proposed in an earlier version.) The c-axis conductivity must be thought of precisely as before,[19] as simply a part of the incoherent background which causes resistivity in the plane.

It is not trivial, really, that the obvious "G_1G_1" result for σ is correct in a strongly-interacting system. It is not at all clear, as we have emphasized, why σ is so large, and requires the spinon mechanism. Very often G_1G_1 does not give the correct answer, as in localizing systems. I feel a plausible explanation of $\rho \propto T$ is vital, but unfortunately can also be obtained for wrong reasons.

One puzzle which still remains and is of great importance is the Hall effect. It is fascinating that the Fermi line is electron-like, almost certainly, yet we see a hole-like Hall effect increasing at low T roughly as $(\ln J/T)^2$. This is, like NMR, an orbital-current phenomenon, and in some way represents a competition between spinon and holon responses; but only that qualitative remark is possible. One very important question is the charge response to phonon or other charge fluctuations. We feel this will encounter the anomalous large holon response, and low to medium frequency phonons may be very heavily damped. This damping will disappear when the gap opens up, possibly explaining the jump in thermal conductitity. Experiment needs to be refined in this area. Perhaps the biggest open question is the nature of the magnetic state below T_c. We cannot rule out either flux phases or weak Néel order.

ACKNOWLEDGMENT

We acknowledge vital discussions with B.S. Shastry, J. Yedidia, A. Georges, D.H. Lee, F.D.M. Haldane, and S. Girvin; also the hospitality of the Aspen Center for Physics and the IBM Yorktown Heights Laboratory.

This work was supported by the NSF, Grant # DMR-8518163 and AFOSR # 87-0392

REFERENCES

1. P.W. Anderson, Proceedings of the Enrico Fermi International School of Physics, *Frontiers and Borderlines in Many Particle Physics*, North-Holland Publ. Co., Varenna, July 1987
2. P.W. Anderson, to be published in Proceedings of *Common Trends in Particle and Condensed Matter Physics* Workshop, Cargese, June 1988, Physics Reports Vol. 184:2–4, Elsevier Science Publ. Co
3. P.W. Anderson, Lecture notes on *Current Trends in Condensed Matter, Particle Physics and Cosmology*, Kathmandu, June 1989
4. E. Stechel, M. Hybertsen, M. Schluter and Jennison, Proceedings of the International Conference – M^{2}S HTSC, Materials and Mechanisms of Super-

conductivity Stanford University, Stanford, CA, July, 1989; also E. Stechel, discussions at the Los Alamos International Conference on the "Physics of Highly Correlated ELectron Systems", Los Alamos, December 1989;

5. C. Varma, Abrahams, A. Ruckenstein, P. Littlewood and Schmitt-Rink, Phys. Rev. Lett. **63**, 1990 (1989), which summarizes many of the ideas of Ref. 2 and 3 in a more available form.
6. Z. Schlesinger and R. Collins, preprint
7. F.D.M. Haldane, J. Phys. C. **14**, 2585 (1981)
8. G. Benfatto and G. Gallivotti, preprint: "Perturbation Theory of the Fermi Surface..."; P.W. Anderson, to be published in Proceedings of *Common Trends in Particle and Condensed Matter Physics* Workshop, Cargese, June 1988, Physics Reports Vol. 184:2–4, Elsevier Science Publ. Co
9. A.M. Luther, Phys. Rev. B **19**, 320 (1979)
10. G.D. Mahan, Phys. Rev. **163**, 612 (1987); P. Noziéres and C. de Dominicis, Phys. Rev. **178**, 1097 (1969)
11. P.W. Anderson, Phys. Rev. **164**, 352 (1967)
12. This is the essence of the procedure of Abrikosov, Gorkov and Dzialoshinskii, (*Methods of QFT in Stat. Mechanics, Sec. 20*, Prentice Hall, 1963) which in the conventional case renormalizes the forward scattering phase shift to $\delta \sim [\ln(\Delta p)]^{-1}$ by use of the Cooper phenomenon. This renormalization procedure cannot work if there are anti-bound states present, I presume, because the assumed "non-singular" parts of the vertex are not harmless, but infinite.
13. T. Hsu and G. Baskaran, unpublished
14. N. Levinson, Kgl. Dansk. Vid. Salsb. Mat. IF **25**, # 9 (1949)
15. J. Friedel, Adv. in Phys. **3**, 446 (1954)
16. C.G. Olson, R. Liu, and D.W. Lynch et al, "High-Resolution Angle-resolved Photoemission Study of the Fermi Surface and the Normal State Electronic Structure of $Bi_2Sr_2CaCu_2O_8$", to be published in Proceedings of the International Conference – M^2S HTSC, Materials and Mechanisms of Superconductivity Stanford University, Stanford, CA, July, 1989; also preprint
17. P.W. Anderson, "Theory of the Excitation Spectrum of the 'Normal' Metal in Cuprate Superconductors: Fit to Experiment and Demonstration of Enhanced Pair Susceptibility", submitted to Nature; P.W. Anderson and Y. Ren, "Theory of the Normal State of Cuprate Superconductors", to be published in Proceedings of MRS Fall Meeting-Symposium M, Boston, Nov. 17, 1989
18. G. Yuval and P.W. Anderson, Phys. Rev. B1, 1522 (1970); P.W. Anderson, "Localized Moments" Lecture notes, Les Houches Summer School, 1968, *Probleme a N Corps Many-Body Physics*, N.Y., Gordon & Breach, pgs. 231-295, C. DeWitt & R. Balian, eds.
19. P.W. Anderson, Z. Zou, Phys. Rev. Lett. **60**, 132 (1988).

Lev P. Gor'kov accepting his honorary doctorate from The Graduate School and University Center in New York City on 5 December 1989.

Award of Doctor of Science *Honoris causa* to Professor Lev P. Gor'kov

by President Harold Proshansky

Lev Petrovich Gor'kov:

One of the first students of the legendary Lev Landau at the Institute for Physical Problems of the Soviet Academy of sciences, you rapidly established your place at the forefront of theoretical condensed matter physics with your extraordinary work connecting the phenomenological theory of superconductivity and the microscopic pairing theory. You went on to formulate an elegant explanation of the effect of minute amounts of magnetic impurities on the superconductive properties of metals and alloys. You were a pioneer in introducing quantum field theory methods in statistical mechanics and condensed matter theory, and your book, co-authored with Alexei Abrikosov and Igor Dzaloshinsky, is, many editions later, still the authoritative text on the subject. Your research on coexisting ferromagnetism and superconductivity broke new ground. Your theory of metallic and superconducting one-dimensional organic solids bridged a gap between chemistry and physics. Your work on the important A-15 superconductors provided a basis for understanding and improving their properties. And recently, you developed, with Volovic, a new symmetry classification for exotic order parameters in superfluids and superconductors. These materials are now under active investigation and your theory may help unravel the riddle of the new high-temperature superconductors.

For your internationally recognized achievements in condensed matter theory,
For the stimulus you have provided as mentor to younger scientists,
For the leadership you have given the scientific community as co-director of the Landau Institute of Theoretical Physics, and
For your principled position on humanitarian issues,

The Graduate School and University Center of the City University of New York is Privileged to award you,

Lev Petrovich Gor'kov,

the degree of Doctor of Science.

TWO BAND MODEL IN HIGH-T_c SUPERCONDUCTORS

L. P. Gor'kov
L. D. Landau Institute for Theoretical Physics,
Academy of Sciences of the USSR,
142432 Chernogolovka, Moscow district, USSR

This presentation discusses the validity of the concept that the conductivity and the other metallic properties in the new high-T_c oxides arise from a doping process. If the concept is correct, the model based on the electron spectrum constructed from a periodic array of localized centers hybridized with a number of itinerant bands (the periodic Anderson model) provides results which may be relevant to the new materials and their properties.

INTRODUCTION

The puzzling properties of the new generation of superconductors first discovered by Bednorz and Müller [1], and especially their high T_c-values, have created numerous proposals concerning the physics underlying the phenomenon. While the phenomenological analyses of their basic superconducting properties leads to the conclusion that new superconductors could be included into the BCS-like scheme (e.g., see in Gor'kov and Kopnin [2]), the apparent interplay between magnetism and superconductivity in the new materials may support the importance of strong Coulomb correlations between electrons. One important circle of theoretical ideas is based on the Hubbard model and its generalizations, as proposed first by Anderson [3]. This approach starts with the Mott insulating state where the mobile charge carriers only appear as the result of some doping procedure (i.e., doping by strontium and oxygen atoms in the La_2CuO_4-based materials, or an intrinsic doping process due to the charge transfer between the CuO-chains and -planes in the YBaCuO-materials). Superconductivity in this picture is due to the additional carriers and arises as the Bose-condensation phenomenon for pairs formed because of a nontrivial ground state symmetry of the system. The strong electron correlations are also responsible for the spin fluctuations in the "spin-bag" mechanism proposed by Schrieffer, although in this picture electrons exhibit a more itinerant character (for recent version see Kampf and Schrieffer [4]).

It is not the goal of this paper to determine the actual superconductivity mechanism in the new superconductors. Therefore we shall not dwell upon the discussion of various possibilities. Instead, in what follows the main emphasis is placed on the possible nature of the normal state of new materials for energy or temperature scales exceeding the T_c-value. We shall also discuss the recent experimental findings concerning the low temperature phase segregation in the oxygen-enriched $La_2CuO_{4+\delta}$ (Jorgensen *et al.* [5], Dabrowski *et el.* [6], Chaillout *et el.* [7]). The choice of La_2CuO_4 as the basic reference material is justified by the idea that, according to the ordinary

Fermi-liquid approach, it would form a noncompensated half-filled band metal , while the genuine stochiometric La_2CuO_4 proves to be an antiferromagnetic insulator. We shall see that these results are of crucial importance for the further interpretation of the superconductivity properties of the new oxides.

THE TWO BAND MODEL

It is now commonly accepted that the additional holes in the conducting CuO_2-planes of the high T_c oxides are situated mainly on the oxygen sites. In the Hubbard model which takes into account nothing but the localized copper sites, the appearance of an additional hole would create the trivalent Cu^{3+} ion (d^8 configuration). This configuration is never seen by different experimental methods. The Extended Hubbard Model, first introduced by Emery [8], is based on the picture of atomic copper *d*-orbitals and oxygen *p*-orbitals. The oxygen orbitals in terms of energy lie higher than the divalent Cu^{2+} (configuration d^9). Cu^{2+} are usually treated as Hubbard centers. Then doping is responsible for the emergence of holes on O^{2-} ions. The transport properties (itinerant states) originate from hybridization between *d*- and *p*-orbitals.

The alternative model (Gor'kov and Sokol [9]) introduces phenomenologically the periodic array of localized Hubbard centers at some energy ε_0 and one or more delocalized bands with dispersion laws $\varepsilon^i(\mathbf{k})$. Therefore this model can be represented in the form of the periodic Anderson Hamiltonian with the hybridization parameter, V:

$$H = \sum_{i,k,\sigma} \varepsilon^i(\mathbf{k}) a^\dagger_{ik\sigma} a_{ik\sigma} + \sum_n \varepsilon_0 d^\dagger_{n\sigma} d_{n\sigma} + \frac{V}{\sqrt{N}} \sum_{nik\sigma} [a^\dagger_{ik\sigma} d_{n\sigma} \exp(i\mathbf{k}\cdot\mathbf{R}_n) + h.c.] \,. \tag{1}$$

Depending upon the relative position of ε_0 and the bottom of the conduction bands three distinct cases are possible, as shown in Fig. 1 (only one band is left for simplicity). In the case of Fig. 1c all localized states are occupied and the ground state is insulating and can be reduced to the Heisenberg spin problem (Gor'kov and Sokol [10,11,12]). At ε_0 above E_f (Fig. 1b) the system would be in the metallic state. Finally, the state shown in Fig. 1c leads to the phase segregation driven mainly by the electronic energy, as discussed in [9]. (Note that in Eq. (1) and in Fig. 1 the energy spectrum is in the hole representation. Inversion of one or more conducting bands also makes it possible to describe the electronic-type superconductors).

PROBLEM OF HEISENBERG SPINS

The physical situation shown in Fig. 1a, as pointed out above, corresponds to the Mott insulating state and by making use of the perturbation expansion in $V/|\varepsilon_0| \ll 1$ can be immediately reduced to the problem of Heisenberg spins interacting via an exchange integral $I(\mathbf{R}_1 - \mathbf{R}_2)$:

$$H_{spin} = \sum_{i>k} I(\mathbf{R}_1 - \mathbf{R}_2)\mathbf{S}_i\mathbf{S}_k \,, \tag{2}$$

where the exchange takes place virtually through the conduction band and $I(\mathbf{R})$ is a positive monotonically decreasing function of the distance R. In other words, $I(\mathbf{R})$ has the antiferromagnetic sign. In the mean field approximation the Neel temperature, T_N, is :

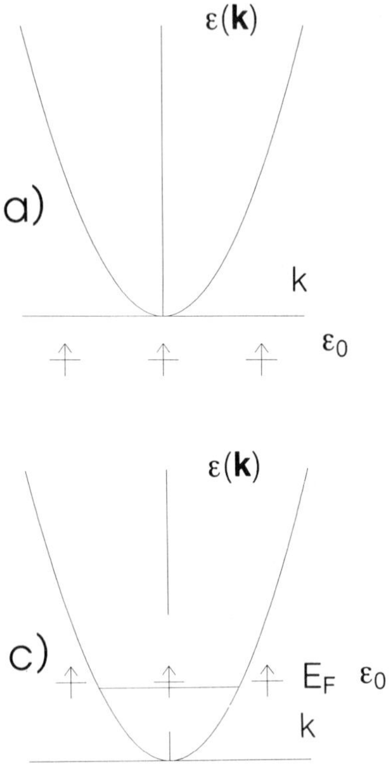

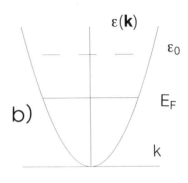

Fig. 1. Three different possibilities for the ground state in the Hamiltonian (1): a) the Mott state; b) the metallic state; c) the phase segregation.

$$T_N \equiv -\frac{1}{2} I(\mathbf{Q}_0) ,$$

where $I(\mathbf{Q})$ is the Fourier transform of $I(\mathbf{R})$ in Eq. (2) for $\mathbf{Q}_0 = (\pi/a, \pi/a)$ and in the 2D-case equals:

$$I(\mathbf{Q}_0) \approx \frac{V^4 a^4 m_B^2}{|\varepsilon_0|} \ln \frac{D}{|\varepsilon_0|} . \tag{3}$$

In Eq. (3) m_B is the conduction band effective mass, a - the lattice parameter and it was already assumed that the level depth $|\varepsilon_0|$ is chosen to be small compared with the bandwidth $D \sim a^{-2} m_B^{-1}$. This conjecture proves to be quite useful. Thus, it is easy to verify that it results in the rise of the long range interaction in Eq. (2) of the asymptotic form:

$$I(R) \approx \frac{V^4 a^4 m_B^2}{2\pi |\varepsilon_0|} \exp\left[-\frac{2R}{R_0}\right] \tag{4}$$

at $R \gg R_0$, where

$$R_0 = (2 m_B |\varepsilon_0|)^{-1/2} \gg a \tag{4'}$$

(see [10-12]. Similar results were also obtained recently by Si *et el.* [13]).

The contribution of second order in V can be presented in the form of the exchange ("s-d") interaction, V_{ex}, between the localized state and a single hole in the conduction band:

$$V_{ex} = \frac{2V^2}{|\varepsilon_0|N}(1/4+\mathbf{S}\vec{\sigma}_{\sigma\sigma'})a^{\dagger}_{p\sigma'}a_{p\sigma} \,. \tag{5}$$

This term starts to play an important role when the carrier concentration in the conduction band becomes finite and the RKKY-interaction needs to be added to the exchange integral in Eq. (2).

When the long range interaction in Eq. (2) has the ferromagnetic sign, has been carefully investigated by Vaks *et el.* [14]. It was shown that the ferromagnetic transition is described in the main approximation by the mean-field approach. For the antiferromagnetic exchange integral the very nature of the genuine ground states far from clear since the system is strongly frustrated. Nevertheless, it is obvious that the temperature of a phase transition, if it exists, retains the same scale T_N as in Eq. (3). Diagram analyses of the generalized magnetic susceptibility

$$\chi_s(\mathbf{q}) = \frac{4\mu_B^2 a^{-2}}{3T} <\mathbf{S_q S_{-q}}> \,, \tag{6}$$

for small $q \le R_0^{-1}$ gives the following result:

$$\chi_s(\mathbf{q}) = \frac{\mu_B^2 a^{-2}}{T + I(\mathbf{q})/2} \,. \tag{7}$$

For the static susceptibility one gets:

$$\chi = \frac{\mu_B^2 a^{-2}}{T + \Theta} \tag{7'}$$

where

$$\Theta \approx \frac{V^4 a^2 m_B}{2\pi\varepsilon_0^2} \sim T_N \frac{D}{|\varepsilon_0|} \gg T_N \,. \tag{8}$$

In other words, Θ exceeds T_N by the factor $(R_0/a)^2 \gg 1$. More generally, it can be proved that the magnetic susceptibility retains its form (7') everywhere below Θ and becomes constant at low temperatures with accuracy $O(a^2 R^{-2})$.

As for the thermodynamic properties, they are determined by the long wave behavior of the correlator in Eq. (6) only at high enough temperatures. In fact, let us rewrite Eq. (2) in the form (for the 2D-case)

$$H^{\lambda}_{spin} = \frac{\lambda V}{(2\pi)^2} \int d^2\mathbf{q}\, I(\mathbf{q}) \mathbf{S_q S_{-q}} \,, \tag{2'}$$

where the parameter λ is introduced to get the thermodynamic identity:

$$\frac{\partial\Omega}{\partial\lambda} = \frac{V}{(2\pi)^2} < \int d^2\mathbf{q}\, I(\mathbf{q}) S_{\mathbf{q}} \mathbf{S}_{-\mathbf{q}} > . \tag{9}$$

(Here and below V is the volume of the system). Making use of Eqs.(6),(7) and (9), one gets:

$$\Omega = \Omega_0 + \frac{3TV}{(2\pi)^2} \int d^2\mathbf{q} \ln\left[1 + \frac{I(\mathbf{q})}{2T}\right],$$

(where $\Omega_0 = -TVa^{-2}\ln 2$.) According to Eq. (8) $I(\mathbf{q})$ is large $(\sim T_N(R_0/a)^2)$ at small $q \le R_0^{-1}$ and is of order of T_N at $q > R_0^{-1}$. Thus, Eq. (9) gives only a small correction of order of $(a/R_0)^2 \ll 1$ to the free spins entropy contribution, Ω_0. In fact, at high enough temperatures the singlet state is favored only for the total sum of all spins in the whole domain with a typical size $\sim R_0$:

$$\sum_i I(\mathbf{R}_i - \mathbf{R}_k)\mathbf{S}_i\mathbf{S}_k \Rightarrow \frac{1}{2}\left[I\left[\sum_i \mathbf{S}_i\right]^2 - I\sum_i \mathbf{S}_i^2\right] . \tag{10}$$

Meanwhile the interaction energy between individual spins responsible for the low temperature ordering is typically of the order of $I \sim T_N$.

The genuine ground state of the magnetic system (2) is not known. Any Neel-type state with the large structure wave vector, $\mathbf{Q}$, would not seriously distort the conduction band in Fig. 1a at $\mathbf{k}=0$. The long-range interaction provides a number of new interesting possibilities including the so-called chiral state (for the discussion see Wen *et el.* [15]). Consider, for instance, the cluster consisting of four spins. For the long-range integral in (2) the interactions between all four spins are identical and the point symmetry group is the tetrahedral group, T_d, which is isomorphic to the permutation group π_4. Therefore, in accordance with Eq. (10), the ground state of this cluster corresponds to the double degenerate singlet representation ^1D. The onset of the long-range chiral order would lift the degeneracy. This results in a force acting on the band electron. This external force can be characterized by a vector Fourier component of the form $\mathbf{k}\times\mathbf{n}$, where $\mathbf{k}$ is responsible for the parity violation and $\mathbf{n}$ (the unit vector, taken perpendicular to the square plane) - for the broken time reversal invariance. One can now construct the single particle Hamiltonian:

$$\hat{h}(\mathbf{k}) = \hat{e}\frac{\mathbf{k}^2}{2m_B} + \lambda([\mathbf{k}\times\mathbf{n}]\sigma) , \tag{11}$$

which splits the band of Fig. 1c into two branches of the form

$$\varepsilon_{1,2}(\mathbf{k}) = \frac{\mathbf{k}^2}{2m_B} \pm \lambda|\mathbf{k}| . \tag{12}$$

The resulting bands of Eq. (12) are schematically shown in Fig. 2.

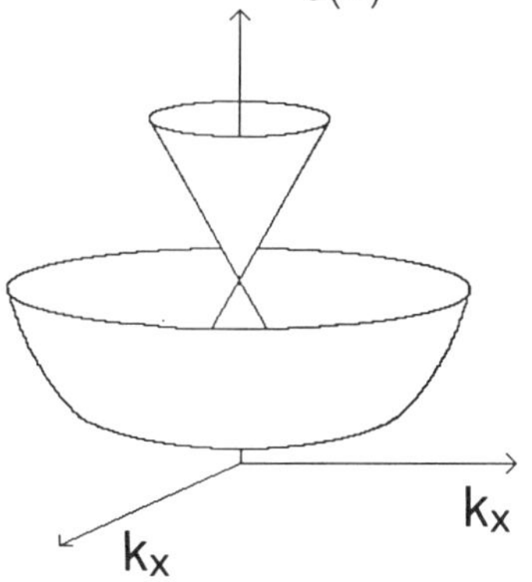

Fig. 2. Splitting of the parabolic conduction band in presence of the chiral order in the magnetic system.

To complete the short overview of some results obtained within the framework of this model let us enumerate the sequence of phenomena which develop in the system with the gradual increase of the number of holes in the conduction band. At very low concentration a single hole tends to homogeneously polarize the surrounding spins via the exchange "s-d" interaction, V_{ex}, from Eq. (5). The corresponding gain in energy is partially compensated by the energy loss due to the antiferromagnetic sign of interaction between spins, according to Eq. (2). This results in the formation of the large size ferromagnetic polaron even for the 2D-case (Gor'kov and Sokol [11,12]). The polaron radius, r_{pol}, is:

$$r_{pol} \sim a \frac{|\varepsilon_0|}{V} \sqrt{\frac{D}{|\varepsilon_0|}} ,$$

and remains large even at $D \sim |\varepsilon_0|$. Therefore their mobility should be very low. The polarons would, probably, play the crucial role in destroying the 3D Neel order.

With the further increase of the hole concentration the polarons start to overlap and the Fermi sea picture for carriers in the conduction band becomes restored. It happens at $p_F r_{pol} \geq 1$. In energy terms this corresponds to the condition that the Fermi energy for holes, ε_F, starts to exceed the strength of the "s-d" exchange field: $\varepsilon_F > V^2 / |\varepsilon_0|$.

The development of the metallic state for holes would provide the background for the appearance of superconductivity in the system. It seems natural to expect that the temperature of the superconducting transition would first increase with the increase of the concentration of holes. In the strontium doped La_2CuO_4 ($La_{2-x}Sr_xCuO_4$), Torrance *et el.* [16] have obtained the T_c concentration dependence at which T_c first increases, reaches a smooth maximum (at $x \approx 0.2$) and then sharply falls down. Above these concentrations the system, being non-superconducting, reveals good metallic properties.

A possible explanation for this type of behavior can be suggested within the scope of the model under consideration (Gor'kov and Sokol [12]). It is based on the important role of the RKKY-interaction at finite hole concentration, especially for the 2D (layered) electronic bands. First, note that in the 2D case the density of states is finite and constant near the bottom of a parabolic band.

Therefore the RKKY-contribution to the Fourier component $I(\mathbf{q})$ of Eqs. (2,2') responsible for the effective spin-spin interaction, is finite and negative at small $\mathbf{q}$. Since the RKKY-interaction is a long range interaction at $p_F a \ll 1$, this contribution appears also in the denominator of Eq. (7), giving rise to an effective increase of the spin pair breaking effects on T_c, in accordance with Eq. (6). This effect would provide a plausible explanation for the diminution of T_c at high enough concentrations.

To conclude this section, let us emphasize that in the above picture the number of holes gradually varies with the increase of dopant concentration. The copper sites are always in the d^9-configuration (Cu^{2+}) possessing the localized moments.

PHASE SEGREGATION AND DOPING

So far we did not discuss the specific mechanism by which additional holes can appear in the conduction band. For clarity, let us keep in mind the simplest case - the solution of oxygen in $La_2CuO_{4+\delta}$. According to Dabrowski *et el.* [6], the low temperature part of its phase diagram has a form schematically shown in Fig. 3.

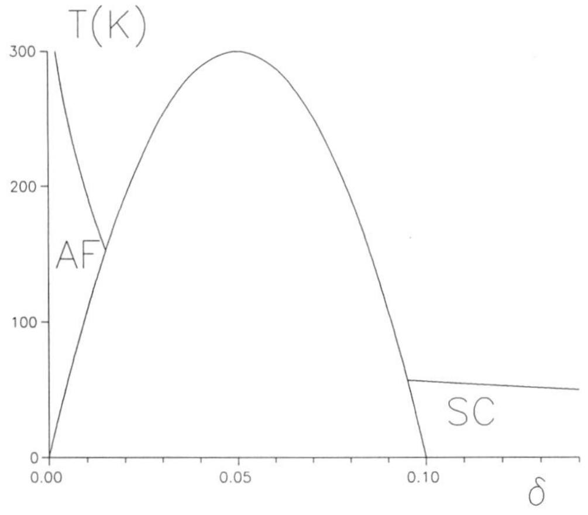

Fig. 3. The schematic view of the phase diagram for $La_2CuO_{4+\delta}$ showing the region of the phase segregation.

Since concentration is the extensive variable the appearance of a miscibility gap would follow from the general thermodynamical arguments. Such a gap manifests the first order transition between two different subphases in the solution. (One of the best known examples is, for instance, the phase separation in the solution of ^3He in ^4He.) Less clear is how the segregation can take place in the solid state at these comparatively low temperatures, since this process must somehow involve the spatial redistribution of oxygen. This, probably, means only that the mobility of the oxygen atom remains rather high. The change of the superconducting volume fraction as measured by the Meissner effect, [depending on a specific treatment] gives an indirect estimate for the typical sizes of domains of single superconducting phase on the scale of the penetration depth (~5000Å). The left hand side of the diagram in Fig. 3 corresponds to the almost stochiometric La_2CuO_4 while for the second (superconducting) phase $\delta \simeq 0.1$, i.e. the excess of oxygen is about 2.5 atomic percent.

Remember in this connection that the doping concept is borrowed from the physics of semiconductors. Dopants (donors or acceptors) in Si or Ge are very shallow Coulomb centers. The effective Bohr orbits considerably exceed the lattice distances

($a_B \approx 10^2 a_0$). Therefore the excess charge is smoothly smeared out and does not introduce large distortions in the surrounding lattice. When the dopant concentration increases the orbits begin to overlap. The impurity band is pushed out from the energy gap into the conduction band. The system manifests the onset of a metallization transition, i.e. carriers become delocalized and can freely move above the homogeneously screened background of the oppositely charged ions. To get this picture one has to combine small values of the effective masses in semiconductors together with a large dielectric constant. Therefore, strictly speaking, there are no reasons to state that similar ideas are applicable to the oxygen ions in La_2CuO_4 at such small concentrations. The oxygen ion may be better described as a lattice defect than a shallow donor center. If this is correct, the transition between two phases could also correspond to a Mott insulator-metal transition driven by a strains induced in the lattice by these defects. Otherwise, at such a transition the moments localized on the copper sites could disappear and all electrons (holes) would be in the conduction band as it is shown in Fig. 1b. In such a case one would have to discuss the superconductivity based on the ordinary Fermi liquid picture. The number of carriers, as measured by the Hall effect, should drastically increase to its nominal stochiometric composition. It is clear that the experimental answer to this question can seriously change our theoretical ideas concerning the new superconductors. It can also be that the numerous observations of the concentration dependence of the transition temperature (e.g. in $La_{2-x}Sr_xCuO_4$, Torrance *et el.* [16]) may be an artifact of the coexistence of two (or more) phases in the sample.

ACKNOWLEDGMENTS

The author is grateful to A. Sokol for many discussions and help with the manuscript. The author thanks Institut für Festkörperphysik , Technishe Hochschule (Darmstadt) for the hospitality and F. Steglich for numerous stimulating discussions of the experimental aspects of the problem. The work was supported by "Sonderforschüngsbereich, 252", Darmstadt/Frankfurt/Mainz.

REFERENCES

1. J.G. Bednorz and K.A. Müller, Z. Phys. B **64**, 188 (1986).
2. L.P. Gor'kov and N.B. Kopnin, Usp. Fiz. Nauk **156**, 117 (1988).
3. P.W. Anderson, Science **235**, 1196 (1987).
4. A. Kampf and J.R. Schrieffer, preprint (1989).
5. J.D. Jorgensen, B. Dabrowski, S. Pei, D.G. Hinks, L. Soderholm, B. Morosin, J.E. Schirber, E.L. Venturini and D.S. Ginley, Phys. Rev. B **38**, 11337 (1988).
6. B. Dabrowski, J.D. Jorgensen, D.G. Hinks, S. Pei, D.R. Richards, H.B. Vanfleat and D.L. Decker, in Proceedings of Stanford Conference, to be published (1989).
7. C. Chaillout, J. Chenavas, S.W. Cheong, Z. Fisk, M.S. Lehman, M. Marezio, B. Morosin, J.E. Schirber, ibid (1989).
8. V.J. Emery, Phys. Rev. Lett. **58**, 2794 (1987).
9. L.P. Gorkov and A.V. Sokol, JETP Lett. **46**, 420 (1987).
10. L.P. Gorkov and A.V. Sokol, JETP Lett. **48**, 547 (1988).
11. L.P. Gorkov and A.V. Sokol, Physica C **159**, 529 (1989).
12. L.P. Gorkov and A.V. Sokol, J. Phys. France **50**, 2823 (1989).
13. Q. Si, J.P. Lu and K. Levin, preprint (1989).
14. V.G. Vaks, A.I. Larkin and S.A. Pikin, JETP **26**, 188 (1968).

15. X.G. Wen, F. Wilczek and A. Zee, Phys. Rev. B **39**, 11413(I) (1989).
16. J.B. Torrance, V. Tokura, A.I. Nazzal, A. Bezinge, T.C. Huang and S.S. Parkin, Phys. Rev. Lett., **61**, 1127 (1988).

Observation of an Hexatic Vortex Glass in Flux Lattices of the High T_c Superconductor $Bi_{2.1}Sr_{1.9}Ca_{0.9}Cu_2O_{8+\delta}$

D. J. Bishop, P. L. Gammel and C. A. Murray
AT&T Bell Laboratories, Murray Hill, New Jersey 07974

D. B. Mitzi and A. Kapitulnik
Stanford University, Stanford, California 94305

We report observation of hexatic order in Abrikosov flux lattices in very clean crystals of the high T_c superconductor $Bi_{2.1}Sr_{1.9}Ca_{0.9}Cu_2O_{8+\delta}$ (BSCCO). Our experiments consist of in situ magnetic decoration of the flux lattice at 4.2 K. Analysis of the decoration images shows that the positional order decays exponentially with a correlation length of a few lattice constants while the orientational order persists for hundreds of lattice constants and decays algebraically with an exponent $\eta_6 = 0.06 \pm 0.01$. Our results confirm recent theoretical speculation that the positional order should be far more sensitive to disorder than the orientational order and that the low temperature ordered phase of the flux lines in these systems might be an hexatic glass.

The behavior of the statics and dynamics of magnetic flux lattices in the high T_c superconductors has triggered considerable experimental and theoretical effort.[1–8] These systems demonstrate a rich and interesting phenomenology. For example, experiments measuring their dynamic response have been interpreted as evidence for glassy response,[1,4] melting[5] and liquid behavior.[6] In addition, the issues raised can be of considerable technological importance because of the possible impact of the lattice on the observed critical currents.[7,8] In this paper we wish to focus on the static properties of the low temperature lattices. In this paper we provide convincing evidence, in agreement with recent theoretical suggestions,[9,10] that the low temperature, low field ordered phase of magnetic flux lattices in BSCCO is a novel phase of condensed matter - the hexatic vortex glass. This phase is characterized by short range (few lattice constant) positional order which decays exponentially and long range bond orientational order which persists for hundreds of lattice constants. We find that the orientational order decays in a power law fashion with a small exponent $\eta_6 = 0.06$ consistent with theory[11] for two-dimensional (2D) equilibrium hexatics and previous 2D simulations.[12] This may indicate a quasi 2D nature of the flux lines upon field cooling. In contrast to what we see here, in the vortex liquid regime seen at higher temperature,[6] one observes blurred flux lines and tracks indicating significant motion of the flux lines on the time scale of the measurement.

The experiments described here consist of the Bitter pattern technique for decorating the magnetic flux lattices in high quality single crystals of BSCCO. The Bitter pattern technique uses very small ferromagnetic particles to sense the field at the surface of a magnetic material. The "smoke" is made by evaporating a magnetic material in a background of inert (helium) gas. The sample is first field cooled to trap

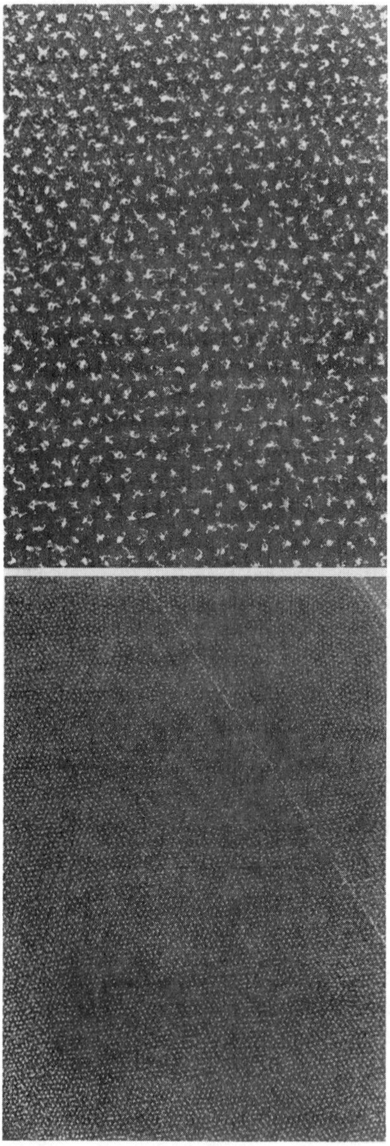

Fig. 1 Shown are two decoration images taken on different regions of a single crystal of BSCCO at a field of 20 Gauss with a decoration temperature of 4.2 K. The field of view in the upper image is 20×30 μm and lower is 80×120 μm.

the flux. Magnetic Ni particles of size ~50Å are then evaporated, thermalized in the gas, and then drift to the sample surface and preferentially decorate regions where there is a magnetic field. Once on the surface, the particles are held immobile by the Van der Waals forces. The sample is then warmed to room temperature and examined using scanning electron microscopy. The technique is limited to low fields and only senses the magnetic structure at the sample surface. Details of the technique as we implement it are given in previous work and references therein.[2]

The samples we have used for this study consist of high quality, single crystals of $Bi_{2.1}Sr_{1.9}Ca_{0.9}Cu_2O_{8+\delta}$ (BSCCO). This material was chosen because of its low pinning as evidenced by μSR experiments[13] and the absence of twinning domains which have been shown to pin and orient the flux lattice in YBCO.[2,18]

The single crystals were grown using a directional solidfication process as described in Ref. 14. The crystals are typically in the form of thin sheets with basal plane dimensions of up to several cm^2 and thicknesses up to 100 μm. The lattice constants, $a_0 = 5.413(2)$ Å, $b_0 = 5.411(3)$ Å, and $c_0 = 30.91(1)$ Å, were determined using a four-circle x-ray diffractometer. As is typical in these materials, an incommensurate periodicity of $4.7(1)\,b_0$ was observed along the *b*-axis. The crystals, as extracted from the melt, demonstrated a large, sharp (< 5K 10-90% transition width) Meissner transition with an onset at 88K, indicative of a homogeneous, bulk superconductor.

Shown in figure 1 are two decoration images taken on a single crystal at different locations on its surface. The crystal surfaces used in these experiments were freshly cleaved in air just before the decoration. The nominal magnetic field used was 20 Gauss applied parallel to the *c* axis of the crystal and the decorations were done at 4.2 K after field cooling. The upper picture in figure 1 has a field of view of 20×30 μm and the lower picture has a field of view of 80×120 μm. Each flux line is marked with a clump of magnetic particles which shows up as a white spot in the photos. As reported previously,[2] the flux quantum as measured here and elsewhere is consistent with $hc/2e$ and excludes the value hc/e.

The decoration images in figure 1 were digitized with a video frame grabber, and the centers of the largest clumps of particles were located by image analysis algorithms described in detail elsewhere.[15] Care was taken to correct for any image distortion introduced by video camera. The positions of the largest clumps, assumed to correspond to the location of the vortex lines, were then used to analyze the translational and orientational order of the vortices. We estimate the accuracy of locating the centers of the vortices is comparable to or better than a radius of a large clump which is $0.14\,a$ where a is an average nearest neighbor spacing of the large clumps.

Shown in figure 2 is a Delaunay triangulation[16] for the upper image presented in figure 1 consisting of 405 centers. In this construction each flux line is represented by a vertex and the lines emanating from it are the bonds to its uniquely defined nearest neighbors. In the figure the nearest neighborhoods of non-six-fold coordinated flux lines are shaded in order to reveal the defect structure.

The most relevant 2D defects for the destruction of long range translational order are free dislocations[11] or isolated pairs of 5- and 7-fold coordinated particles, interstitials that exist as 4-fold coordinated particles with a pair of 7-fold coordinated neighbors and vacancies which often show up as 8-fold coordinated centers with two 5-fold coordinated centers. We observe in the large scale image containing 6763 vortices an approximately equal interstial and vacancy density of roughly 0.4% each and free dislocation density of approximately 1.5%, counting clusters with non-zero local Burger's vector. The apparent waviness of the lattice rows in figure 2 could be caused by inherent inaccuracy in the decoration technique and in locating the centers of vortices or by genuine random pinning distortions, but does not destroy the long range bond orientational order, or contribute significantly to the fast decay of translational order with distance.

The net concentration of dislocations observed is much larger than that which could be caused by a gradient in the field (or density of vortices). For our images this

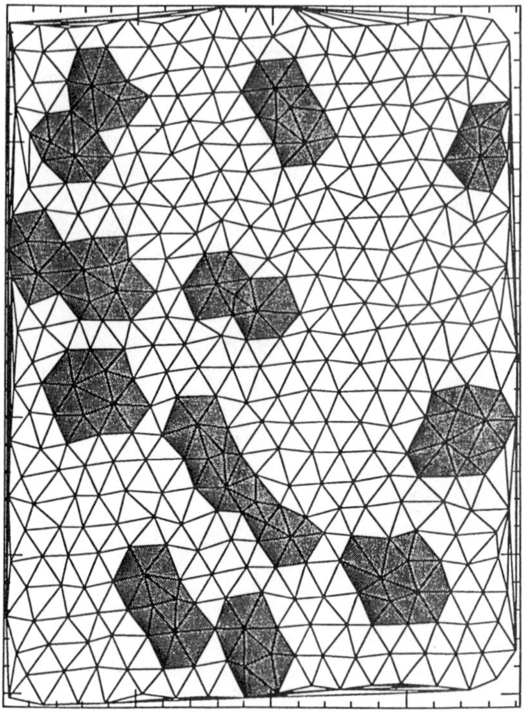

Fig. 2 Shown is a Delaunay triangulation for the upper, smaller scale image in figure 1. In this construction, each flux line is represented by a vertex and the lines emanating from it are the bonds to its nearest neighbors. In the figure the nearest neighborhoods of the non-sixfold coordinated flux lines are shaded in order to reveal the defect structure. An isolated dislocation is located in the upper right, and a cluster with net Burger's vector is located at the left, just above the center of the image.

gradient is measured by counting vortices and is found to be < 1(1)% in the short image direction and < 2.5(1)% in the long image direction. This measured density gradient will introduce a density of unbound dislocations that is ~0.04% with an average separation of ~40*a*. The intrinsic free dislocation density average separation is ~10*a*, which is quite close to the measured translational correlation length.

The image in figure 2, and a similar defect analysis of the larger scale image of figure 1, are strikingly reminiscent of that obtained for 2D colloidal hexatics.[17] If one sights along rows of particles, one can clearly see that there exists long-range orientational order, while the dislocations, interstitials, and vacancies in the defective neighborhoods limit the positional order to only a few nearest neighbors. The analysis can be made quantitative by extracting from the images shown in figure 1 the orientational and positional correlation functions.

Figures 3 and 4 are the central result of this paper. Shown in figure 3 is a plot of the static pair correlation function $g(r)$ and the static orientational correlation function $g_6(r)$ for the upper image in figure 1. The pair correlation function is defined as the quantity:

$$g(r) = \frac{1}{2\pi N} \int \int_A \rho(\mathbf{r}')\rho(\mathbf{r}' - \mathbf{r}) d^2\mathbf{r}' d\phi \qquad (1)$$

for the continuum limit, where $\rho(\mathbf{r})$ is the vortex density at $\mathbf{r}$, ϕ is the azimuth angle,

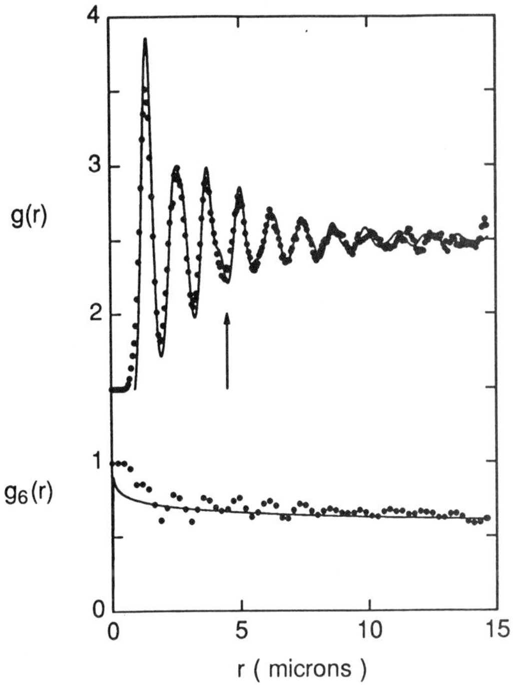

Fig. 3 Shown is a plot of the static pair correlation function $g(r)$ (top) and the static orientational correlation function $g_6(r)$ (bottom) for the upper, smaller scale image in figure 1. An estimate of the orientational averaged translational correlation length ξ is labeled by the arrow.

and N is the total number of vortices in the image area A. This function gives the probability that, given a vortex line at the origin, one will locate another line at any distance r (in the $x-y$ plane, $\perp$ to c). The orientational correlation function $g_6(r)$ used here is the correlation function of the orientational order parameter:[11]

$$\psi_6(\mathbf{r}) \equiv \frac{1}{n_j} \sum_j e^{i6\theta_j} \qquad (2)$$

for a vortex located at $\mathbf{r}$ where θ_j is the angle with respect to the x axis of the bond to the jth nearest neighbor, summed over all n_j nearest neighbors, determined by a Voronoi polyhedron analysis.[16] It is defined as the quantity:

$$g_6(\mathbf{r}) \equiv \frac{1}{2\pi g(\mathbf{r})} \int\int_A \psi_6(\mathbf{r}' - \mathbf{r}) \psi_6(\mathbf{r}') d^2\mathbf{r}' d\phi \qquad (3)$$

Its decay measures directly the fall off of bond-orientational order of the system.

An estimate of the translational correlation length ξ (marked by the arrow on figure 3) was determined by a fit to $g(r)$ using an exponential decay envelope multiplied by the pair correlation function of a broadened perfect triangular lattice. The fits to an exponential are good, and for this image and direction we obtain $\xi \sim 3.6\,a$. As will be discussed later, there is a small anisotropy in the real space lattices with a corresponding anisotropy in the correlation lengths.

The plot of $g_6(r)$ shown in figure 3, is indicative of slowly decaying, quasi-long range orientational order. Clearly the orientational order persists throughout the

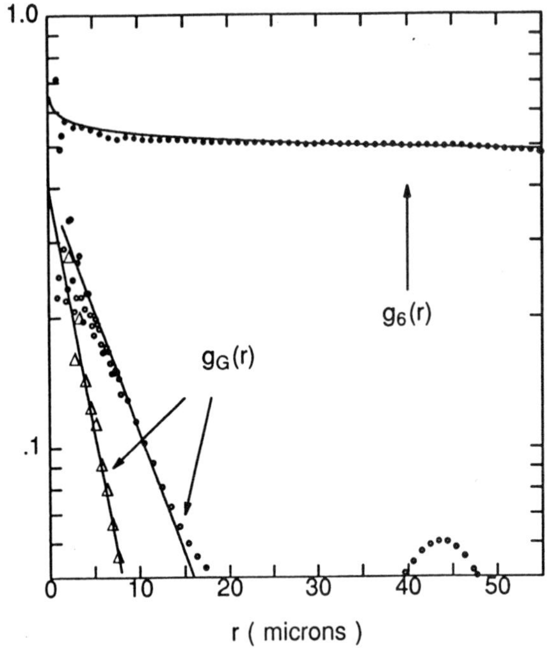

Fig. 4 Shown are two extreme translational correlation functions $g_G(r)$ and the orientational correlation function $g_6(r)$ on a semi-long plot for the lower, larger scale image in figure 1.

entire image. In order to examine the decay of this order over longer length scales, the lower image in figure 1 was analyzed. Shown in figure 4 is a plot of the orientational correlation function $g_6(r)$. Also shown for comparison is the translational correlation function $g_G(r)$ for two different directions in the lattice, also for larger scale image. The translational correlational function $g_G(r)$ measures the decay of the translational order parameter:

$$\psi_t(\mathbf{r}) = e^{i\mathbf{G}\cdot\mathbf{r}} \tag{4}$$

at a location **r**, where **G** is any reciprocal lattice vector.[11] It's definition is identical to Eq. (3) above, with ψ_t substituted for ψ_6. Its decay directly measures the fall off of translational order of a specific wavelength. The orientational order is sufficiently long range in these images that positions of the first reciprocal lattice vectors $|G_0|_i = (4\pi)/(a_i 3^{1/2})$ were easily located to an accuracy better than 0.05% by performing a Fourier transform of the data. Here a_i is the nearest-neighbor separation in one of three symmetry directions of $i = 1, 2, 3$ of the lattice. We have plotted in figure 4 the extreme cases for G_0 with the shortest and longest correlation lengths.

We find a slight anisotropy of 9(1)% and 8(1)% in the *ab* plane of the lattice for the smaller and larger scale image respectively, slightly smaller than that previously observed on the YBCO *ab* plane.[19] The translational correlation function is a better measure of the decay of translational correlations than $g(r)$ as it is effectively a one-dimensional correlation and automatically takes the lattice anisotropy into account. We see for three different directions, a fast exponential decay with translational correlation lengths of 5(1), 6(1), and 8(1) average nearest neighbor spacings in the three directions for the smaller area image and 3.6(5), 4.6(5) and 7(1) average nearest neighbor spacings for the larger area image.

The longer translational correlation lies in the direction of the longest reciprocal lattice vector, which is rotated 90° from the longest real space vortex spacing. In the larger image of figure 1, the longest reciprocal lattice vector is in a direction that is 31.6° counterclockwise from the obvious line defect in the larger scale image in the lower right. As the ratios of the effective masses range between $m_a/m_b \approx 1.5—2.2$ for this material[20] we expect a distortion of the perfect hexagonal real space lattice by the ratio of penetration depths $\lambda_b/\lambda_a = (m_a/m_b)^{1/2} = 1.2—1.5$, with the long direction along the b axis.[19,21] We have X-rayed the sample and find that the BSCCO crystal a axis is rotated 6.1 ° counterclockwise from the defect with the flux lattice real space direction rotated 2.5 ° counterclockwise from the defect. Then the b axis is 83.9 ° rotated clockwise from the line defect. We conclude the defect is not obviously a twin boundary, for which the angle should be 45 °. Of three different reciprocal lattice directions, the direction of the shortest translational correlation length is 6.7 ° clockwise from the b direction of the lattice. The closest real space direction of the average lattice is 2.5 ° removed in angle from the defect as well, so that it's effect on the order of the lattice seems to be quite local.

Figure 4 suggests several things. The first is that, while the translational order is decaying exponentially on a scale of 4–10 a, the orientational order is decaying only very slowly. A fit to an exponential decay for this curve yields an effective orientational correlational length of 250a. The second point is that the data are consistent with a power low decay where $g_6(r)$ varies as $r^{-\eta_6}$ with the exponent η_6 having the value 0.06(1). We have analyzed in a similar fashion, four regions on three different crystals. All gave similar results to that quoted above. For η_6 we have found the values of 0.06(1) and 0.06(2) from different places on sample 1 at 20 Gauss (Figs. 3 and 4), 0.05(2) from sample 2 at 80 Gauss and 0.30(5) from sample 3 at 10 Gauss. For ξ we find 2.5(5), 3.2(5), 3(1) and 3.2(5)nearest neighbor lattice constants for the four sample studied, as determined from the angularly averaged $g(r)$. The larger value of η_6 found for 10 Gauss data suggest that as H_{c1} is approached, the lattice is more disordered with a more liquid like response with shorter range orientational order.

The present work needs to be compared and contrasted with previous magnetic decorations on high T_c single crystals. Our early work[2] on both YBCO and BSCCO observed flux lattices which showed a tendency for triangular order but both the positional and orientational order for those lattices was found to decay approximately exponentially with correlation lengths of a few lattice constants. Such an early image is shown in figure 5. Shown is a decoration image taken on BSCCO at a field of 20 Gauss at 4.2K. What is seen is both the orientational and positional order decaying over a few lattice constants. In the present work, on much cleaner samples, the positional correlation length is similar to the previous work but the orientational length has increased dramatically. This suggests that for our best materials we are now in a intermediate regime of disorder for the lattices in these materials. Dolan, *et al* [18] have also seen very high quality flux lattices in both YBCO and BSCCO single crystals with real space images similar to those presented here, suggesting long range orientational order, although the type of quantitative analysis we have presented here has not yet been done on those images.

After the idea of an hexatic phase was first introduced by Nelson and Halperin,[11] the issue of disorder and decay of positional and orientational order was

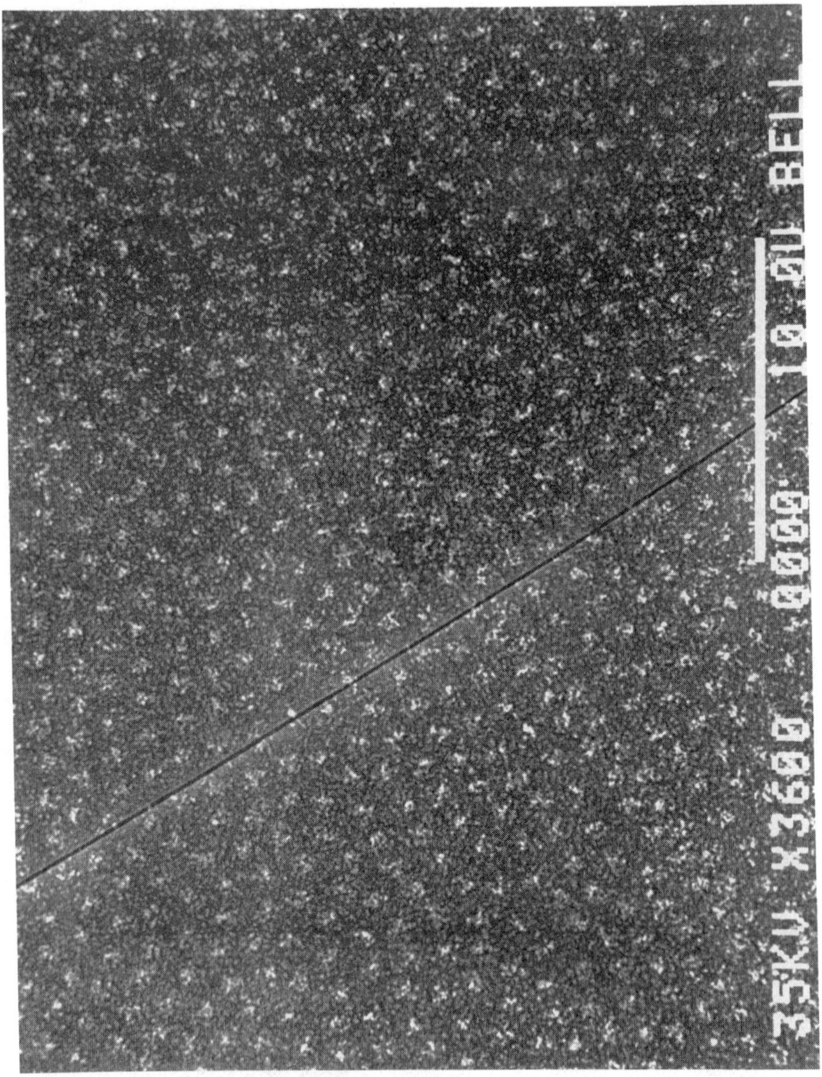

Fig. 5 Shown is a decoration on an early BSCCO crystal at a field of 20 Gauss at a temperature of 4.2K. Both the orientational and the positional order decay over a few lattice constants.

discussed by a number of workers in a variety of contexts. Nelson and co-workers[12] discussed the effects of disorder on the orientational and positional correlations in their simulations on planar arrays of spheres. They were the first to point out that random disorder couples much more strongly to positional than to orientational correlations.

More recently many of these idea have been reexamined in the context of 3D superconducting flux lattices. Fisher[22] has argued that, in 3D, the low temperature vortex phase dominated by strong pinning disorder will be a glass. Wordenweber and Kes[23] and Brandt[24] have suggested that in the regime of strong pinning, with quenched impurity disorder, that one would find a glassy state consisting of entangled flux lines. Nelson and co-workers[3,10] have suggested the possibility of a 3D fluid of entangled flux lines which is driven by large thermal fluctuations in the high T_c materials rather than by the quenched in disorder. They have suggested a phase diagram with a low temperature ordered phase, an intermediate entangled hexatic flux liquid and a high temperature isotropic entangled liquid phase. There are a variety of experiments which are consistent with various aspects of this phase diagram with both melting[5] and an intermediate hexatic phase[25] having been observed.

Nelson and Halperin's theory[11] for 2D hexatics with thermal disorder, as opposed to those with the quenched-in disorder found here, indicate that this exponent η_6 should be 0.25 at the hexatic to fluid transition, decreasing to zero at the hexatic to crystal transition. This is consistent with observations in 2D colloidal[17] systems with thermal disorder. Simulations in systems of 2D arrays of hard spheres[12] with trapped random dislocations, find a value of $\eta_6 = 9c/\pi$, where c is the fraction of dislocation cores. This relation was obtained through a cumulant expansion of $g_6(r)$ within continuum elastic theory. In our case if we take as an estimate of c the fractional area of Voronoi polyhedra in which there is net Burgers vector (~ 1.5%), then $\eta_6 \sim 0.04$, which is close to our observed result. This simple analysis shows that the dislocations are responsible for the large part of both the short range positional order as well as the quasi-long range orientational order.

The detailed nature of the low temperature ordered phase is the subject of a recent paper by Chudnovsky.[9] In that work he suggests that in the limit of strong pinning disorder, that the in plane positional order of vortices should be only short ranged while the orientational order should be long ranged (infinite) leading to an hexatic vortex glass. However in that work, dislocations have not been explicitly been put in. In work done simultaneously and independently by Marchetti and Nelson,[10] the effects of dislocation loops in the 3D vortex lattice of combination screw and edge character have been included. They suggest that, similar to Chudnovsky, the positional order should fall off exponentially and the orientational order will be long ranged. Within the framework of the present theoretical understanding, there are at least two way one can understand our experimental observations. In the first scenario, the hexatic vortex glass, which we see at low temperatures is a vestige of high temperature hexatic liquid state which then gets quenched-in as the temperature is lowered due to the low mobility of the edge dislocations (or vortices). In the second scenario, favored by Chudnovsky, the hexatic glass is not a non-equilibrium vestige of a high temperature state but is the true low temperature ground state which is produced by the competition between disorder in the system and the repulsive interactions between the lines.

In the first scenario, the apparent 2D-like algebraic decay of orientational order we observe could come about if the screw dislocations, which have no Peierls stress,[24] were expelled upon cooling, leaving behind rigid, long edge dislocations possibly piercing the entire length of the sample. The long edge dislocations do experience Peierls tress, so would be quenched in. If the vortex lines are sufficiently rigid at low fields, they would effectively be a 2D system. If the low mobility of the dislocations were, on the other hand, caused by trapped entanglements in the bulk of the sample, the system would remain effectively 3D with long range orientational order.

In the second scenario, in which random pinning is important for generating the glass, dislocations, interstitials, and vacancies will presumably be required to explain the observed algebraic decay of orientational order. At the moment it is not clear to us which scenario is the appropriate one.

In conclusion we have presented data which we feel provides convincing evidence for the observation of an hexatic vortex glass in superconducting flux lattices in the high T_c superconductor BSCCO. In this state the positional order falls off exponentially with a correlation length of a few lattice constants while the orientational order persists for hundreds of lattice constants and has algebraic long range order.

The authors would like to thank G. J. Dolan, D. Fisher, M. P. A. Fisher, D. Huse, P. Littlewood, and D. Nelson for many stimulating discussions. Two of the authors would like to acknowledge support received from AT&T (D.B.M) and the NSF (A.K.). This work was supported in part by the Joint Services Electronic Program under grant No. N00014-84-K-0327 and by the Stanford Center for Materials Research through the NSF Department of Materials Research.

REFERENCES

1. A. C. Mota, D. Visani, K. Muller and J. G. Bednorz, Phys. Rev. B36, 4011 (1987).
2. P. L. Gammel, D. J. Bishop, G. J. Dolan, J. R. Kwo, C. A. Murry, L. F. Schneemeyer, and J. V. Waszczak, Phys. Rev. Lett. 59, 2592 (1987).
3. D. R. Nelson, Phys. Rev. Lett. 60, 1973 (1988); D. R. Nelson and S. Seung, Phys. Rev. B39, 9153 (1989).
4. Y. Yeshurun and A. P. Malozemoff, Phys. Rev. Lett. 60, 2202 (1988).
5. P. L. Gammel, L. F. Schneemeyer, J. V. Waszczak, and D. J. Bishop, Phys. Rev. Lett. 61, 1666 (1988).
6. R. N. Kleiman, P. L. Gammel, L. F. Schneemeyer, J. V. Waszczak, and D. J. Bishop, Phys. Rev. Lett. 62, 2331 (1989).
7. M. A. Dubson, S. T. Herbert, J. J. Calabrese, D. C. Harris, B. R. Patton, and J. C. Garland, Phys. Rev. Lett. 60, 1061 (1988).
8. B. van Dover, L. F. Schneemeyer, E. M. Gyorgy, and J. V. Waszczak, Applied Phys. Lett. 52 1910 (1988).
9. E. M. Chudnovsky, Phys. Rev. B40, 11357 (1989).
10. M. C. Marchetti and D. R. Nelson, Phys. Rev. B41, 1910 (1990).
11. B. I. Halperin and D. R. Nelson, Phys. Rev. Lett. 41 121 (1978); D. R. Nelson and B. I. Haperin, Phys. Rev. B19, 2457 (1979).
12. D. R. Nelson, M. Rubinstein and F. Spaepen, Phil. Mag. A46, 105 (1982).
13. D. R. Harshman, R. N. Kleiman, D. Mitzi and A. Kapitulnik, to be published.
14. D. B. Mitzi, L. W. Lombardo, A. Kapitulnik, S. S. Laderman, and R. D. Jacowitz, Phys. Rev. B (1990, accepted for publication).
15. D. H. Van Winkle and C. A. Murry, J. Chem. Phys. 89, 3385 (1988).
16. F. F. Preparata and M. L. Shamos, "Computational Geometry, an introduction",

(Springer Verlag, New York, 1985).
17. C. A. Murray, D. H. Van Winkle and R. A. Wenk, "Phase Transitions" in press; C. A. Murray, W. O. Sprenger and R. A. Wenk, to be published.
18. G. J. Dolan, G. V. Chandrasekar, T. R. Dinger, C. Field, and F. Holtzberg, Phys. Rev. Lett. 62, 827 (1989).
19. G. J. Dolan, F. Holtzberg, C. Field, and T. R. Dinger, Phys. Rev. Lett. 62, 2184 (1989).
20. S. Martin, A. T. Fiory, R. M. Fleming, L. F. Schneemeyer and J. V. Waszczak, Phys. Rev. Lett. 60, 2194 (1989); S. Martin, private communication.
21. L. J. Campbell, M. M. Doria and V. G. Kogan, Phys. Rev. B38, 2439 (1988).
22. M. P. A. Fisher, Phys. Rev. Lett. 62, 1415 (1989).
23. R. Wordenweber and P. Kes, Phys. Rev. B34, 494 (1986).
24. E. Brandt, Phys. Rev. B34, 6514 (1986).
25. T. Worthington, F. H. Holtzberg and C. A. Field, to be published.

MAGNETIC ORDER OF THE QUASI-2D-ANTIFERROMAGNET WITH FRUSTRATING IMPURITIES ($La_{2-x}Sr_xCuO_4$)

L. I. Glazman
Institute of Problems of Microelectronics Technology and High-Purity Materials, USSR Academy of Sciences, Chernogolovka, Moscow district 142432, USSR.

A.S.Ioselevich
L. D. Landau Institute for Theoretical Physics, USSR Academy of Sciences, Kosygina street 2, Moscow 117940, USSR.

The effect of impurity-induced states on the long range order in a lamellar antiferromagnet (AF) is studied; and the magnetic phase-diagram of lightly doped $La_{2-x}Sr_xCuO_4$ is proposed. It is shown that long-range magnetic perturbations and the layered structure cause the shrinkage of AF domains on the phase diagram and lead to the reentrant AF transition. A nonmonotonous dependence of the correlation length ξ_{2D} on temperature T is obtained; the dependence $\xi_{2D}(x)$ is exponential for high T, and $\xi_{2D} \sim x^{1/2}$ for low T.

Introduction

The antiferromagnetism of high-T_c copper oxides is caused by the superexchange between localized copper spins. The high anisotropy of these compounds provides the small value of an interplane exchange integral I' as compared with the in-plane one I [1]. Doping creates holes in the p-shells of the oxygen ions of CuO_2 -planes [2]. At low temperatures and low doping level x these holes are bound to acceptors. This is clear from the Mott type of conductivity [3].

The localized holes produce magnetic defects, which frustrate the order in AF matrix [4]. In fact, the AF order in $La_{2-x}Sr_xCuO_4$ is destroyed by a surprisingly small doping: $x_2 = 0.02$ [1]. We show in this paper that the strong doping-sensitivity of magnetic order may be due to the long range of magnetic perturbations combined with the strong anisotropy.

A localized hole and a number of adjacent copper-spins form a magnetic defect [4,5] of size D. The interaction of a defect with the outer spins produces a perturbation in AF matrix which drops off as r^{-1}. This kind of perturbation was considered by Villain [6] in connection with a spin-glass problem and later was applied to LaSrCuO by Aharony et al. [4]. It is the long-range interaction and the small parameter I'/I that makes it possible [7] to obtain the boundary of the 3D-AF-region on a T-x diagram (Fig. 1), i.e. to determine the x-dependence of Neel temperature $T_N(x)$. The long-range order is greatly distorted by acceptors only if $x > x_1 \sim I'/I$. For $x > x_2 \sim [D^2 \ln(I/I')]^{-1}$ the long-range order is destroyed at all temperatures. Since the domain size $D \geq 1$ and $I/I' \gg 1$, the value $x_2 \gg x_1$. In a wide interval $x_1 < x < x_2$ a reentrant transition arises; its origin is the magnetic defect-defect correlation enhanced at low T.

The impurity-induced magnetic phase transition is *not of the percolation type:*

$x_2 \ll x_c$ (where $x \sim D^{-2}$ is a threshold for the percolation through the defects). At this point our concept of the transition differs from that of Aharony et al. [4].

The defect-induced disorder modifies the dependence $\xi_{2D}(T)$. Under the finite x this dependence becomes nonmonotonous. Such behavior is of particular importance, because it forms the source for the reentrant transition. At low temperatures one has $\xi_{2D} \sim x^{-1/2}$, which agrees with the experiments [1]. At relatively high temperatures (above the AF transition) the function $\xi_{2D}(x)$ is exponential.

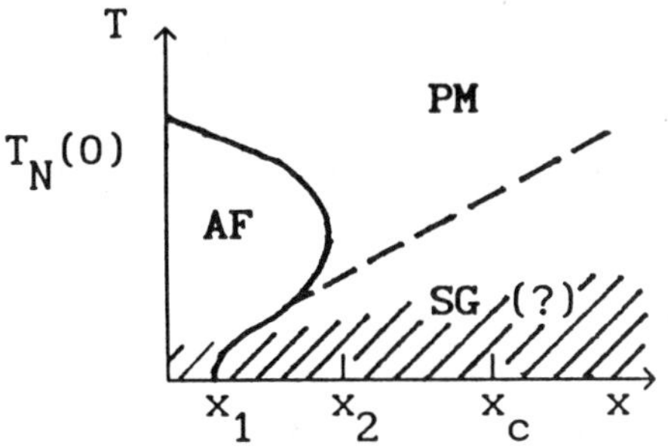

Fig. 1. The magnetic phase-diagram of a lightly doped lamellar antiferromagnet. AF: antiferromagnetically long-range-ordered phase of a matrix; PM: paramagnetic phase, x_c is a percolation threshold. The region of strong fluctuations ($\xi_{2D} \sim x^{-1/2}$) lies below the broken line; the supposed region of impurity spin-glass (SG) is shadowed.

THE EFFECTIVE HAMILTONIAN OF THE MAGNETIC SYSTEM

To determine the in-plane correlation length ξ_{2D} we have to derive a Hamiltonian of the long-wave-length 2D-fluctuations of $\mathbf{n}(\mathbf{r})$ ($\mathbf{n}$ is the unit vector of antiferromagnetism). For a pure AF, this Hamiltonian has a well-known form:

$$H_0 = \int d^2\mathbf{r}(\nabla\mathbf{n})^2 \ , \tag{1}$$

where $\rho \sim I$ is a spin stiffness.

The magnetic defect carries a finite spin $\mathbf{M}$. The value of $\mathbf{M}$ (as well as D) can be obtained from a microscopic model calculation [5, 8]. The interaction energy of a defect and AF matrix is minimal when $\mathbf{M} \perp \mathbf{n}_0$ ($\mathbf{n}_0 = \mathbf{n}(r \to \infty)$) [6]. The perturbation of vector $\mathbf{n}$ at large r is:

$$\delta\mathbf{n} \equiv \mathbf{n}(\mathbf{r}) - \mathbf{n}_0 = (\mu/2\pi)\mathbf{m}(\mathbf{er})/(r^2) \ , \tag{2}$$

which corresponds to the field of a dipole in two dimensions. Here $\mathbf{m} = \mathbf{M}/M$, $\mathbf{e}$ is the unit vector, parallel to one of the tetragonal axes ($\mathbf{a}$ or $\mathbf{b}$). Four possible directions of $\mathbf{e}$ correspond to the four nonequivalent positions of a defect in the magnetic lattice. The asymptotics (2) and the randomness of $\mathbf{e}$ are universal and do not depend on the specific form of a microscopic Hamiltonian. The frustrating dipolar moment modulus μ is the only model-dependent parameter, it is determined by a defect structure. For the simplest model [4] this structure is shown in Fig. 2. In the general case the existence of nonzero vector $\boldsymbol{\mu} = \mu\mathbf{e}$ is governed by the symmetry of a defect. For the case of Sr^{2+} ion, substituting La^{3+}, the real symmetry demands $\mu = 0$ in the Tetragonal phase and $\mu \neq 0$ in the Orthorhombic one. Moreover, it is very likely that large local distorsions and, therefore, large μ can arise from the Jahn-Teller effect, independent of the macroscopic state of the crystal [8].

The 2D-law (2) is valid on the scale $D \ll r \ll r_{3D}$; for $r \geq r_{3D} = (I/I')^{1/2}$ the

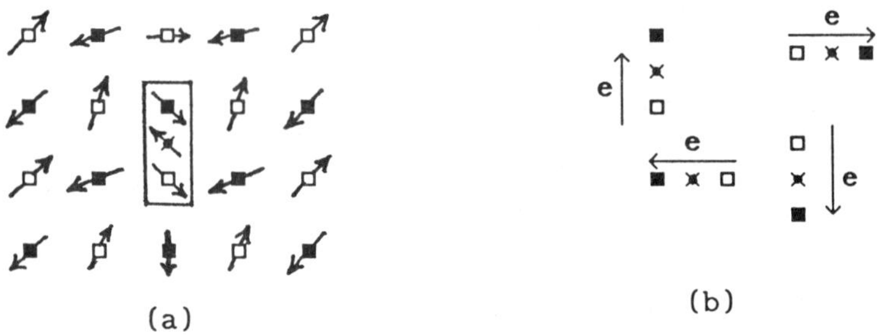

Fig. 2. The magnetic structure of a defect in the model of Aharony et al. [4]. a) Magnetic frustration; ❑ and ■ are the Cu sites in two sublattices; × is the O site, where the hole is localized; the spin-directions are shown. b) four types of defects, differing in the **e**-direction.

interplanar exchange transfers the perturbation onto the adjacent planes and leads to a 3D-law: $\delta\mathbf{n} \sim r^{-2}$.

Now we shall construct a long-wave-length Hamiltonian of the **n**–**m** interaction, which produces the same perturbation $\delta\mathbf{n}$, as Eq. (2):

$$H = H_0 + H_{\text{int}}; \quad H_{\text{int}} = \rho\mu\sum_i(\mathbf{e}_i\nabla)(\mathbf{n}(\mathbf{r}_i)\mathbf{m}_i) , \tag{3}$$

where $\mathbf{r}_i$ are the positions of the acceptors. It can be easily shown that Eq. (2) minimizes the energy Eq. (3) in the case of single acceptor. Eq. (3) holds for the case of a finite acceptor concentration x if the average distance $r_a \sim x^{-1/2}$ between acceptors satisfies the condition: $D \ll r_a \ll r_{3D}$.

Thus the fluctuations of unit vectors $\mathbf{n}(\mathbf{r})$ and $\mathbf{m}$ are governed by the Hamiltonian Eq. (3) with randomly distributed $\mathbf{r}_i$ and $\mathbf{e}_i$. It is well-known that in a 2D-theory with a Hamiltonian Eq. (1) the fluctuations of **n** diverge logarithmically in the thermodynamic limit: $<\delta\mathbf{n}^2> \sim \ln L$, where L is the size of the system. We shall demonstrate that the defect-induced fluctuations are also logarithmically divergent. The calculation up to the second order in H_{int} yields:

$$<\delta\mathbf{n}^2> = <\delta\mathbf{n}^2>_0 + <\delta\mathbf{n}^2>_2; \quad <\delta\mathbf{n}^2>_0 = (\mathrm{T}/\pi\rho)\ln L ,$$

$$<\delta\mathbf{n}^2>_2 = (\mu/2\pi)^2<[\textstyle\sum_i(\mathbf{e}_i\mathbf{r}_i)r_i^{-2}\mathbf{m}_i]^2> \approx x\mu^2/4\pi\ln(L/r_a) . \tag{4}$$

The distances are measured in lattice-spacing units. Indices 0 and 2 denote the orders of the perturbation theory; the first-order term is obviously zero. The result for the second order is valid for $r_a \ll L$, when the sum may be substituted by the integral. The crossing terms in $(\sum\ldots)^2$ are neglected because they are of order x^2. The result can be rewritten in the form:

$$<\delta\mathbf{n}^2> \approx (\mathrm{T}/\pi\rho_{\text{eff}})\ln L; \quad \rho_{\text{eff}} = \rho(1-x\mu^2\rho/4\mathrm{T}) . \tag{5}$$

So, at least for the small values of x, the influence of magnetic defects on $<\delta \mathbf{n}^2>$ is manifested as a small renormalization of a spin-stiffness ρ.

Our idea is that an appropriate renormalization of ρ in the simplest Hamiltonian Eq. (1) is sufficient for taking into account the impurity effects on $\mathbf{n}$ at arbitrary x. According to this idea we shall integrate out the fluctuations of $\mathbf{n}$ with the spatial scales $r<L_0$, where $r_a \ll L_0 \ll r_{3D}$, ξ_{2D}. Applying the standard procedure to the Hamiltonian Eq. (3) we obtain after renormalization:

$$H_1 = H_0 + H_{\text{int}} + H\{\mathbf{m}\}; \quad H\{\mathbf{m}\} = (\rho\mu^2/2)\sum_{i \neq j}(I_{ij}\mathbf{m}_i\mathbf{m}_j);$$

$$I_{ij} = r_{ij}^{-2}[(\mathbf{e}_i\mathbf{e}_j) - 2(\mathbf{r}_{ij}\mathbf{e}_i)(\mathbf{r}_{ij}\mathbf{e}_j)r_{ij}^{-2}] \, . \tag{6}$$

The term $H\{\mathbf{m}\}$, which is generated by the short-wave-length part of H_{int} in Eq. (3), describes the $\mathbf{m}$–$\mathbf{m}$ interaction via the $\mathbf{n}$-field. This term is L_0-independent due to the condition $r_a \ll L_0$. The characteristic value of the random $\mathbf{m}$–$\mathbf{m}$ exchange in $H\{\mathbf{m}\}$ is $U = \rho\mu^2 x/4$. The renormalizations of ρ and μ in H_0 and H_{int} are negligible since $\ln L \ll \ln \xi_{2D}$.

After the above procedure $\nabla\mathbf{n}$ becomes a small and slowly varying function in $\mathbf{r}$-space. After that one can easily perform the integration over $\mathbf{m}_i$. In zeroth order in $\nabla\mathbf{n}$ this integration gives the free energy $F_0(\mathrm{T})$ of a random magnet with a Hamiltonian $H\{\mathbf{m}\}$. The smallness of $\nabla\mathbf{n}$ enables us to consider H_{int} in the framework of the linear response formalism and to obtain a contribution to the free energy of the random magnet quadratic in $\nabla\mathbf{n}$:

$$F = F_0 - (\chi/2)\int d^2\mathbf{r}(\nabla\mathbf{n})^2 \, , \tag{7}$$

where $\nabla\mathbf{n}$ appears as a generalized external field and χ is a corresponding susceptibility:

$$\chi = (\rho/V)(\rho\mu^2/4\mathrm{T})\sum_{i,j}<(\mathbf{e}_i\mathbf{e}_j)<(\mathbf{m}_i\mathbf{m}_j)>_{\mathrm{T}}>_c \, . \tag{8}$$

Here V is the volume, and $<\ldots>_{\mathrm{T}}$ and $<\ldots>_c$ denote the thermodynamic (with the Hamiltonian $H\{\mathbf{m}\}$), and the configurational averaging; respectively. Eq. (7) is just a formal expansion of a free energy F in small "external field" $\nabla\mathbf{n}$. It does not require any restrictions on the Hamiltonian $H\{\mathbf{m}\}$. In particular, the assumption of the Gaussian type of fluctuations is not needed. All the problems, connected with the non-Gaussian field $\mathbf{m}$ are reduced to the two unknown functions $F_0(\mathrm{T})$ and $\chi(\mathrm{T})$.

The spatial dispersion of χ occurs on the scale r_a and can be neglected since $L_0 \gg r_a$. Random $\mathbf{m}$–$\mathbf{m}$ interactions in $H\{\mathbf{m}\}$ do not break the isotropy either in spin, or in real space; hence χ is a scalar function. As follows from Eq. (6) and Eq. (8), χ depends on x and T only via ratio $U/\mathrm{T} = \rho\mu^2 x/4\mathrm{T}$; so one has $\chi = \rho[1-\phi(\rho\mu^2 x/4\mathrm{T})]$, where $\phi(z)$ is a universal dimensionless function.

After the integration over $\mathbf{m}_i$, the free energy Eq. (7) replaces the last two terms in H_1 (Eq. (6)). Thus, as we have just expected, the effective Hamiltonian of long-wave-length fluctuations of $\mathbf{n}$ coincides with H_0, except for the renormalization of the spin-stiffness:

$$H_{\text{eff}} = (\rho_{\text{eff}}/2)\int d^2\mathbf{r}(\nabla\mathbf{n})^2;\quad \rho_{\text{eff}} = \rho\phi(\rho\mu^2 x/4\text{T})\ . \tag{9}$$

The lowest order perturbation theory results in Eq. (4) and Eq. (5) coincide with the first two terms of the expansion for $\phi(z)$ in Eq. (9).

THE 2D-CORRELATION LENGTH

The correlation length for the 2D-**n**-field with the Hamiltonian Eq. (1) or Eq. (9) is well-known [9]:

$$\xi_{2\text{D}} \sim \exp(2\pi\rho_{\text{eff}}/\text{T})\ . \tag{10}$$

Thus, a problem of x- and T-dependences of ξ_{2D} is reduced to that of ϕ shape.

At large z a function $\phi(z)$ follows the Curie-Weiss law:

$$\phi(z) = 1 - z\ . \tag{11}$$

This limit corresponds to the case at relatively high temperatures (T exceeds the **m–m** interaction U). The corrections to Eq. (11) can be obtained from the cluster expansion of $\tilde{\chi}$. It can be easily shown that the first nonzero correction to Eq. (11) is of order of z^3.

The shape of $\phi(z)$ at $z \sim 1$ is determined by the qualitative behavior of the impurity spin system at intermediate temperatures $\text{T} \sim U$. The properties of a 2D-planar magnet with random dipolar interactions are not known exactly so far. However, there are strong arguments (see [10]) in favor of a spin-glass-transition at zero temperature for such a system. Hence one can expect the magnetic susceptibility χ to increase monotonously without a saturation under lowering T. The generalized susceptibility $\tilde{\chi}$ differs from χ by a factor $(\mathbf{e}_i\mathbf{e}_j)$ in Eq. (8). However, there is no reason for a qualitative difference between shapes of χ and $\tilde{\chi}$ in the paramagnetic phase. So we assume the monotonous increase of $\tilde{\chi}(\text{T})$ or, in other words, the monotonous decrease of $\phi(z)$.

At large T the renormalization of ρ_{eff} can be neglected and $\xi_{2D}(\text{T})$ increases with lowering T. On the other hand ρ_{eff} goes to zero and $\xi_{2D}(\text{T})$ decreases when T tends to U/z_0 (where z_0 satisfies an equation: $\phi(z_0)=0$). So $\xi_{2D}(\text{T})$ is a nonmonotonous function. The asymptotics Eq. (11) enables one to write an analytical expression

$$\xi_{2\text{D}}(\text{T}, x) \sim \exp[(2\pi\rho/\text{T}) - (\pi\rho^2\mu^2 x/2\text{T}^2)]\ , \tag{12}$$

that reveals the exponential decrease of ξ_{2D} with doping. Eq. (13) is exact at $z \ll 1$ (when the second term in Eq. (12) is much less, than the first one). For $z \sim 1$ Eq. (12) provides a qualitative illustration of nonmonotonous temperature dependence of ξ_{2D}.

There is an instability ($\rho_{\text{eff}} < 0$) in the Hamiltonian Eq. (9) for $z > z_0$. Formally, it means the divergence of fluctuations, and the shorter the wave-length, the stronger the divergence. Physically, the divergence should be cut at the shortest relevant scale, i.e. at the mean interdefect spacing r_a. Hence, the correlation length $\xi_{2D} \sim r_a \sim x^{-1/2}$ at $T < x\rho\mu^2/4z_0$ (i.e. below the broken line in Fig. 1). A square root x-dependence of ξ_{2D} and its weak temperature variations at low T and $x > 0.02$ were found in the experiments [1,11].

The above study of ξ is related to a strictly 2D AF. In the real case of 3D AF with a strong anisotropy the system undergoes a phase transition. The results of this section are relevant to the experiment only outside the region of AF-phase. However, the results for ξ_{2D} and the condition $I/I' \ll 1$ enable us to determine the phase boundary.

THE 3D-NEEL TEMPERATURE

The Neel temperature T_N of an undoped quasi-2D-antiferromagnet is defined approximately by a relation: $T_N \sim I'\xi_{2D}^2(T_N)$. In a strictly-2D-AF an infinitesimal concentration of quenched dipolar defects destroys the long-range order [6] even at $T=0$ (i.e. $\xi_{2D}(T=0)$ is finite). So, in a doped quasi-2D-AF one can expect the suppression of long-range order for rather small x. The above relation for T_N can be easily generalized for this case: $T_N \sim I'\xi_{2D}^2(T_N, x)$, or in the leading logarithmic approximation, we may obtain $T_N(x)$ from the equation:

$$\xi_{2D}(T_N, x) = r_{3D} \equiv (I/I')^{1/2} . \tag{13}$$

Using Eq. (9) and Eq. (10), one can rewrite Eq. (13) in the form:

$$\phi(y/\tau) = \tau , \tag{14}$$

where $y = x/x_0$, $\tau = T_N/T_N(0)$ are normalized concentration and transition temperature; $T_N(0) = 2\pi\rho/\ln r_{3D}$, $x_0 = 8\pi/\mu^2 \ln r_{3D}$. For low concentrations $y \ll 1$ Eq. (11) is valid and Eq. (14) can be solved:

$$\tau = 1 - y - y^2 . \tag{15}$$

The last term in Eq. (15) is within the accuracy of calculation because of the absence of z^2 terms in the $\phi(z)$ expansion.

We have no accurate analytical expression for $\phi(z)$ at $z \sim 1$, but the shape of the $\tau(y)$-function can be qualitatively established with the help of graphic analysis of Eq. (15) (see Fig. 3). Above the critical concentration y^* Eq. (14) has no solutions (AF ordering is impossible for all T's); below y^* there are two solutions that correspond to the reentrant behavior of phase transition. In dimensional variables a critical concentration is $x_2 = 8\pi y^* / \mu^2 \ln r_{3D}$. So the reentrant transition is a direct consequence of non-monotonous temperature dependence of ξ_{2D}.

The physical origin of the phenomena discussed lies in the **m**–**m** correlation enhancement at low temperatures. However, there is a limitation on the concentration x, for which these correlations occur: $r_a < r_{3D}$, i.e. $x > x_1$. At $x < x_1$ it is only 3D-tails of perturbations that overlap. Hence there is no divergence in the impurity-induced fluctuations, no ρ-renormalization and no reentrant transition for $x < x_1$.

We can estimate a value of x_2, by extrapolating the Curie-Weiss law Eq. (11) onto the $z \sim 1$ region. The solution of Eq. (14) with $\phi(z) = 1 - z$ gives:

$$\tau(y) = [1 \pm (1-4y)^{1/2}]/2 , \quad y^* = 1/4 . \tag{16}$$

Using the values $T_N(0) = 300$ K, $2\pi\rho = 1200$ K [12] one can see, that a value $x_2 = 0.02$

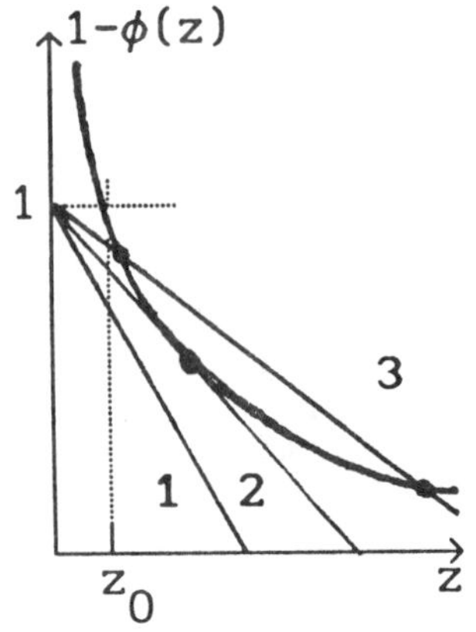

Fig. 3. A graphical solution of Eq. (15). The three stright lines shown are defined by the equation $\phi=y/z$ with: 1) $y>y^*$(no solutions), 2) $y=y^*$ (one solution), 3) $y<y^*$ (two solutions). It is clear, that the proposed picture holds for any shape of $\phi(z)$ as long as a finite value z_0 exists ($\phi(z_0)=0$).

corresponds to $\mu\sim 10$. The estimate $\mu\sim 6D$ [5] shows, that the experimentally observed x_2 can be interpreted in terms of defects with realistic size D of 1-2 lattice spacings.

There is no direct experimental evidence for a reentrant transition in LaSrCuO so far. But several neutron-scattering experiments (especially the suppression of the sublattice magnetization at low T) [1], as well as resistivity measurements [10] indicate the presence of a reentrant behavior.

The paramagnetic state of the defect spins is preserved even on the AF background. Hence the impurity contribution χ to the magnetic susceptibility may be comparable to that of AF-matrix despite the low doping.

True long-range order is destroyed below the line of the reentrant transition. However the 3D-interactions and the magnetic anisotropy, which are inevitably present in the crystal, can result in 3D-Ising-like behavior at the lowest T's. Consequently, a 3D-spin-glass transition can arise.

REFERENCES

1. Birgeneau, R.J., Shirane G. In: "Physical Properties of High Temperature Superconductors", D.M.Ginsberg, ed., (World Scientific Publishing, 1989).
2. Emery V.J., Phys.Rev.Lett., 58, 2794, (1987).
3. Kastner M.A. et al, Phys.Rev.B37, 111, (1988).
4. Aharony A. et al, Phys.Rev.Lett., 60, 1330, (1988).
5. Glazman L.I., Ioselevich A.S., unpublished.
6. Villain J.V., Z.Phys. B33, 31, (1979).
7. Glazman L.I., Ioselevich A.S., Pis'ma v ZhETF, 49, 503, (1989); and Z.Phys.B, in press.
8. Gogolin A.O., Ioselevich A.S., Pis'ma v ZhETF, 50, # 11 (1989),in press.
9. Polyakov A.M., Phys.Lett., B59, 79, (1975).
10. Binder K., Young A.P., Rev.Mod.Phys., 58, 801, (1986).
11. Birgeneau R.J., et al, Phys. Rev. B38, 6614, (1988).
12. Chakravarty S., et al, Phys.Rev.Lett., 60, 1057, (1988).
13. Thio T., et al, Phys.Rev. B38, 905, (1988).

ANDERSON MODEL FOR LOCALIZATION WITH LONG-RANGE TRANSFER AMPLITUDES: DELOCALIZATION OF VIBRATIONAL MODES CAUSED BY THEIR ELECTRIC DIPOLE INTERACTION

L.S.Levitov
L.D.Landau Institute for Theoretical Physics, GSP-1, Kosygin st., 2
Moscow, 117940, USSR

Phonons in disordered dielectric materials at T=0 are always delocalized owing to the electric dipole interaction (no matter how weak it is). Real-space renormalization of the dipole coupling is studied. A renormalization group equation for the distribution of dipole parameters is derived and its similarity with the Boltzmann kinetic equation is revealed. Conservation laws are found, the H-theorem is proven. Stationary distributions form a six-parameter manifold of fixed points. Energy transport is critical, slower than diffusive.

1. INTRODUCTION

Here I study the Anderson model for localization with transfer amplitudes of a long-range character. The most general form of the Hamiltonian is

$$H(\psi)_i = V_i \psi_i + \sum_{i \neq j} T_{ij} \psi_j, \tag{1.1}$$

where i, j label sites, V_i is a random on-site potential, $T_{ij} \approx |\mathbf{r}_i - \mathbf{r}_j|^{-\sigma}$ are transfer amplitudes, ψ_i is the amplitude of a quantum particle at the site $\mathbf{r}_i$. It is known [1] that all eigenstates of the Hamiltonian (1.1) are localized if the following two conditions are fulfilled simultaneously: a) $\langle V_i^2 \rangle \gg \langle T_{ij}^2 \rangle$; b) $\sigma > D$, where D is space dimension. Therefore, there are two basic ways to come out of the localized regime; one may violate either one of the inequalities a), b). Usually only the possibility a) is discussed in connection with delocalization of electrons. Here I consider the second possibility: $\sigma = D$, while condition a) remains fulfilled. In this case the diverging number of resonances destroys localized states. Problems of this type show a new critical delocalization behavior and, probably, form a universality class of Anderson transition different from the usual one studied for electron transport in disordered metals. (The main difference is in the long-range character of the interaction which is absent in metals owing to screening.)

For dimension 3, which is the most important one, we need $\sigma = 3$. Many systems exist for which cubic r^{-3} interactions are important: localized optical phonons in disordered dielectric materials coupled by the electric dipole interaction, two-level systems in glassy solids coupled by the r^{-3} elastic interaction, magnetic impurities in metals coupled by the r^{-3} RKKY interaction, etc. Of course, the first problem of delocalization of vibrational modes is the simplest one, since it is linear: one has to find normal modes for a large (but quadratic) Hamiltonian. Although only this problem is discussed below, one can expect that delocalization of an excitation in other systems mentioned above is similar to that studied here.

Thus, I am going to demonstrate how the long-range electric dipole interaction of localized vibrational states kills their localization. It was shown [2] that in the presence of this interaction localized states cannot exist, no matter how weak the interaction is. The main reason for this conclusion is that the number of resonances diverges (Sec.2). The divergence is logarithmic, so the effect of delocalization is weak. We will see that this enables one to apply the renormalization group technique and study the effect within its framework (see Sec.3). The renormalization group equation derived there for dipole parameters is extremely similar to Boltzmann kinetic equation. This equation is studied and its stationary solutions (analogous to Maxwell distribution) are found (Sec.4). "Entropy" is introduced and the H-theorem is proved. In Sec.5 a scaling analysis of energy transport by vibrational modes is presented. The dynamics is critical: instead of diffusive transport, $<|\mathbf{r}|^2> \approx t$, a slower behavior is derived, $<|\mathbf{r}|^2> \approx t^{2/3}$. Energy -energy correlations are shown to have a scale invariant form $K(r,t) = t^{-1}F(t^{-1/3}r)$ at $|\mathbf{r}| \to \infty$, $t \to \infty$, where $F(x)$ is a universal function of x depending also on six parameters.

2. BASIC MODEL: DIVERGENCE OF THE NUMBER OF RESONANCES.

We are interested in the part of the spectrum corresponding to states localized in the absence of the long-range interaction [2]. The Hamiltonian can be written as

$$H(p,q) = \sum_i \frac{1}{2}(p_i^2 + \omega_i^2 q_i^2) + \sum_{i<j} q_i q_j T_{ij}, \tag{2.1.i}$$

where

$$T_{ij} = \frac{\mathbf{a}_i\cdot\mathbf{a}_j - 3\mathbf{a}_i\cdot\mathbf{n}_{ij}\,\mathbf{a}_j\cdot\mathbf{n}_{ij}}{|\mathbf{r}_i - \mathbf{r}_j|^3}, \qquad \mathbf{n}_{ij} = \frac{\mathbf{r}_i - \mathbf{r}_j}{|\mathbf{r}_i - \mathbf{r}_j|}, \tag{2.1.ii}$$

p_i, q_j are scalar canonical variables corresponding to localized modes. The first part of (2.1.i) describes bare non-perturbed localized modes, while the second part stands for their electric dipole interaction. The positions $\mathbf{r}_i$ of localized modes randomly and uniformly fill the space with concentration n. Vectors $\mathbf{a}_i$ are defined by

$$\mathbf{d}_i = \mathbf{a}_i\, q_i \tag{2.2}$$

where $\mathbf{d}_i$ is the electric dipole caused by the displacement $\mathbf{q}_i$ of the i-th oscillator. We call the $\mathbf{a}_i$ "dipole parameters". Squared frequencies ω_i^2 are non-correlated random numbers uniformly filling a wide interval $[\Delta_-^2, \Delta_+^2]$, so their distribution function is

$$\nu(\omega) = 2\nu\omega \quad \text{for} \quad \Delta_- < \omega < \Delta_+, \quad 0 \quad \text{otherwise}, \tag{2.3}$$

where $\nu = (\Delta_+^2 - \Delta_-^2)^{-1}$. Dipole parameters $\mathbf{a}_i$ are random non-correlated vectors with some distribution $f(\mathbf{a})$: $dP = f(\mathbf{a})\, d^3\mathbf{a}$.

Since we are going to treat the second term of (2.1.i) as a perturbation let us impose the condition

$$\lambda = \langle \mathbf{a}^2 \rangle \nu n \ll 1, \tag{2.4}$$

where $\langle \mathbf{a}^2 \rangle = \int \mathbf{a}^2 f(\mathbf{a}) d^3\mathbf{a}$. The parameter λ plays a role of a coupling constant in this problem. When λ is not small the decomposition of the interaction into short-range and long-range parts (essential for the derivation of (2.1)) becomes inconsistent. The smallness of λ is a very important assumption used systematically below.

Simple qualitative arguments can be given to show that normal modes of the problem (2.1) cannot be localized [3]. Consider two oscillators having frequencies ω_i, ω_i , positions $\mathbf{r}_i$, $\mathbf{r}_j$ and dipole parameters $\mathbf{a}_i$, $\mathbf{a}_j$. They are resonant if

$$|T_{ij}| \geq |\omega_i^2 - \omega_j^2| \tag{2.5}$$

If condition (2.5) is true, then normal modes of the problem $H(q,p) = \frac{1}{2}(p_i^2 + \omega_i^2 q_i^2) + \frac{1}{2}(p_j^2 + \omega_j^2 q_j^2) + T_{ij} q_i q_j$ are not localized on one oscillator but they are essentially non-zero at both places $\mathbf{r}_i$, $\mathbf{r}_j$. One can estimate the number of resonances and show that it diverges.

The quantity to be calculated is n_V, the average number of oscillators falling in resonance with a given one. We easily find an estimate $n_V = \int n P(\mathbf{r}) d^3\mathbf{r}$, where the integral is taken over the sphere of volume V centered at $\mathbf{r}_i$, $P(\mathbf{r})$ is the probability that two oscillators separated by a distance $|\mathbf{r}|$ are in a resonance. After estimating $P(\mathbf{r})$ as $\nu \langle \mathbf{a}^2 \rangle |\mathbf{r}|^{-3}$ obtain

$$n_V \approx \frac{4\pi}{3} n\nu \langle \mathbf{a}^2 \rangle \ln(V) \approx \lambda \ln(V) \tag{2.6}$$

The divergence of n_V with V indicates delocalization. However, the divergence is logarithmic, so the effect of delocalization is weak. This suggests that we apply the renormalization group ideas.

3. RENORMALIZATION EQUATION

Why is it necessary to do something else after we established delocalization in Sec.2? Three types of states are known in disordered systems: localized, critical and extended. According to Sec.2 localized states are absent in this problem. So we have to choose between critical and extended ones. This can be done by real space renormalization analysis of the "effective coupling" of the modes (crudely speaking, given by λ). According to Sec.2 the coupling of bare modes is weak. If λ grows as a function of scale and becomes ≈ 1 at some scale R_c , then the system is driven to the strong coupling regime and normal states are extended at sufficiently large scales ($\gg R_c$). However, our analysis leads to a completely different result: starting with small bare λ one is left with small λ at all scales. Since the system remains in the weak coupling regime at all scales the eigenstates turn out to be critical. The criticality yields an unusual law for an excitation transport in this system. When normal states are extended the transport is diffusive. When they are critical the transport is usually described by non-standard scaling laws. Since the second case takes place we have to study space and time dependence of the energy-energy correlation function and find its scaling exponent. All this work is done step by step in Secs.3-5.

First we mention one property of resonance oscillators that is basic for our approach. Simple qualitative analysis presented in [3] shows that resonance oscillators are mainly coupled in pairs, i.e. triplet and other higher-order resonances occur very rarely compared with pairs. Of course, resonance pairs are in their turn coupled in resonance pairs at larger scales, so the overall structure is hierarchical. This picture emerges from the estimate (2.6) which tells that the resonances are rarely distributed in the "logarithmic" space. This property gives a basis for our method. Let us truncate the $\mathbf{r}^{-3}$ -interaction at some R_0 , i.e. put $T_{ij}=0$ for all pairs (i, j) such that $|\mathbf{r}_i-\mathbf{r}j|>R_0$. Find exact normal modes for this truncated Hamiltonian (denote them R_0 -modes). Then replace R_0 by R_1 such that

$$R_1 \gg R_0, \qquad \text{but} \quad \lambda \log_2(\frac{R_1}{R_0}) \ll 1. \tag{3.1}$$

Find R_1-modes and consider them as linear combinations of R_0-modes. According to the above remark R_1-modes are either single R_0-modes or resonance pairs of R_0-modes (we ignore by triple and all other multi-oscillator resonances). When considering R_1-modes as a result of the interaction (resonance) of R_0-modes one can assume that interacting R_0-modes are far apart compared with their localization radius (this assumption is self-consistent [3]). This enables us to proceed in the following manner. For R_1-modes calculate effective dipole parameters $\tilde{\mathbf{a}}_i$ assuming that $\mathbf{a}_i$ for R_0-modes are known. Then use for the coupling of R_1-modes the dipole-dipole interaction of the form (2.1.ii) with $\tilde{\mathbf{a}}_i$ instead of $\mathbf{a}_i$.

Find all necessary quantities for two oscillators ($\mathbf{R}_0$-modes) having numbers i, j. They interact according to

$$H_{ij}(q,p)=\frac{1}{2}(p_i^2+\omega_i^2 q_i^2)+\frac{1}{2}(p_j^2+\omega_j^2 q_j^2)+T_{ij}q_i q_j \tag{3.2}$$

Normal modes q^+, q^- are defined by

$$q^+ = \cos\theta\ q_i + \sin\theta\ q_j, \qquad q^- = -\sin\theta\ q_i + \cos\theta\ q_j$$

$$ctg\, 2\theta = \frac{\omega_i^2 - \omega_j^2}{2T_{ij}} \tag{3.3}$$

Oscillation frequencies $\omega_\pm$ of the modes (3.3) are given by

$$\omega_\pm^4 - (\omega_i^2+\omega_j^2)\omega_\pm^2 + \omega_i^2\omega_j^2 - |T_{ij}|^2 = 0 \tag{3.4}$$

and the electric dipole $\mathbf{d}$ of the modes i,j can be expressed as

$$\mathbf{d} = \mathbf{a}_i\, q_i + \mathbf{a}_j\, q_j = \mathbf{a}^+ q^+ + \mathbf{a}^- q^-, \tag{3.5}$$

where

$$\mathbf{a}^{+} = \cos\theta\, \mathbf{a}_i + \sin\theta\, \mathbf{a}_j, \quad \mathbf{a}^{-} = -\sin\theta\, \mathbf{a}_i + \cos\theta\, \mathbf{a}_i \tag{3.6}$$

Expression(3.5) shows that any mode (say, the k-th one) that falls in a resonance with the (+)-mode or the (-)-mode at some next step of the renormalization procedure interacts with it according to (2.1.i), where $D_{k\pm}$ is given by (2.1.ii) and $\mathbf{a}^{\pm}$ are taken from (3.6).

Subsequent resonances (interactions) of the modes can be considered as non-correlated [3] (we mean correlations at different moments of the renormalization "time" $\xi=\ln(R)$). Thus we have only to find the distribution of $\mathbf{a}_i$ for R_1-modes assuming it known for R_0-modes, i.e. derive a renormalization equation for $f(\mathbf{a})$. This derivation can be found in [3], here I quote only the result:

$$\frac{\partial f(\mathbf{a})}{\partial \xi} = n\nu \int d\tau d^3\mathbf{a}_1 d^3\mathbf{a}_2\, f(\mathbf{a}_1) f(\mathbf{a}_2)\, Q(\mathbf{a}_1, \mathbf{a}_1)\, [\delta(\mathbf{a}-\mathbf{a}^{+})+\delta(\mathbf{a}-\mathbf{a}^{-})-\delta(\mathbf{a}-\mathbf{a}_1)-\delta(\mathbf{a}-\mathbf{a}_2)], \tag{3.7}$$

where

$$Q(\mathbf{a}_1,\mathbf{a}_2) = \int d\Omega\, |\, \mathbf{a}_1\cdot\mathbf{a}_2 - 3\mathbf{a}_1\cdot\mathbf{n}\, \mathbf{a}_1\cdot\mathbf{n}\, |. \tag{3.8}$$

Here $\xi=\ln(R)$, $\tau = ctg\, 2\theta$ (see (3.3)), $\mathbf{a}^{\pm}$ are the functions (3.6) of θ (or τ), $d\Omega$ is a unit sphere area element. Important properties of the function $Q(\mathbf{a},\mathbf{b})$ are: $Q(\mathbf{a},\mathbf{b})=4\pi|\mathbf{a}|\,|\mathbf{b}|Q(\phi)$, where ϕ is the angle between the vectors $\mathbf{a},\mathbf{b}$; $Q(\phi\pm\pi)=Q(\phi)=Q(-\phi)$; $\max[Q(\phi)]=0.77..$, $\min[Q(\phi)]=0.63..$. Thus $Q(\phi)$ is almost constant: $|(Q(\phi)-Q^{*})/Q^{*}|<0.1$ for all $\phi[-\pi,\pi]$, where $Q^{*}=0.7$.

Concerning Eq.(3.7) our main task is to find and study its solutions $f(\mathbf{a},\xi)$ such that $f(\mathbf{a},0)=f(\mathbf{a})$, where $f(\mathbf{a})$ is the microscopic distribution of the vectors $\mathbf{a}_i$. Of special interest are its asymptotics at $\xi\to\infty$ related with important dynamic characteristics of the problem (see Sec.5).

4. SOLUTIONS OF THE RENORMALIZATION EQUATION.

Our analysis of Eq. (3.7) will be strongly motivated by its analogy with the Boltzmann equation. First consider "Integrals of the motion " for Eq. (3.7). There are six important quantities conserved under Eq. (3.7). Consider 3 components of the vector $\mathbf{a}=(a_x,a_y,a_z)$. Using the orthogonality of the transformation (3.6) one easily finds that each of the quantities

$$\langle a_x^2\rangle,\ \langle a_y^2\rangle,\ \langle a_z^2\rangle,\ \langle a_x a_y\rangle,\ \langle a_y a_z\rangle,\ \langle a_z a_x\rangle \tag{4.1}$$

is conserved when $f(\mathbf{a},\xi)$ satisfies Eq. (3.7) [3]. We remark: An important question is whether the conservation of the quantities (4.1) is an exact or an approximate result. One might suspect that the conservation fails when not only interacting pairs but also multi-oscillator resonances are taken into account. However, it is shown in [3] that such resonances do not destroy the conservation of $\langle a_\alpha a_\beta\rangle$ ($\alpha, \beta=x, y, z$).

Further results concerning Eq. (3.7) can be obtained only for its approximate version which we get by replacing $Q(\mathbf{a},\mathbf{b})\to Q^{*}|\mathbf{a}|\,|\mathbf{b}|$ (an error introduced by this replacement is of order of 10%). So the modified Eq. (3.7) (denote it "Eq. (3.7M)") allows an exact treatment. The properties of Eq. (3.7M) are very similar to those of the Boltzmann equation.

I [Integrals of motion] The quantities (4.1) are invariants of Eq.(3.7M), not only of Eq.(3.7).

II [H-theorem] Let $f(\mathbf{a},\xi)$ satisfy Eq.(3.7M). Define entropy as

$$H[f] = -\int \ln(|\mathbf{a}| f(\mathbf{a},\xi))\, f(\mathbf{a},\xi)\, d^3\mathbf{a} \qquad (4.2)$$

The function $H(\xi)=H[f(\mathbf{a},\xi)]$ grows monotonously: $\frac{\partial}{\partial\xi}H(\xi)\geq 0$.

III [Stationary distributions] All stationary solutions of Eq. (3.7M) are given by

$$f_G(\mathbf{a})=\frac{A}{|\mathbf{a}|}\exp(-a_\alpha G_{\alpha\beta} a_\beta), \qquad (4.3)$$

where G is a positively defined symmetric 3×3 matrix.

IV [Maximum of entropy] The entropy $H[f]$ (4.2) reaches its maximal value for the distributions (4.3). More precisely, consider all functions $f(\mathbf{a})$ such that

$$f(\mathbf{a})\, d^3\mathbf{a}=1, \qquad \int a_\alpha a_\beta f(\mathbf{a})\, d^3\mathbf{a}=G_{\alpha\beta} \quad (\alpha,\beta=1,2,3) \qquad (4.4)$$

Then $H[f]\leq H[f_G]$, where $f_G(\mathbf{a})$ is defined by (4.3); $H[f]=H[f_G]$ only if $f(\mathbf{a})=f_G(\mathbf{a})$.

We see that the asymptotic properties of solutions of Eq. (3.7M) are very simple: any solution converges to one of the stationary distributions $f_G(\mathbf{a})$. The parameters $G_{\alpha\beta}$ are completely determined by second moments of the initial distribution $f(\mathbf{a},\xi{=}0)$.

The closeness of Eqs. (3.7) and (3.7M) enables us to deduce the main features of Eq. (3.7):

a) Every solution $f(\mathbf{a},\xi)$ of Eq.(3.7) converges to a stationary distribution as $\xi\to\infty$.

b) Stationary distributions form a 6-parametric set (generally speaking, not given by (4.3)), where the parameters can be chosen as $G_{\alpha\beta}=\int a_\alpha a_\beta f(\mathbf{a},\xi{=}0)\, d^3\mathbf{a}$ ($\alpha,\beta=1,2,3$, $\alpha<\beta$).

Thus we come to the main conclusion of this section: The renormalization equation (3.7) has a 6-dimensional manifold of fixed points parameterized by symmetric positively defined 3×3 matrices.

5. SCALING ANALYSIS OF ENERGY TRANSPORT

In this section some implications of the renormalization group analysis of Secs.3 and 4 for dynamics of the system are studied. We are particularly interested in the energy-energy correlation function $K(\mathbf{r},t)=\langle h(\mathbf{r},t)\, h(0,0)\rangle/\langle h(0,0)\rangle^2$. Here $h(\mathbf{r},t)$ is the energy density.

Now let us show that the dynamics is critical and find the critical exponents. Consider the process of propagation of an excitation in the system. Let one of the oscillators be set in motion at $t=0$. After some time T it sets in motion its nearest resonant neighbor. According to Sec.3, $T\approx|\omega_+-\omega_-|^{-1}\approx\nu^{-1/2}\langle\mathbf{a}^2\rangle|\mathbf{r}_1-\mathbf{r}_2|^3$, where $\mathbf{r}_1$, $\mathbf{r}_2$ are the oscillators positions. After a period $\approx T$ the energy of the excitation is localized not only at the place $\mathbf{r}$ but at both places $\mathbf{r}_1$ and $\mathbf{r}_2$, since both two modes (linear combinations of q_1 and q_2) are excited. Later, after some larger time T' ($T' \gg T$) new oscillators are involved in the energy transport. Each of the two excited modes sets in motion other modes which in their turn can be considered as a result of interaction of several oscillators during the time period $\ll T'$. The speed of further propagation of the excitation is determined by effective coupling of these complex modes. But the interaction of the complex modes is given by the dipole form (2.1.ii) with effective dipole parameters $\mathbf{a}$ (see Sec.3). Since the distribution $f(\mathbf{a})$ has a limit at $\xi \to \infty$, the interaction of the complex modes at distance L scales as L^{-3} ($\xi = ln(L)$).

Therefore, after a long time T the excitation spreads over a region of size $L(T) \approx T^{1/3}$. An important implication of this result is that the dynamics is (asymptotically) invariant under rescaling

$$(\mathbf{r}, t) \to (Z\mathbf{r}, Zt), \tag{5.1}$$

where Z is an arbitrary number. Consequently, the correlation function $K(\mathbf{r}, t)$ must be invariant under the rescaling (5.1). Thus we obtain

$$K(\mathbf{r}, t) = t^{-1} F(t^{-1/3}\mathbf{r}) \quad at\ large \quad t, |\mathbf{r}|, \tag{5.2}$$

where $\int F(x) d^3x = 1$ due to the conservation of energy. The function $F(x)$ in (5.2) depends on 6 parameters $\langle a_\alpha a_\beta \rangle$ in some universal way. The non-diffusive dynamics with the exponent 1/3 instead of the usual 1/2 follows directly from (5.2):

$$\langle |\mathbf{r}|^2 \rangle = \int |\mathbf{r}|^2 K(\mathbf{r},t)\, d^3\mathbf{r} = t^{2/3}\ const, \tag{5.3}$$

where $const = \int |x| F(x)\, d^3x$.

6. CONCLUSION

We studied delocalization of vibrational modes using a real space renormalization method. A "Kinetic" renormalization equation is derived for the distribution of dipole parameters. This equation has non-trivial fixed points forming a six-dimensional manifold. The absence of the dependence of the limiting distribution on the renormalization cutoff parameter leads to scale independent coupling of renormalized modes and, therefore, critical dynamics with the exponent 1/3 which is slower than the standard diffusive one (having exponent 1/2).

REFERENCES

1. P. W. Anderson, Phys. Rev. **109,** 1492 (1958)
2. L. S. Levitov, Europhys. Lett. **9(1),** 83 (1989)
3. L. S. Levitov, submitted to Phys. Rev. Lett.

THEORY OF THE MULTIPHONON TRAPPING RATE

A. S. Ioselevich and E. I. Rashba
L. D. Landau Institute for Theoretical Physics,
Academy of Sciences of the USSR,
117940, Moscow, V-334, USSR

A theory of the self-trapping (ST) rate $w(T)$, and also of the rate of the multiphonon trapping of charge carriers and excitons by defects is developed in the semiclassical approximation. The preexponential factor in $w(T)$ is estimated in both low-temperature (instanton) and high-frequency (Arrhenius) limits.

INTRODUCTION

The mechanisms of non-radiative multiphonon transitions are among the fundamental problems of condensed matter physics. Their elucidation is of primary importance for the theoretical description of numerous phenomena. Initially, attention was focused on non-radiative transitions in local centers. In the course of the last two decades it has shifted to the non-radiative multiphonon capture of particles (charge carriers, excitons) from free (F) states to different states with a strong lattice relaxation. The latter may be ST states in the perfect lattice, and also local states at defects with a strong electron-phonon coupling. The phenomena, for which the non-radiative capture is of primary importance, are recombination in semiconductors (including the effect of persistent photoconductivity [1]), ST of excitons in alkali halides, rare gas solids and organic crystals [2, 3], radiation induced defect production and defect reactions, etc.

We briefly outline here fundamentals of the formalism [4,5a,b,c] based on the semiclassical approximation which seems to be the most adequate for describing non-radiative capture. Then we present the basic results. The probability of the capture may be written as:

$$w(T) = \bar{\omega} \nu B(T) \exp(-S) . \quad (1)$$

Here $\bar{\omega}$ is the characteristic phonon frequency, $\nu = 1$ for intrinsic ST and equals the concentration of defect sites ($\nu < 1$) for extrinsic trapping, the exponent $S(T)$ is the classical Hamiltonian action, the enhancement factor is $B(T) \gg 1$. The probability $w(T)$ is exponentially small since surmounting of the ST barrier by thermally activated tunneling (instanton mechanism) in the low-T region and by the Arrhenius process in the high-T region is part of the ST process. The theory succeeds in estimating the value of $S(T)$ and $B(T)$, and in determining their temperature dependence in the most important limiting cases.

ADIABATIC POTENTIALS AND CONTINUUM APPROXIMATION

The two basic types of adiabatic potential surfaces (APS) for the problem of non-radiative trapping are shown in Fig.1. Sheet 1 is the total energy when a particle is at the bottom of the band, and Sheet 2 is the adiabatic potential for a trapped particle. APS of the Fig. 1a type is the only possible surface for intrinsic ST; it also describes extrinsic ST by defects. APS of the Fig. 1b type is typical of recombination centers (normal centers). The lowest saddle point W at APS is an analytical point in the first case, but in the second case it is positioned at the branching surface (br); hence, it turns into a singular crest point (cr).

The ST state always has the scale of the lattice spacing a. But the barrier state W has a macroscopic scale $b \approx \Lambda a \gg a$ if the coupling constant $\Lambda \equiv J/E_B \gg 1$, where E_B is the half-width of the band, and J is the ST energy. In this case the ST barrier height is $W \sim E_B/\Lambda^2$ [2]. When $\Lambda \gg 1$, the nonpolar electron-phonon interaction may be described in the continuum approximation and chosen as linear in phonon amplitudes. For acoustic phonons $\omega(q) = cq$, with coupling constant $\gamma(q) = \gamma q$, for optic ones $\omega(q) = \omega_0$, $\gamma(q) = \gamma_0$.

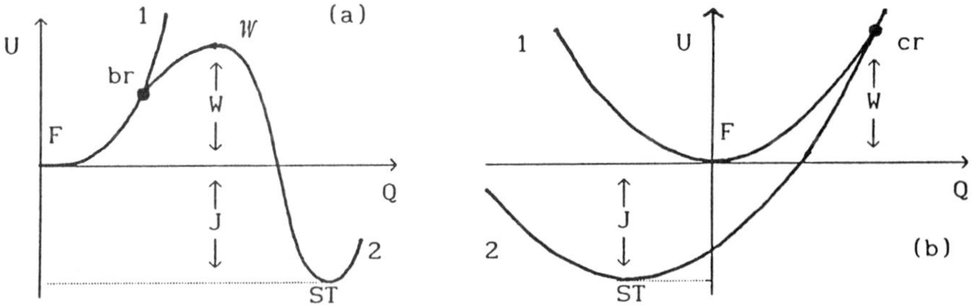

Fig. 1. Configurational diagrams for ST (a) and normal centers (b). W is the height of the ST barrier, J is the ST energy.

GENERAL FORMALISM. EXPONENTIAL APPROXIMATION

The probability $w(\mathbf{k}, T)$ of the transition from the free state with the momentum $\mathbf{k}$ to the ST state in a time $t_1 - t_2$ may be expressed as a path integral in the space of lattice variables $Q(t)$:

$$(t_2 - t_1) w(\mathbf{k}, T) = \int DQ\, G^R_{\{Q\}}(\mathrm{ST}, t_2; \mathbf{k}, t_1) G^A_{\{Q\}}(\mathbf{k}, t_1, \mathrm{ST}, t_2) \exp\{ i \int L_0(Q)\, dt \}, \quad (2)$$

Here $G^{R,A}$ are (retarded, advanced) electron Green's functions for a given phonon trajectory $Q(t)$, L_0 is the Lagrangian of a free lattice. Integration over t is performed along the path Γ (Fig. 2). $Q(t)$ is periodic in t, the period equals to $i\beta$, $\beta = 1/T$. The fast time dependence may be factored out from $G^{R,A}$ and incorporated into the last factor of (2). Then $L_0 \to L$, L is the total Lagrangian, and the exponential factor becomes equal to $\exp(iS(t))$, $S(t)$ being the total Hamiltonian action. The semiclassical approximation implies that this factor is treated by the method of steepest descent, $\delta S = 0$. The method is applicable when $\mathrm{Im}\, S \gg 1$, this contribution to S comes from the integration over imaginary time τ, $|\tau| \leq \beta/2$. The elimination of lattice variables yields

$$S = \min_\psi \frac{1}{2} \int\int d\mathbf{r}\, d\tau \{ |\nabla\psi(\mathbf{r}, \tau)|^2/m + V(\mathbf{r}, \tau)\, |\psi(\mathbf{r}, \tau)|^2 \}, \quad (3)$$

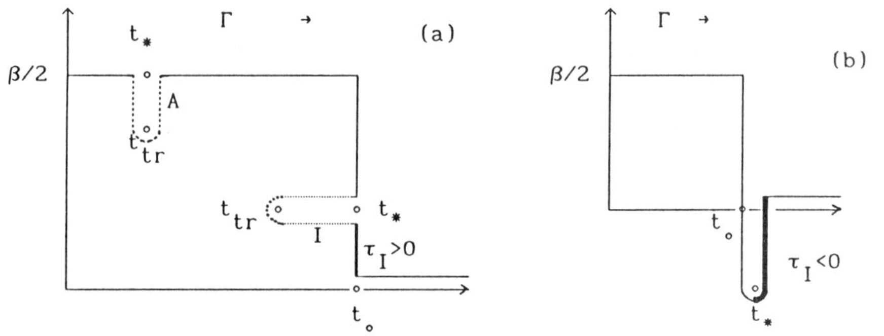

Fig. 2. Paths in the upper half of the complex t-plane. a) Self-trapping. Solid line: the path for trapping of thermal particles. Appendages: for trapping of hot particles in the Arrhenius process (A) and in the instanton process (I). The instanton is shown by the heavy line. b) Normal centers.

where m is the mass of a particle, V is defined as

$$V(\mathbf{r},\tau) = \int\int d\mathbf{r}' d\tau' K(\mathbf{r}-\mathbf{r}', \tau-\tau')\, |\psi(\mathbf{r}',\tau')|^2, \tag{4}$$

and K is the kernel

$$K(\mathbf{r},\tau-\tau') = \int d\mathbf{q}\, e^{i\mathbf{q}\cdot\mathbf{r}}\, \gamma(\mathbf{q})^2 D(\mathbf{q},\tau-\tau')/(2\pi)^3 . \tag{5}$$

Here $D(\mathbf{q},\tau)$ is a free phonon Green's function, and the electron wave function $\psi(\mathbf{r},\tau)$ is determined by the Schrödinger equation

$$\{-\Delta/2m + V(\mathbf{r},\tau)\}\, \psi(\mathbf{r},\tau) = E(\tau)\, \psi(\tau) \tag{6}$$

with a time-dependent eigenvalue $E(\tau)$.

Eq. (6) has time-dependent solutions, i.e., instantons (I), both for $T=0$ [6] and $T\neq 0$ [5a], having the duration $\tau<\bar{\omega}_I^{-1}$. They correspond to loaded motion of the lattice with a particle in a local state $\psi(\mathbf{r},\tau)$. Free tunneling begins at $t=t_0+i\beta/2$; at $t_*=t_0+i\tau_I$ the local level appears in the potential $V(\mathbf{r},\tau)$, and a particle is trapped at this level. The instanton solutions describe the tunneling of the compound system (the particle + the lattice) through the ST barrier; at $T\neq 0$ the tunneling is thermally activated.

Eq. (6) also has a time-independent solution $\psi(\mathbf{r})$ [4], but only for APS of Fig. 1a. It corresponds to the configuration W The action $S_A=\beta W$ and corresponds to the Arrhenius (A) process.

For APS of Fig. 1a $S(\beta)$ reaches its minimum at low T for the I-solution, and at high T for the A-solution (Fig. 3). The curve $S_I(\beta)$ may touch the S_A-line on $\beta=\beta_c=2\pi/\omega_1$ (Fig. 3, curve a), ω_1 being the frequency of vibrations in imaginary time near the point W [7, 5b]. In this case the dependence $S(\beta)$ is smooth. It may also intersect the line $S_A(\beta)$ [8,5b] (curve b), in this case a kink in $S(\beta)$ at $\beta_c < 2\pi/\omega_1$ must exist. Hence, at the temperature $T_c = 1/\beta_c$ the dominating mechanism of the process is changed. For APS of Fig. 1b only the I-solution exists, and $S_I(\beta) < \beta W$, for $\beta \to 0$ the curve $S_I(\beta)$ touches βW (curve c, Fig. 3) [9,5b]. So, for high T the process is either of the Arrhenius type (APS of Fig. 1a), or of the I-type (APS of Fig. 1b) but similar to the A-type.

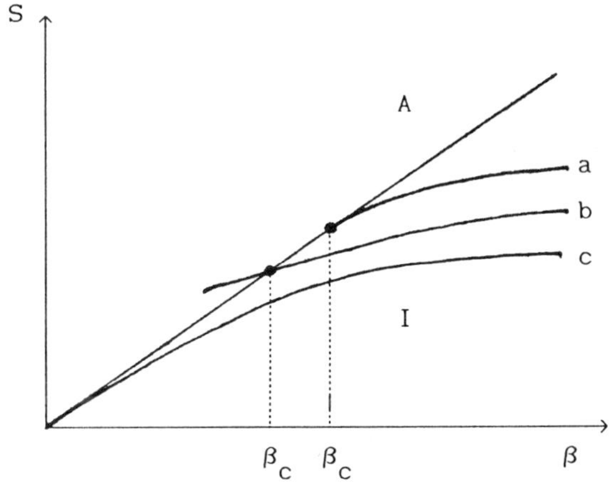

Fig. 3. Dependence $S(\beta)$ for I (curves a, b and c) and A processes.

The contribution of long-wave-length acoustic phonons to $S_I(T)$ always dominates at low T. In this limit $S(\infty) - S(\beta) \propto T^4$ and $\propto T^2$ for the deformation and piezoelectric interaction, respectively.

For optic phonons instantons have the same size b as the ST barrier, the tunneling occurs near it, and $\tau_I \sim \omega_0^{-1}$, $S \approx 6.2\, W/\omega_0$. For acoustic phonons the instanton size r_I is either $r_I \sim a$ or $r_I \sim \Lambda^{1/5} a$; i.e., $r_I \ll b \approx \Lambda a$. In the second case $\tau_I \sim 1/\bar{\omega}\Lambda^{1/5}$, $T_c \approx 2.4\,(W/E_B)(c/a)$ and $S_I(\beta = \infty) \approx 2.8\, c\rho/m^2 C^2 \approx 0.06\, Wb/c$. For $T \to 0$ the tunneling process occurs far from the ST barrier in the part of APS having an almost sheer back wall. So, the pattern of tunneling is quite different for these groups of phonons.

PREEXPONENTIAL FACTOR B(T) - LOW TEMPERATURES

Calculation of the prefactor $B(T)$ in (1) is most controversial problem in the theory of non-radiative transitions. While calculating the $B(T)$ for impurity centers by the method of multi-phonon transitions (or, which is the same, by the method of generating functions), the correct order of magnitude of $B(T)$ may be found only in the non-Condon approximation. As a result, the consideration becomes intricate. The necessity to include the non-adiabatic stage of the process connected with the trapping of a particle from a F-state at the emerging local level produces additional difficulties. In our formalism $B(T)$ may be decomposed into a product of several multipliers, each of which may be estimated independently.

For $T < T_c$ the instanton mechanism dominates, calculations are based on Eq. (2).

1. Free electron temperature factor

The Lagrangian L includes the energy of a free particle, $E(\mathbf{k})$, when

$|\tau_I| < |\tau| < \beta/2$. As a result, the averaging of (2) over the Maxwellian distribution produces the multiplier:

$$p_e(T) = (1 - 2\tau_I T)^{-3/2} . \quad (7)$$

The T dependence brought about by $p_e(T)$ dominates in $w(T)$ at low temperatures. Since τ_I has different signs for ST and for normal centers, the opposite T dependence of $w(T)$ can be expected in the low T limit.

2. Electronic transition

A slow free particle is trapped by the increasing potential $V(\mathbf{r}, t)$ after a local level appears at $t = t_0$. The trapping is completed when the level becomes adiabatic; its depth at this moment is $\Omega \sim (\bar{\omega}^2 w)^{1/3}$, and the radius of the quantum state $r_{tr} \sim (m\Omega)^{-1/2}$. The volume effective in the trapping $v_{tr} \sim r_{tr}^3$, brings in the multiplier

$$w_{tr} \sim v_{tr}/V \sim (m\Omega)^{-3/2}/V , \quad (8)$$

where V is the normalization volume.

3. Number of extremals

Trapping may occur along different extremal paths since a particle may be trapped at different lattice sites. This results in the multiplier

$$\nu(V/v) , \quad (9)$$

$v \approx a^3$, $\nu = 1$ for intrinsic ST and equals the dimensionless concentration of defect sites in other cases.

4. Time-type zero mode

Since the moment when the trapping begins is arbitrary, there exists a degeneracy resulting in the existence of a zero frequency mode. As usual [10], this generates the multiplier

$$\bar{\omega} S_I^{1/2} \quad (10)$$

in the result of the integration over different extremal paths in (2).

All together these multipliers yield $\bar{\omega}\nu B(T)$, hence

$$B(T) \sim p_e(v_{tr}/v) S_I^{1/2} \sim p_e(E_B/\bar{\omega})^{3/2} . \quad (11)$$

Eqs. (9) and (11) hold when $r_I \sim a$, and different extremal paths are well separated. In the opposite case, when $r_I \gg a$, there are three soft modes in the expansion of $S[Q(t)]$ near each extremal path. In the continuum limit these modes transform into three space-type zero modes, they introduce in $B(T)$ an additional multiplier $(a/r_I)^3 S_I^{3/2}$. Hence, Eq. (11) must be replaced by

$$B(T) \sim p_e(W/\bar{\omega})^3 , \quad (12)$$

in the continuum limit for extrinsic ST.

The enhancement parameter $B(T)$ in Eqs. (11) and (12) is always found to be much greater than one because it varies to a power greater than one of the large adiabatic parameter, $E_B/\bar{\omega}$ or $W/\bar{\omega}$, the ratio of a characteristic electronic energy to the vibrational quantum energy. The effect of reemission [11] is incorporated in (8) and (11). The evaluation of the numerical coefficient in $B(T)$ is a very complicated problem, and the coefficient has not so far been found rigorously for any model yet, for specific models it may be estimated by perturbation theory provided the non-adiabaticity is small [9].

PREEXPONENTIAL FACTOR B(T) HIGH TEMPERATURES

For $T > T_c$ the Arrhenius mechanism dominates for the APS of Fig. 1a. For $T > T_c$ the total flux includes a classical component and besides an appreciable part of it passes by tunneling near the top of the barrier. The analogy with the Arrhenius surmounting of a barrier by a nonlinear multimode system becomes clear if (2) is expressed not via electron $G^{R,A}$ but via phonon $D^{R,A}$ Green's functions, the latter include the factor $\exp(iS_0)$. For a nonlinear system the prefactor $B'(T) \sim 1$ [7]. However, in the trapping problem, there is also an additional factor connected with free-bound electronic transitions. The final expression is:

$$\nu B(T) \sim B'(T)(v_{tr}/V)(\nu V/v)(\Omega/T)^{3/2} \sim \nu (E_B/T)^{3/2}. \tag{13}$$

In (13) two of the multipliers coincide with (8) and (9). The factor $(\Omega/T)^{3/2}$ is proportional to the phase volume in the **k**-space inside which the matrix element of the transition is large (it decreases rapidly for $kr_{tr} \gg 1$, cf. Eq. (15)). Eq.(13) differs from (11) by the absence of p_e and by the substitution $\bar{\omega} \to T$. The same holds for (12) in the continuum limit. So, in the Arrhenius region

$$B(T) \sim (E_B/T)^{3/2} \quad \text{or} \quad B(T) \sim (W/T)^3 \tag{14}$$

for the discrete and continuum limits, respectively. Besides, a slower dependence on T for $T < \bar{\omega}$ comes from B' because of the change in the vibrational frequencies at ST.

TRAPPING OF HOT PARTICLES

The initial electron energy E can in principle facilitate the surmounting of the ST barrier by a system. This is the case only when the energy is transferred from the electron to the lattice coherently. However, the probability of a such energy transfer is [5c, 12]

$$w_{tr}(E) \sim (v_{tr}/V)\exp\{-(4/3)(E/\Omega)^{3/2}\}. \tag{15}$$

The trapping rate w_{tr} (cf. Eq. (8)) is strongly suppressed at large E and may become a bottle-neck in the process.

For $E < W$, in the Arrhenius regime the action is [5b,c]

$$S_A(E, T) = (W - E)/T + (4/3)(E/\Omega)^{3/2}. \tag{16}$$

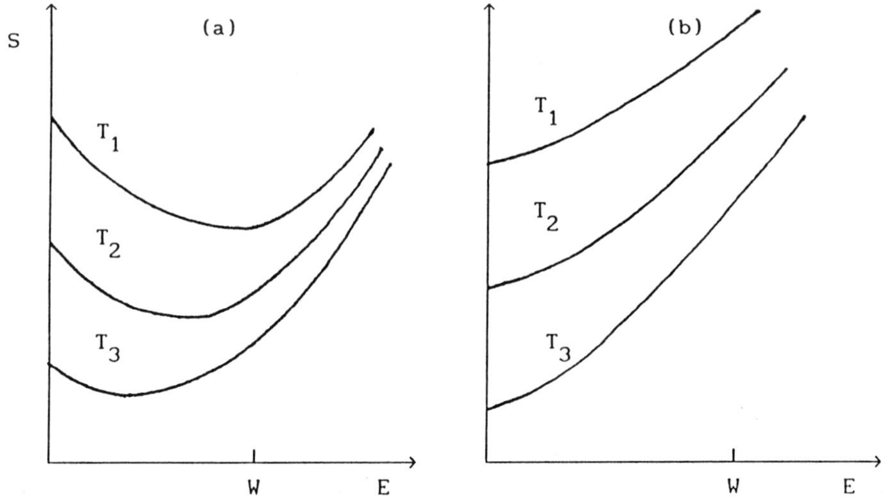

Fig. 4. The E-dependence of S at various lattice temperatures ($T_1 < T_2 < T_3$). a) ST, b) normal centers.

The first term is the energy deficit which is covered by heat bath. The last term comes from the integration along the appendage A (Fig. 2a), the moment of trapping $t_{tr}(E)$, being determined by the condition $E_{ST}(t) = -\Omega^3(t-t_*)^2 = E$, $E_{ST}(t)$ is the depth of the energy level. In the instanton regime

$$S_I(E, T) = S_I(T) - 2\tau_I E\,. \tag{17}$$

The integration along the appendage I (Fig. 2a) makes only an imaginary contribution into S which may be omitted. For $E \gg w$, S increases monotonously with E. The dependence $S(E)$ is schematized in Fig. 4, for $E \ll w$ it has an opposite slope for ST and for normal centers [5b,c,8].

Eqs. (15) and (16) describe the E and T dependence of $w(T, E)$ in the exponential approximation.

DEFECT PRODUCTION

The ST process can result in production of lattice defects (vacancies and interstitials). The self-consistent theory of defect production must be microscopic, but the stage of the process preceding the emergence of the lattice defect can be described by the continuum theory. After the system exits from below the barrier, at $t > t_0$ (Fig. 2), the electron wave function is contracting and the potential well is getting deeper. At the moment $t_c((t_c - t_0) \sim \overline{\omega}^{-1})$ the system collapses; for $t \to t_0$ it is described by the self-similar solution

$$\psi(\mathrm{r}, t) = a^{-3/2}(t)\psi_c(r/a(t)), \quad a(t) \propto (t_c - t)^{\alpha}\,. \tag{18}$$

The magnitude of $\alpha > 0$ depends of the type of electron-phonon coupling. As a result, a finite part of the energy J is released on a microscopic scale as the kinetic energy of a few atoms. This concentration of the kinetic energy conditions the appearance of a lattice defect.

REFERENCES

1. C. H. Henry, D. V. Lang, Phys. Rev. B15, 989 (1977).
2. E. I. Rashba, in "Excitons" (North Holland, Amsterdam, 1982), p. 543.
3. M. Ueta et al., Excitonic Processes in Solids (Springer, Berlin, 1986).
4. A. S. Ioselevich, Sov. Phys. JETP 54, 800 (1982).
5. A. S. Ioselevich, E. I. Rashba, (a) Sov. Phys. JETP 61,1110 (1985), (b) ibid. 64, 1137 (1986), (c) Solid State Commun. 55, 705 (1985).
6. S. V. Iordanskii, E. I. Rashba, Sov. Phys. JETP.47, 975 (1978).
7. I. Affleck, Phys. Rev. Lett. 46, 388 (1981).
8. S. V. Meshkov, Sov. Phys. JETP 62, 1000 (1985).
9. V. N. Abakumov et al., Sov. Phys. JETP. 62, 853 (1985).
10. J. S. Langer, Ann. Phys. 41, 108 (1967).
11. T. N. Morgan, Phys. Rev. B28, 7141 (1983).
12. Yu. N. Demkov, Sov. Phys. JETP. 19, 762 (1964).

IS THE WORLD AT THE BORDER OF CHAOS?

Per Bak
Physics Department, Brookhaven National Laboratory, Upton NY 11973

Dissipative dynamical systems with many degrees of freedom naturally evolve to a self-organized critical state with fluctuations (avalanches) extending over all length- and time-scales [1]. The systems operate at the border of chaos, with zero Lyapunov exponent and algebraic growth of initial deviations. This picture has support from numerical and analytical model calculations, and from experiments by Held on sandpiles [2], by Babcock and Westervelt [3], and by Che and Suhl [4] on magnetic domain patterns, and by God on earthquakes. Applications to turbulence, biology, and economics have been suggested.

THE SELF-ORGANIZED CRITICAL STATE

Many dynamical phenomena in nature are characterized by long-range spatial and temporal fluctuations. In particular, the signal emitted from sun-spots, quasars, river flow, traffic, resistors, and economics etc may have a power-law "1/f" spectrum extending over many decades. Also, it has been pointed out that dissipation in turbulence, and the bright matter in the universe, and earthquake epicenters are distributed on fractal sets with power-law spatial correlations [5]. Despite the ubiquity of these phenomena, no general explanation has been put forward until recently. In 1987, I pointed out in a collaboration with C. Tang and K. Wiesenfeld that extended, interactive, dissipative dynamical systems may evolve to a critical attractor with no characteristic time scale [1]. The dynamics arises from a distribution of avalanches of all sizes limited only by the size of the system. The "1/f" noise should be seen as the natural signal from extended dynamical systems since it is essentially deterministic in nature, and does not originate from external random noises. The attractor can be viewed as critical in three different senses: first, the temporal and spatial correlation functions are power-laws as for equilibrium thermodynamic systems precisely at the critical point for second order phase transitions; second, the response to a single perturbation propagates as a critical chain reaction where the branching of activity is balanced by the deaths of activity; third, the attractor is critical with respect to chaotic behavior, with zero Lyapunov exponent. Numerical and analytical calculations by us, and by others [6-9], on discrete and continuous, deterministic and random, models have been carried out, and unequivocally support the existence of the self-organized critical state.

The prototypical example of self-organized critical phenomena is a pile of sand onto which grains are added slowly. As the pile grows, there will be bigger and bigger avalanches. Eventually the slope will settle around the "angle of repose" and there will be avalanches with a wide range of sizes limited only by the size of the pile. The most striking feature of the self-organized critical state is its resilience. The system responds to any perturbation, randomness etc. by returning to the critical state after a transient period. If "snowscreens" are added in an effort to prevent falling sand, the pile simply builds up to a steeper critical state, thus completely counteracting any effort to take the system away from criticality. It is this resiliency which makes the self-organized critical state a viable candidate for the phenomena mentioned above.

In the following, the emergence of the self-organized critical state will be demonstrated in the context of a few simple model systems. One model may represent the dynamics of earthquakes, [10,11] or a squeezed metal rod. Condensed matter physics generally deals with "interactive dissipative dynamical systems", so (not surprisingly) model systems for the study of self-organized criticality are abundant here. Motion of dislocations in resistors and pinned flux lines in superconductors come to mind. Two very recent experiments on self-organized criticality in domains in magnetic recording tape, and in magnetic garnet films will be briefly reviewed.

EARTHQUAKE DYNAMICS OR PLASTIC FLOW

The most direct application of the idea of self-organized criticality might be to earthquakes. Consider two techtonic plates grinding against each other. The surface of contact is continuously being pushed towards an instability where a part of one plate slips relative to the other plate. The number of earthquakes at which an energy E is released is given by the famous Gutenberg-Richter law, $N(E) \approx E^{-a}$, where a is a constant between 1.2 and 1.7 [12]. Thus, earthquakes of all sizes are possible, although large earthquakes are less likely than small ones.

The situation that we want to describe can be visualized in terms of the set-up shown in fig.1. Two segments of material are slowly squeezed against each other causing them to slip along their interphase. The experiment has actually been done by Bobzov and Lebedkin [13] who used aluminum and niobium rods. A "fault" region was generated as the pressure increased, causing a transition from elastic flow to plastic flow. They measured voltage pulses across the sample generated by motion of dislocations. They indeed observed "earthquakes" along the fault with a power-law distribution independent on the material and mechanism (believed to be different for the two materials) for the slip. In the present context, the blocks are techtonic plates grinding against each other along a fault, or a fault system. Now and then parts of the plates will slip relative to each others; these slips are the ruptures of the crust at earthquakes.

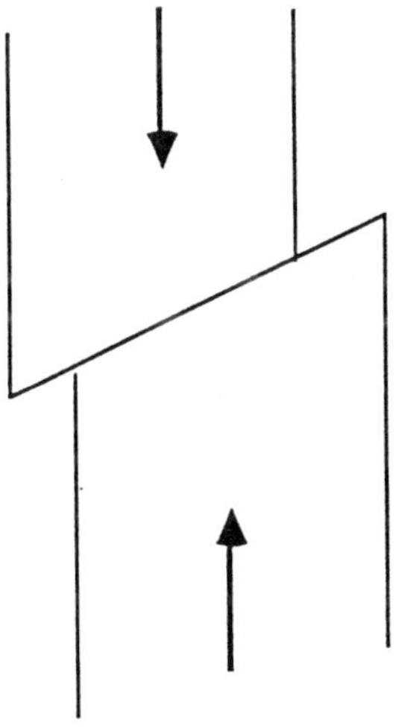

Figure 1. Slip region generated by squeezing a metal rod or techtonic plates.

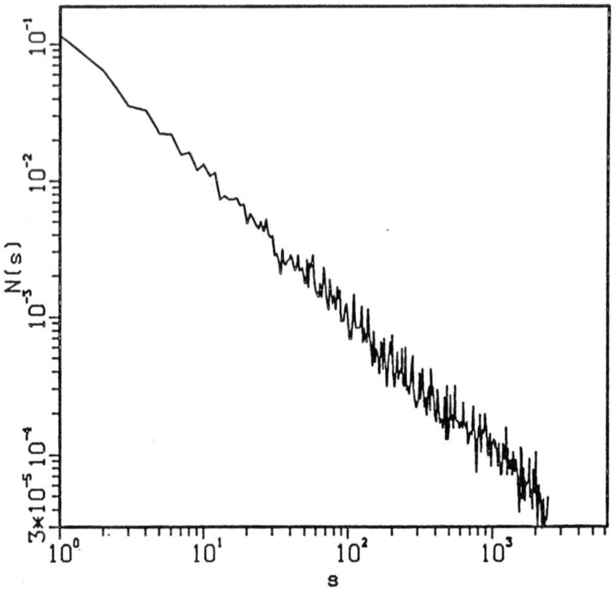

Figure 2. Size distribution of earthquakes for 50 ×50 system [11].

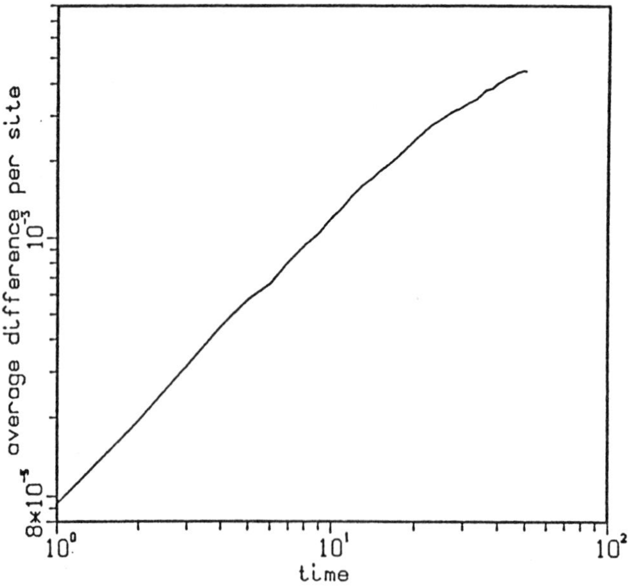

Figure 3. Power-law evolution of small perturbation. The system is "at the border of chaos".

For simplicity one plate is assumed to be rigid, and the other to be an elastic medium represented by a 2d array of blocks, at positions (i,j), $0<i,j<N$ connected by springs. The forces represented by the springs by no means have to be harmonic; they may for instance be hard core interactions. In general, the force is an antisymmetric function of the distances to the neighbors. The blocks interact with the rigid plate by static and dynamic friction forces. We assume that the array is open at one end and extends infinitely in the other direction. Whenever the spring force on a particular block exceeds the critical static friction force, it slides until the interaction forces have been reduced to the critical dynamical friction; then it stops. In the aluminum rod experiment, the process may be dislocation motion caused by shift of an atomic bond. During this process, potential energy is first converted to kinetic energy, and then dissipated (radiated) when the blocks are decelerated by the frictional forces.

Of course, since the blocks are at rest between the slips the total force on each block is zero, so the spring forces exactly balance the friction forces (Newton's third law). Hence, the reduction in friction force is transferred to the spring forces, and thus to neighbor friction forces. For simplicity, it is assumed that the force is redistributed evenly among the nearest neighbors. Note that while the forces are conserved, the density of blocks is not; this distinguishes the model from leap spring models where the average distance fixed by the connection of leap springs to rigid parts of the plates. The model is driven by slowly pushing the rigid surface relative to the other surface. The time-scale set by the pushing is a geological one and can be viewed as infinitely large compared with any realistic observation time; this is essential for the generation of power-laws and fractal scaling for spatial and temporal correlation functions.

Let us monitor the friction force z_i (= - the spring force) on the i'th block. Initially a random distribution of sub-critical forces are chosen. The z_i's grow at a small rate p until somewhere z reaches the critical value and a slip event takes place [11]. Without loss of generality the critical friction force is chosen to be an integer Z_{cr}, and the reduction of friction force is taken to be 2d units, so

$$z_i \rightarrow Z_{cr} - 2d\,, \quad z_{nn} \rightarrow z_{nn}+1 \text{ whenever } z_i > Z_{cr}.$$

This equation may simulate a general nonlinear transport situation. With a little imagination it can be viewed as the stop-and-go process for highway traffic flow, one of the phenomena believed to show 1/f-like behavior.

The process initiated by a single slip event transfers force to the neighbors causing them to be more unstable, allowing for a chain reaction. This chain reaction is the earthquake in our model. As the process continues, the forces z_i generally increase, causing bigger and bigger earthquakes. Eventually, the system will be pumped-up to a statistically stationary state where the increase in force is precisely balanced by the slips. We assume that the crust of the earth has had sufficient time to reach the stationary state, so we shall generally be concerned with this state only.

Let us convince ourselves that the system is indeed critical. Figure 2 shows the distribution of earthquakes found by simulation of the model in two dimensions. The straight line in the log-log plot indicates that the energy distribution is indeed a power law,

$$N(s) = s^{-a}\,, \qquad a \approx 1.1$$

in accordance with the Gutenberg-Richter law. Geophysicists have told us that the crust of the earth (at least for moderate earthquakes) should be viewed as a three-dimensional system. In 3d we find a ≈ 1.3, in somewhat better agreement with observations. For large earthquakes extending over more than the thickness of the crust one might expect a cross-over from 3d to 2d exponents.

We have studied the evolution of two configurations of the critical state which differ by a very small random perturbation, and followed the evolution of the difference. In chaotic systems this difference increases exponentially $\approx e^{-\lambda t}$ where λ is the Lyapunov exponent. Figure 3 shows that λ is zero, and that the difference increases algebraically $\approx t^c$ where c is a positive exponent: our turbulent system is at the border of chaos. The fact that the power is positive indicates that the uncertainty of the state of the system grows, albeit much less dramatically than for chaotic systems. The situation for predicting earthquakes is less desperate than for fully chaotic systems, although one has the added complexity of having to deal with very many degrees of freedom. We denote such systems as "weakly chaotic". Since many dynamical systems are expected to be self-organized critical, we expect weak chaos to be quite ubiquitous in nature.

Once the existence of the self-organized critical state has been established it is not so difficult to derive other exponents, such as the fractal dimension, characterizing different correlation functions. In particular, Ito and Matsuzaki [14] have generalized our model by adding a random disturbance to sites which have just been subjected to an earthquake. They obtained a clustering of epicenters with a fractal dimension of 1.1. They also were able to obtain a power law distribution of aftershocks (Omori's law). Sornette and Sornette [15] were able to show the existence of 1/f noise in the time gap between large earthquakes. Carlson and Langer have found self-organized criticality in a one-dimensional continuous stick-slip model with inertia [16].

THE GAME OF LIFE

The "Game of Life", a simple cellular automaton showing complex static and dynamic configurations [17]. The game simulates the rise, fall, and alterations of a society of living organisms. The main interest in "Life" has been on the generation of complexity in local configurations, which has been thought of as mimicking a general scenario of the emergence of complexity in nature. In contrast to the "sandpile" model (but in accordance with most realistic situations) "Life" has no local conservation laws. We have shown that local configurations in the "Game of Life" self-organize into a critical state with the critical density of live individuals $p_c \approx 0.03$ [18].

The "Game of Life" is defined on a square lattice. There are two states on each lattice site, representing the presence or absence of a live individual. The rules for the evolution of "Life" are very simple:

i) The fate of a live individual depends on its nearest neighbors; it will die at the next time step if there are less than 2 (over-exposure) or more than 3 live neighbors (over-crowding); it will remain alive otherwise.

ii) At a dead site, a new individual will be born at the next time step only if there are exactly 3 live neighbors.

In order to elucidate the collective behavior of the society we study the following process: Starting with a random distribution of live sites, the system evolves according to the rules i) and ii) until it comes to "rest" in a simple periodic state with a distribution of local still life and simple cyclic life. There is no propagating activity. The system is then perturbed at a randomly chosen local site, for instance by adding a live individual, and is allowed to evolve according to the rules until it comes to rest again. As the process is repeated, the system evolves into a statistically stationary state.

We measure the number, s, of births (or deaths) following a single perturbation in the stationary state. Indeed the distribution of clusters of size s averaged over thirty thousand perturbations is a power law, $D(s) \propto s^{-\tau}$, $\tau \approx 1.4$, and so is the distribution of the durations of the perturbations $D(T) \propto T^{-b}$, with $b \approx 1.6$. The fact that the activity does not decay, or explode exponentially (becomes chaotic) indicate that life and death are highly correlated in time and space to allow the activity to continue indefinitely: the system has evolved into a critical state.

If life (and its environment) is indeed at a critical state, the concept of a stable equilibrium of evolution is meaningless. Nature is ever changing along consecutive configurations of the critical state. The disappearance of a large group of species, such as the extinction of the dinosaurs, does not require a large external force (meteor) but may be an intrinsic evolutionary phenomenon.The apparent logical connectivity does not indicate that Nature is in balance. In analogy with the sand model, any effort to stop the clock and trying to "conserve Nature" is a meaningless and certainly losing battle. The more we try to place "snowscreens" around us to preserve ourselves, or the more we try to restructure Nature, the bigger will be the apocalypse when the configuration of nature changes into one where we are no longer present.

A FOREST-FIRE MODEL

The phenomenon of turbulence is essentially not understood. On the geometric aspects, it has been suggested that in the turbulent state, energy is dissipated on a fractal set, with dimension slightly higher than two [5]. Some phenomenological models have been proposed, in which the fractal set is pre-assumed [19]. It is, however, essential to understand the dynamical mechanism which generates the fractal itself. We have studied a simple "forest-fire" model [20]. Specifically, we focus on the spatial distribution of dissipation (fire) and its dependence on the driving force (tree growth). The mechanism for fractal dissipation is demonstrated explicitly. Our lattice model is defined in any dimension, with the following simple rules:

i) Trees grow with a small probability p from empty sites at each time step;
ii) Trees on fire will burn at the next time step.
iii)The fire on a site will spread to trees at its nearest neighbor sites at the next time step.

There is only one parameter in the model, namely the growth rate of trees. We are interested in the limit where trees grow infinitely slowly. We find that in this limit the fire is characterized by a correlation length $\xi(p) \approx p^{-\upsilon}$. The critical point is at p=0. Because of the finite size of our lattices, the probability p has to be non-zero (such that the correlation length is smaller than the system) in order to prevent accidental extinction of the fire. Starting from a homogeneous distribution of trees and fires, the forest fire evolves to a stationary state. Figure 4 shows a snapshot of the forest on fire, taken in the stationary state after the initial transient period. Note the large coherent domains of trees separated by a fires, indicating that the system is operating near a critical point. By measuring the number distribution D(r) of fire at a distance r from a chosen site on fire, we obtain the fractal dimension: $D = 1.0 \pm 0.2$ in 2d and $D=2.5 \pm 0.2$ in 3d. This value of D agrees with experimental observations for turbulence. Of course, this could be accidental. If one prefers the language from traditional equilibrium critical phenomena, the fire-fire correlation function $G(x)=\langle f(x')f(x'+x)\rangle$ decays as $G(x) \propto x^{2-d-\eta}$, $\eta = 2-D \approx 1.0$ in 2d; $\eta \approx -0.5$ in 3d. The scaling has been confirmed by means of a Monte Carlo Renormalization Group type calculation [20].

The model may be rather directly applied to spreading of diseases, or propagation of chemical activity, such as real fire. We believe that the model is simple enough to allow for explicit theoretical analysis, for instance renormalization group theories based on expansions around the upper critical dimension. The limitation of the present model when applied to turbulence is obvious. It is a lattice model, thus the "Kolmogorov" length for energy dissipation is a fixed length scale, namely the lattice spacing. The dynamics is simple; for instance, the coherent domains are structureless. However, our study on a specific dynamical model shows explicitly that certain principles are viable: (a) Driven non-equilibrium systems may operate near critical points. This is NOT low-dimensional chaos and the fractal nature of turbulence can not be described in terms of a strange attractor. The dynamics operate at the border of chaos with zero Lyapunov exponent [21].

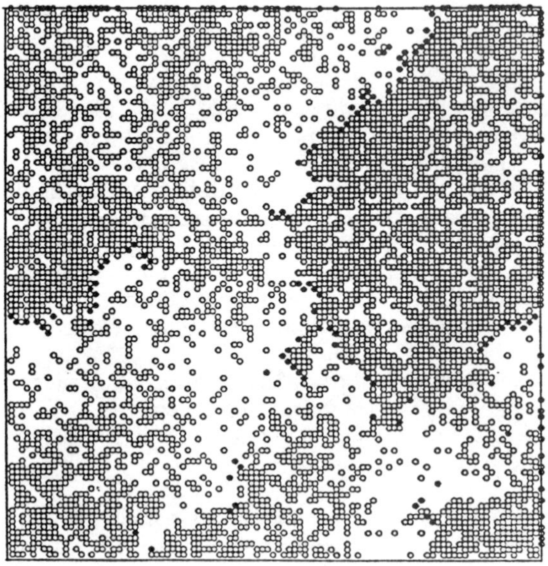

Figure 4. Forest-fire at the critical state. Open circles represent live trees, filled circles represent burning trees [20].

EXPERIMENTS

In view of the deterioration of funding for fundamental science, experiments on sandpiles may be the way of the future. Indeed, several groups have undertaken investigations on the dynamics of sandpiles. Glen Held [2] has measured the distribution of avalanches in a not-so-low cost computerized sandpile, where single grains of aluminum sand (or sand from Smith Point Beach on Long Island) were deposited on a flat plate, and the distribution of avalanches were measured by monitoring the mass fluctuations of the pile. Babcock and Westervelt [3] observed avalanche-like, topological rearrangements of cellular domain patterns in magnetic garnet films. The avalanches, triggered by variations in magnetic field, showed self-organized criticality where the lifetimes T and the sizes s obey power law distributions $D(T) \approx T^{-2.23}$ and $D(s) \approx s^{-2.30}$. Che and Suhl [4] have studied arrangements of domain walls separating regions of opposite magnetization on magnetic recording tape and argue that they are arranged in self-organized critical states as discussed above. But of course, since the criticality is self-organized we do not have to set up experiments in the laboratory; it is enough to watch the dynamics of naturally occurring dynamical systems, as was done for earthquakes.

ACKNOWLEDGMENTS

The research reviewed here was performed in collaboration with Chao Tang, Kurt Wiesenfeld, and Kan Chen. The work was supported by the Division of Materials Science, US Department of Energy, under contract DE-AC02-76CH00016.

REFERENCES

1. P. Bak, C. Tang, and K. Wiesenfeld, Phys. Rev. Lett. **59**,381(1987); Phys. Rev. A **38**, 364 (1988); C. Tang and P. Bak, Phys. Rev. Lett. **60**, 2347 (1988); C. Tang and P. Bak, J. Stat. Phys. **51**, 797 (1988); P. Bak and C. Tang, Phys. Today **42**, S27 (1989).
2. G. Held, Private communications
3. K. L. Babcock and R. M. Westervelt, Phys. Rev. Lett. (in press)
4. X. Che and H. Suhl, Phys. Rev. Lett. (in press)
5. B. Mandelbrot, J. Fluid Mech. **62**, 331 (1974); *The Fractal Geometry of Nature* (W. H. Freeman, San Francisco, 1982).
6. S. P. Obukhov, in "*Random Fluctuations and Pattern Growth: Experiments and Models*" edited by H. E. Stanley and N. Ostrowsky (Kluwer Academic, Dordrecht, Boston, London, 1989), p 336.
7. L. P. Kadanoff, S. R. Nagel, L. Wu, and S. Zhou, Phys. Rev. A, **39**, 6524 (1989).
8. Y-C. Zhang, Phys. Rev. Lett. **63**, 470 (1989).
9. T. Hwa and M. Kardar, Phys. Rev. Lett. **62**, 1813 (1989).
10. P. Bak and C. Tang, J. Geophys. Res. **94**, 15635 (1989).
11. P. Bak and K. Chen, "Dynamics of Earthquakes", to be published in "*Fractals and their Application to Geology*" (Geological Society of America, Denver, 1990)
12. B. Gutenberg, and C. F. Richter, Ann. di Geofis. **9**, 1 (1956).
13. W. S. Bobzov and M. Lebedkin, (1989); their results were kindly communicated to us by S. Obukhov.

14. K. Ito and M. Matsuzaki, submitted to J. Geophys. Res.
15. A. Sornette and D. Sornette, Europhys. Lett. **9** 192 (1989)
16. J. M. Carlson and J. S. Langer, Phys. Rev. Lett. **62**, 2632 (1989).
17. E. R. Berlekamp, J. H. Conway, and R. K. Guy, *Winning Ways* (Academic Press, 1985).
18. P. Bak, K. Chen and M. Creutz, Nature, 342, **780** (1989)
19. U. Frisch, P. Sulem, and M. Nelkin, J. Fluid Mech. **87**, 719 (1978); R. Benzi, G. Paladin, G. Parisi, and A. Vulpiani, J. Phys. A **17**, 3521 (1984)
20. P. Bak, K. Chen, and C. Tang, submitted to Phys. Rev. Lett.
21. M. H. Jensen, P. Bak, and K. Chen, submitted to Physics Letters.

The submitted manuscript has been authored under contract DE-AC02-7600016 with the Division of Material Sciences, U. S. Department of Energy. Accordingly, the U. S. Government retains a nonexclusive, royalty-free license to publish or reproduce the published form of this contribution, or allow others to do so, for U.S. Government purposes.

HYDRODYNAMICS OF CLASSICAL AND QUANTUM LIQUIDS WITH FREE SURFACES

I. M. Khalatnikov*
Theoretische Physik,
Eidgenössische Technische Hochschule-Hönggerberg, 8093 Zürich, Switzerland
*on leave from L.D. Landau Inst. for Theoretical Physics, USSR Academy of Sciences

1. HYDRODYNAMICS OF A CLASSICAL LIQUID

Let us start our study with the simplest case of the potential isentropic motion of liquids. As is known, it is then possible to introduce the potential of the velocity $+\alpha$ so that the velocity $\mathbf{v}$ is defined by the relation:

$$\mathbf{v} = \nabla\alpha \tag{1.1}$$

and the momentum corresponding to a unit volume of the liquid by

$$\mathbf{j} = \rho\,\nabla\alpha\,. \tag{1.2}$$

As has been shown in our paper [1] Formula (1.2) is a particular case of the general expression for momentum density

$$\mathbf{j} = -\sum p\nabla q \tag{1.3}$$

where p and q are canonically conjugate variables (momentum p with the coordinate q) following from the general symmetry properties of the system. In the liquid to the group of translations the generator ∇ corresponds, and its density, according to (1.3), determines the momentum. Thus, in accordance with (1.2) the variables $-\rho$ and α are the momentum and coordinate, respectively. For the general case of non-isotropic and non-potential motion, according to [1] for the density and momentum we have

$$\mathbf{j} = \rho\,\nabla\alpha + s\,\nabla\beta + f\,\nabla\gamma\,, \tag{1.4}$$

where f and γ are the Clebsch variables and β is a coordinate, conjugate to the momentum $-s$. In (1.4) there is an extra function related to a certain hidden gauge group (for details see [2]).

Coming back to the simplest case of potential motion, we write out the action for our system. Having in mind a further investigation of the interaction of surfacial: and bulk modes in the liquid, we shall take into account gravitational and capillary forces.

Recall the formula, determining the Hamiltonian

$$H = \sum p\dot{q} - L \, . \tag{1.5}$$

Then for the action we have

$$A = \int dt \int^{\zeta(x,y,t)} dV\,(-H + \sum p\dot{q}) = \int dt \int^{\zeta} dV\,(-\frac{\rho \mathbf{v}^2}{2} - \varepsilon(\rho) - \rho g z - \rho\dot{\alpha}) - \int dt\, \kappa \int dS. \tag{1.6}$$

κ is the surface tension, dS is an element of the surface of the liquid, $\varepsilon(\rho)$ is the energy density of the liquid where it is assumed that the liquid fills the half-space $(z = \xi,\ z = -\infty)$

Calculating the variation of the action (1.6) with (1.1) taken into account, we get

$$\delta A = \int dt \int^{\zeta(x,y,t)} dV \{[-\frac{(\nabla\alpha)^2}{2} - \frac{\partial\varepsilon}{\partial\rho} - \dot{\alpha} - gz\,]\,\delta\rho + (\dot{\rho} + div\,\rho\mathbf{v})\delta\alpha\}$$

$$+ \int dt \int dx\,dy\,\{-\rho v_N\sqrt{1+(\nabla\zeta)^2}\,\delta\tilde{\alpha} + \dot{\zeta}\rho\,\delta\tilde{\alpha} - \delta\zeta\,[\rho\dot{\tilde{\alpha}} + \frac{\rho}{2}(\nabla\tilde{\alpha})^2 + \varepsilon + g\zeta - \kappa\, div\,\frac{\nabla\zeta}{\sqrt{1+(\nabla\zeta)^2}}]\,\}_{z=\zeta} = 0. \tag{1.7}$$

Here

$$\upsilon_N = \frac{\upsilon_z - \upsilon_x\zeta_x - \upsilon_y\zeta_y}{\sqrt{1+(\nabla\zeta)^2}}$$

∇ and Δ are $2d$ operators in the x, y plane

$$\tilde{\alpha} = \alpha\,|_{z=\zeta} \, .$$

Equating variations of the bulk part of the action to zero, we get an equation of motion

$$-\frac{(\nabla\alpha)^2}{2} - \mu - \dot{\alpha} - gz = 0\,, \quad \mu = \frac{\partial\varepsilon}{\partial\rho} \tag{1.8}$$

and a continuity equation

$$\dot{\rho} + div\,\rho\mathbf{v} = 0 \qquad \mathbf{v} = \nabla\alpha\, . \tag{1.9}$$

Similarly, from the surfacial part of the action, we find

$$\dot{\zeta} = \upsilon_N\sqrt{1+(\nabla\zeta)^2} \tag{1.10}$$

$$\rho\dot{\tilde{\alpha}} + \rho\frac{(\nabla\tilde{\alpha})^2}{2} + \varepsilon + \rho g\zeta + \kappa\, div\,\frac{\nabla\zeta}{\sqrt{1+(\nabla\zeta)^2}} = 0\, . \tag{1.11}$$

It follows from (1.11) and (1.8) that at $z=\zeta$ on the surface of the liquid we have

$$p+\kappa\, div \frac{\nabla\zeta}{\sqrt{1+(\nabla\zeta)^2}} = 0. \tag{1.12}$$

where the pressure is

$$p = -\varepsilon + \mu\rho \tag{1.13}$$

Before passing over to the Hamiltonian form of the equations, recall that for the bulk part of the liquid a canonical pair is $-\rho$ (momentum) and α (coordinate). For the surfacial part the problem of choice is more difficult, since $\dot{\zeta}$ first emerged not in the expression for the action A but in expression (1.7) for its variation. We think it naturally follows from (1.7) how to obtain an appropriate ζ and momentum ψ; the momentum being determined by its variation

$$\delta\psi = \rho\delta\tilde{\alpha}\,. \tag{1.14}$$

This fact can be interpreted as manifestation of non-holonomity. There is no other explanation of this fact but the correctness of the choice is confirmed by the fact that eqs. (1.10) and (1.11) are obtained by us in the Hamiltonian form.

Actually, the Hamiltonian of the system (equ. (1.6)) equals

$$H = \int^{\zeta} dV\,(\rho\frac{(\nabla\alpha)^2}{2}+\varepsilon+\rho g z)+\kappa\int dS \tag{1.15}$$

Write down variations of H

$$\delta H = \int^{\zeta} dV\,[\delta\rho\,(\frac{(\nabla\alpha)^2}{2}+\mu+gz)+\delta\alpha\,(-div\,\rho\nabla\alpha)]$$

$$+\int dx\,dy\,[\delta\zeta\,(\rho\frac{(\nabla\tilde{\alpha})^2}{2}+\varepsilon+\rho g\zeta)+\rho\upsilon_N\sqrt{1+(\nabla\zeta)^2}\,\delta\tilde{\alpha}-\kappa\, div\frac{\nabla\zeta}{\sqrt{1+(\nabla\zeta)^2}}\,\delta\zeta]\,. \tag{1.16}$$

υ_N is the projection of the velocity onto the normal to the surface

$$\upsilon_N = \frac{\upsilon_z-\upsilon_x\zeta_x-\upsilon_y\zeta_y}{\sqrt{1+(\nabla\zeta)^2}}$$

The Hamiltonian equations ensuing from (1.16), for the bulk part are

$$\dot{\alpha}=\frac{\delta H}{\delta(-\rho)} = -(\frac{(\nabla\alpha)^2}{2}+\mu+gz) \tag{1.8'}$$

$$-\dot{\rho} = -\frac{\delta H}{\delta \alpha} = div\,(\rho \nabla \alpha) \tag{1.9'}$$

and for the surfacial part are

$$\dot{\zeta} = \frac{\delta H}{\rho \delta \tilde{\alpha}} = upslion_N \sqrt{1+(\nabla \zeta)^2} = \upsilon_z - \upsilon_x \zeta_x - \upsilon_y \zeta_y \tag{1.10'}$$

$$\dot{\psi} = \rho \dot{\tilde{\alpha}} = -\frac{\delta H}{\delta \zeta} = -(\rho \frac{(\nabla \tilde{\alpha})^2}{2} + \varepsilon + \rho g \zeta - \kappa\, div \frac{\nabla \zeta}{\sqrt{1+(\nabla \zeta)^2}}). \tag{1.11'}$$

Eqs. (1.8')-(1.11') exactly coincide with eqs. (1.8)-(1.11) which virtually confirms the correct choice of the pairs of canonically conjugate variables.

For an arbitrary non-isentropic and non-potential motion, as is known, it is necessary to introduce 3 pairs of canonically conjugate variables (see (1.4)).

$$(p\ q), \quad (\rho\ \alpha), \quad (s\ \beta), \quad (f\ \gamma)\,.$$

And by analogy with the given derivation, one can repeat all calculations and obtain an overall system of hydrodynamic equations. For the surfacial part the canonically conjugate variables are the coordinate ζ and momentum ψ, determined by their variations

$$\delta \psi = \tilde{\rho}\ \delta \tilde{\alpha} + \tilde{s}\ \delta \tilde{\beta} + \tilde{f}\ \delta \tilde{\gamma} \tag{1.17}$$

(the tilde labels the values of the functions on the surface $z = \zeta$.)

To study nonlinear phenomena, e.g., processes of transformation of bulk waves into surfacial waves one should write down the expansion of the Hamiltonian up to cubic terms with respect to the canonical variables.

2. HYDRODYNAMICS OF QUANTUM SUPERFLUID LIQUID

The problem of the canonical formalism of the equations of hydrodynamics of superfluid liquid has been studied in detail in [1].

Recall, that the equations of hydrodynamics are obtained from the Hamiltonian

$$H = \int^{\zeta} dV(\frac{\rho v_s^2}{2} + \mathbf{p}\cdot\boldsymbol{\upsilon}_s + \varepsilon(\rho, s, \mathbf{p}) + \rho g z) + \kappa \int dS \tag{2.1}$$

where the superfluid velocity υ_s is determined by the potential α

$$v_s = \nabla \alpha$$

the momentum with respect to the motion of normal and superfluid parts is determined by the potentials β, f, γ

$$\mathbf{p} = s\, \nabla \beta + f \nabla \gamma \tag{2.2}$$

and, consequently, the total momentum **j** is

$$\mathbf{j}=\rho\,\nabla\alpha+s\,\nabla\beta+f\,\nabla\gamma\,, \tag{2.3}$$

the energy density is determined by the identity

$$d\varepsilon=Tds+\mu d\rho+(\mathbf{v}_n-\mathbf{v}_s)\cdot d\mathbf{p} \tag{2.4}$$

In the Hamiltonian (2.1) we have taken into account gravitational and capillary forces on the surface. According to (2.3) the pairs of canonically conjugate variables are

$$(-\rho,\ \alpha),\quad (-s,\ \beta),\quad (-f,\ \gamma). \tag{2.5}$$

To study acoustic and surfacial waves it is not necessary to take into account the Clebsch variables $(f,\ \gamma)$. Write out the hydrodynamic equations in the form of the Hamiltonian ignoring (f γ)

$$\dot{\alpha}=-\frac{\delta H}{\delta\rho}=(\frac{\upsilon_s^2}{2}+\mu+gz) \tag{2.6}$$

$$\dot{\rho}=+\frac{\delta H}{\delta\alpha}=-div\,(\rho\mathbf{v}_s+\mathbf{p}) \tag{2.7}$$

$$\dot{\beta}=-\frac{\delta H}{\delta s}=-T-\mathbf{v}_n\cdot\nabla\beta \tag{2.8}$$

$$\dot{s}=+\frac{\delta H}{\delta\beta}=-div\,(\mathbf{v}_n s)\,. \tag{2.9}$$

To describe surface motion we introduce canonical variables ζ and ψ, where the momentum ψ is determined by its variation similarly to (1.14).

$$\delta\psi=\tilde{\rho}\,\delta\tilde{\alpha}+\tilde{s}\,\delta\tilde{\beta} \tag{2.10}$$

(note that the tilda points to the fact that all functions are taken at $z=\zeta$)

$$\rho\dot{\tilde{\alpha}}+s\dot{\tilde{\beta}}=-\frac{\delta H}{\delta\zeta}=-(\frac{\rho\upsilon_s^2}{2}+\mathbf{p}\cdot\mathbf{v}_s+\varepsilon+\rho gz)|_{z=\xi}+\kappa\,div\,\frac{\nabla\zeta}{\sqrt{1+(\nabla\zeta)^2}} \tag{2.11}$$

$$\dot{\zeta}=\frac{\delta H}{\delta\psi}=\upsilon_{nN}\sqrt{1+(\nabla\zeta)^2}=\frac{jN}{\rho}\sqrt{1+(\nabla\zeta)^2} \tag{2.12}$$

where the subscript N labels the component of υ_n, normal to the surface.

Note we do not take into account the interaction of liquid with vapor in the boundary conditions (2.12). To do so would have considerably complicated the results without changing them qualitatively.

It follows from eqs. (2.6) and (2.11) that at $z=\zeta$ there is a constraint

$$p+\kappa\, div \frac{\nabla\zeta}{\sqrt{1+(\nabla\zeta)^2}}=0 \tag{2.13}$$

3. EQUATIONS OF HYDRODYNAMICS FOR SUPERFLUID VELOCITY WITH THE SURFACIAL WAVES TAKEN INTO ACCOUNT IN THE QUADRATIC APPROXIMATION

As independent variables, let us take the pressure p and temperature T. Then from the definition of the pressure

$$p=-\varepsilon+Ts+\mu\rho+\mathbf{p}(\cdot\mathbf{v}_n-\mathbf{v}_s) \tag{3.1}$$

in the approximation, quadratic with respect to velocities, we get

$$\rho\delta\mu=-s\,\delta T+\delta p-(\mathbf{p}\cdot\delta(\mathbf{v}_n-\mathbf{v}_s))=-s\,\delta T+\delta p-\frac{\rho_n}{2}(\mathbf{v}_n-\mathbf{v}_s)^2. \tag{3.2}$$

δ denotes deviation of the values of the functions from the equilibrium values.

Combining eqs. (2.6) and (2.8) and the boundary conditions (2.13), we get $(\sigma=s/\rho)$ at $z=\zeta$

$$-(\dot{\tilde{\alpha}}+\sigma\dot{\tilde{\beta}})=\frac{\upsilon_s^2}{2}-\frac{\rho_n}{2\rho}(\mathbf{v}_n-\mathbf{v}_s)^2+\frac{\rho_n}{\rho}\mathbf{v}_n\cdot(\mathbf{v}_n-\mathbf{v}_s)-\frac{\kappa}{\rho}\Delta\zeta+g\zeta$$

$$+\frac{1}{2}\frac{\partial\sigma}{\partial T}(\delta T)^2-\frac{\partial\sigma}{\partial p}\delta p\,\delta T+\frac{1}{2}\frac{\partial}{\partial p}\frac{1}{\rho}(\delta p)^2\ . \tag{3.3}$$

It is clear that the terms quadratic in velocity are equal to the kinetic energy

$$(\rho_s\frac{\upsilon_s^2}{2\rho}+\rho_n\frac{\upsilon_n^2}{2\rho})\ .$$

In the approximation when one can ignore the thermal expansion in superfluid liquid $(\partial p/\partial T=0)$ there occurs a complete division of the first sound (δp) and second sound (δT). For the first sound from (3.3) we find

$$-\dot{\tilde{\alpha}}=\frac{1}{2}(\nabla\tilde{\alpha})^2-\frac{1}{2c_1^2}\dot{\tilde{\alpha}}^2-\frac{\kappa}{\rho}\Delta\zeta+g\zeta. \tag{3.4}$$

Here $c_1^2=\frac{\partial p}{\partial\rho}$ is the velocity of the first sound.

We average this equation for a time exceeding the time of the acoustic oscillation transition and we find that

$$\bar{\zeta}=0$$

i.e., the average oscillations of the surface in the running acoustic wave have order higher than 2. This is a well-known result [L. Landau, E. Lifschitz "Mechanics of Continuous Media"].

For the second sound from (3.3) we get

$$-(\dot{\tilde{\alpha}}+\sigma_0\dot{\tilde{\beta}})=\frac{\rho_s}{2\rho_n}\sigma^2(\nabla\tilde{\beta})^2-\frac{1}{2}\frac{\partial\sigma}{\partial T}\dot{\tilde{\beta}}^2-\frac{\kappa}{\rho}\Delta\zeta+g\zeta\,. \tag{3.5}$$

Averaging this expression for a time much larger than the period of oscillations in the second sound and taking into account that the velocity of the second sound c is defined by the relation

$$c_2^2=\frac{\rho_s}{\rho_n}\frac{\sigma^2}{\frac{\partial\sigma}{\partial T}} \tag{3.6}$$

Then we again see that

$$\bar{\zeta}=0 \tag{3.7}$$

in the running wave of second sound.

This assertion holds also for calculations performed with the accuracy up to the the terms quadratic in $\gamma=(\frac{\partial\rho}{\partial T}\frac{T}{\rho})$.

The statement that $\zeta=0$ naturally holds only for the running wave. In the standing wave one can from (3.5) find a non-zero value of ζ. But for second sound we can make a stronger statement. For second sound $\tilde{\alpha}+\sigma_0\dot{\tilde{\beta}}$ is identically equal to zero $(j=\rho\nabla\alpha+s\nabla\beta=0)$. Therefore from (3.5) follows that instead of (3.7) we have

$$\zeta=0$$

for running wave of the second sound.

This result differs qualitatively from the result of R. Sorbello [3]. (See also [4].)

4. STANDING SURFACE WAVES, EXCITED BY SECOND SOUND

Let us briefly dwell upon the analysis of the experimental results of J. Olsen et al. [4], who have observed excitation of standing surfacial waves by the second sound in superfluid helium. To carry out this analysis we need equations of hydrodynamics with accuracy up to the second order terms. Then we shall retain only nonlinear terms generated by the second sound; these terms will play the role of the driving force in the linear equations. Introduce the designations

Then the boundary conditions on the surface according to (2.12) and (3.5) are written with the necessary accuracy as

$$-\dot{\tilde{\phi}} = \frac{1}{2}\frac{\partial\sigma}{\partial T}\,[u_2^2(\nabla\tilde{\beta})^2 - \dot{\tilde{\beta}}^2] - \frac{\mathcal{H}}{\rho}\Delta\zeta + g\zeta \tag{4.1}$$

$$\dot{\zeta} = \frac{\partial\tilde{\phi}}{\partial z} \tag{4.2}$$

Then the equations of motion in bulk, we need, are found from (2.6)-(2.9)

$$\dot{\phi} + \frac{1}{\rho_0}c^2\delta\rho + \frac{1}{2}\frac{\partial\sigma}{\partial T}\,[u_2^2(\nabla\beta)^2 - \dot{\beta}^2)] + \frac{1}{2}\gamma u_2^2(\nabla\beta)^2 = 0 \tag{4.3}$$

$$\dot{\rho} + \rho_0\Delta\phi - \rho_0\frac{\partial\sigma}{\partial T}div\,(\dot{\beta}\nabla\beta) = 0 \tag{4.4}$$

(where $\gamma = \frac{\partial\sigma}{\partial T}\frac{\partial}{\partial\ln\rho}\ln(\rho_s/\rho_n)$).

For simplicity, consider a linear standing wave, dependent on one coordinate x (in experiment [4] there has been cylindrical symmetry). The standing wave of the second sound is then defined by the expression for the potential

$$\beta = \beta_0\cos(kx)\cos(\omega t), \qquad \omega = u_2 k \tag{4.5}$$

Remember that the potential β is related to the temperature variation as

$$\dot{\beta} + \delta T = 0 \tag{4.6}$$

From Eqs. (4.1) and (4.9), if we neglect the bulk motion (which can be done since the velocity of surfacial wave is much smaller than the velocity of the second sound), after inserting (4.5) into them, we get

$$\zeta = \zeta_0\cos(2kx) \tag{4.7}$$

$$\tilde{\phi} = -\phi_0\,\frac{\sin(2\omega t)}{2\omega} \tag{4.8}$$

where the amplitudes ζ_0 and ϕ_0 equal

$$\zeta_0 = \frac{1}{4}\frac{\partial\sigma}{\partial T}\beta_0^2\omega^2\,\frac{1}{\Lambda(2k)}, \quad \Lambda(2k) = g + \frac{\mathcal{H}}{\rho}(2k)^2 \tag{4.9}$$

$$\phi_0 = \frac{1}{4}\frac{\partial\sigma}{\partial T}\beta_0^2\omega^2 \tag{4.10}$$

This result differs considerably from the results obtained in [3]. We see that in the main approximation the relief of the surface is time-independent. Therefore, the picture of the relief on the surface, obtained in [4] by photographing with a long exposure is not averaged over time, but is stationary and time-independent. Formula (4.8) is also very significant. Thus it follows from this formula that the potential $\tilde{\phi}$ on the surface oscillates with time but, naturally, does not depend on the coordinate (since the second sound in this approximation cannot excite the motion of the liquid as a whole for which the gradient of the potential of $\tilde{\phi}$ is responsible). Oscillation of *pis* somewhat resembles the Josephson effect. Uncomplicated calculations, taking into account the bulk equations (4.3) and (4.4), yield a non-stationary addition to the expression (4.7) for ζ, which has the order of u_s^2/u_2^2, where

$$u_s(k) = k^{-1}\Lambda(k) \tag{4.11}$$

Let us present without derivation the result for this non stationary addition to the

$$\zeta_1 = -\frac{1}{8}\beta_0^2 \frac{\partial\sigma}{\partial T} \cos(2\omega t)\cos(2kx) \tag{4.12}$$

These problems have been investigated in more detail in the article written with V. L. Pokrovsky and submitted for publication.*

ACKNOWLEDGEMENT

I wish to thank Prof. T. M. Rice for his hospitality during my stay at ETH and Prof. J. L. Olsen for very stimulating discussions.

* The author is sincerely thankful to S. Korshunov for useful discussion of the derivation of Formula (4.12).

REFERENCES

1. I. M. Khalatnikov, Zh. Eksp. Teor. Fiz. **23,** 169 (1952).
2. V. L. Pokrovskii and I. M. Khalatnikov, Pis'ma Zh. Eksp. Teor. Fiz. **23,** 653 (1976); [JETP Lett. (1976)];**23,**599 Zh. Eksp. Teor. Fiz. **71,** 1974 (1976); [JETP, (1976)].**44,**1036
3. R. S. Sorbello, J. Low Temp. Phys. **23,** 411 (1976).
4. J. Olsen, Journ. of Low Temp. Phys. **61,** 17 (1985).

TEMPORALLY-PERIODIC STATES OF EXTENDED NONEQUILIBRIUM SYSTEMS

G. Grinstein, IBM Research Division, T.J. Watson Research Center, Yorktown Heights, NY 10598

We summarize recent results concerning the stability of spatially-coherent, time-periodic states in noisy, classical, discrete-time, many-body systems with short-range interactions. Generic stability of periodic k-cycles with $k > 2$ can be achieved only by rules carefully constructed to exploit lattice anisotropy and so suppress droplet growth. For ordinary rules which do not utilize spatial anisotropy in this way, periodic k-cycles with periods $k > 2$ are metastable rather than stable under generic conditions, losing spatial coherence through nucleation and growth of droplets.

INTRODUCTION

In this talk we review some recent progress[1] in elucidating the conditions under which temporally-periodic states occur in noisy, classical, spatially-extended nonequilibrium systems. Such states, characterized by equal-time correlation functions which oscillate periodically in time, cannot of course occur in equilibrium, where only stationary states with time-independent equal-time correlations obtain.[2] It is obvious that periodic states will always result in systems driven by a periodic external force. We consider here only the more interesting situation wherein the equations of motion describing the systems's time evolution are time-translation invariant, and the system must spontaneously break this time-translation symmetry to produce a periodically-oscillating state. Given that we treat only "fully probabilistic" systems, i.e., ones in which any microscopic state is accessible from any other with nonzero probability, this broken symmetry can, strictly speaking, only occur in the thermodynamic limit.[2] We therefore need consider only very large systems.

From the experimental viewpoint, the existence of coherent time-periodic states is completely obvious: They occur ubiquitously, in systems such as Rayleigh-Benard convection cells,[3] surface wave experiments,[4] Taylor vortex flows,[5] and scores of others. We argue here that producing genuinely stable periodic states is, nonetheless, a subtle and rather difficult business, and hypothesize that many of the coherent periodic oscillations observed experimentally may represent long-lived <u>metastable</u> states, rather than genuinely stable ones.

WHY STABILIZING TEMPORALLY-PERIODIC STATES IS DIFFICULT

Let us first restrict ourselves to systems (like probabilistic cellular automata) which evolve in discrete time according to local, dynamical rules with (discrete) time-translation invariance. We shall try to make clear why, in the presence of any noise, stabilizing collective periodicity is difficult. Imagine that we have succeeded in constructing a (discrete-time) dynamical rule which has broken its time-translation invariance by producing, say, a stable 3-cycle. That is, imagine that some fourier amplitude, say the amplitude of the k=0 mode (the spatial average, M, of the system's dynamical variables), assumes three distinct values, M_1, M_2, and M_3, say, in regular periodic fashion

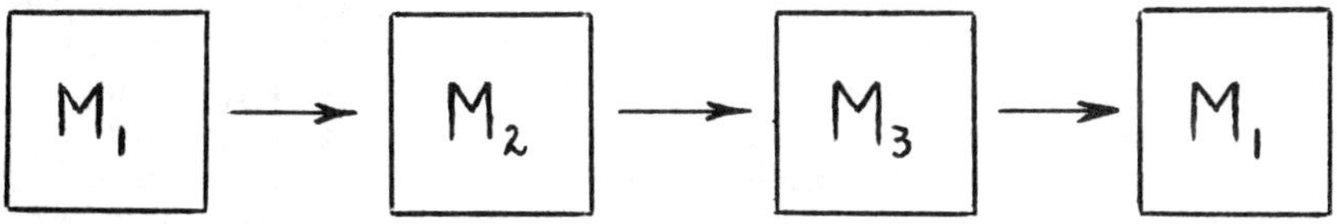

Fig. 1. A periodic 3-cycle state at four successive time steps. At the first, second, and third time steps, most of the system variables assume the values M_1, M_2, and M_3, respectively. At the fourth step, M_1 occurs again.

at consecutive time steps (fig. 1). (The arguments that follow apply equally well under the assumption that the fourier amplitude of a mode with finite k undergoes the temporal oscillations. We choose k=0 only for simplicity.) The rule is assumed noisy, so that no single variable will assume this precise, repeating sequence of values, M_1, M_2, M_3; only the spatial average (or, equivalently, the noise-averaged value of any given variable) in the thermodynamic limit is truly periodic.

Suppose that we start the system off in an M_1 initial state, and monitor its evolution at every third time step thereafter. If the evolution is genuinely periodic, then the system viewed at three-step intervals should look stationary, i.e., the M_1 state should persist indefinitely. In the presence of any (Gaussian or other unbounded[6]) noise, the system will of course nucleate droplets of M_2 in the infinite sea of M_1. The stability of the periodic time dependence is determined by considering the dynamics of such droplets, whose radius R (monitored at every third step), evolves in time, t, according to the same phenomenological equation, viz.,[7]

$$\partial R/\partial t \sim -\sigma/R + h \quad , \tag{1}$$

that describes droplet evolution in equilibrium systems such as the Ising model.[8] The term on the right side of (1) proportional to the droplet's curvature, $1/R$, expresses the tendency of the droplet to shrink and so reduce the length of the domain wall between the states M_1 and M_2; σ represents the analogue, for nonequilibrium systems, of surface tension. The second term, h, is the average velocity with which an infinite, flat domain wall between these two states translates in the course of three steps. This term is a measure of the inequivalence of the states M_1 and M_2; it plays a role analogous to an applied magnetic field in the domain growth kinetics of the equilibrium Ising model.[8] The main point is that the generic situation is the one where there is no symmetry guaranteeing the equivalence of M_1 and M_2, i.e., where h is nonzero in Eq. (1). While it is possible to construct rules with a special symmetry which ensures that walls do not translate on average (an example is given in ref. 1), the typical case, and certainly the overwhelmingly probable one encountered in practice, is the one where the states are inequivalent. (It is easy to check, e.g., for the periodic states studied numerically in the coupled map lattices of ref. 9, that no such symmetry is present.[10])

Let us assume for the moment that h is positive, so that M_2 is favored over M_1. Eventually, owing to the noise, the system nucleates droplets of M_2 with radii R larger than the critical radius, $R_c \equiv \sigma/h$. For such droplets, the right side of (1) is positive, implying that the droplets expand with time, ultimately replacing the state M_1 by M_2. After sufficiently long time, therefore, the state M_1 fails to occur at every third time step. In other words, the assumed temporally- periodic state of the system is not stable, but <u>metastable</u>. One might think that the M_2 state which supplants M_1 simply persists indefinitely (still viewed at each third time step), thereby implying that the system is still in a stable 3-cycle, but shifted in phase from the initial condition. This is not the case: Since in one time step M_1 turns into M_2, and M_2 turns into M_3, the instability of M_1 with respect to droplets of M_2 implies the instability of M_2 with respect to droplets of M_3 (and also the instability of M_3 to droplets of M_1). Thus, starting as we did from a state of pure M_1, one finds that large droplets of M_2 begin to grow and supplant M_1 after many 3-step cycles; but M_2 itself then starts being supplanted by M_3, etc. After many cycles, therefore, one expects to find a mixture of equal parts of M_1, M_2, and M_3 in the system at any given time. In other words, the overall spatial average, M, of the variables in the system ceases to vary with time, and the behavior of the system is stationary, not periodic.

Note that the instability of the periodic state does not depend on our assumption of positive h in Eq. (1) for the evolution of drops of M_2 in a sea of M_1. If h is in fact negative, so that M_1 is favored over M_2, then droplets of M_1 immersed in a sea of M_2 evolve according to Eq. (1) with h positive; we can simply repeat the arguments above for droplets of this type, with the same result, viz., the destabilization of the assumed periodic state. Thus the nucleation and growth mechanism of Eq. (1) destroys collective periodicity under all generic conditions. It is clear from (1), however, that a periodic state can readily be made metastable, its lifetime roughly given by the time (a function of the noise, h, and σ), required for the nucleation of droplets of critical size. This time grows exponentially with R_c. If the velocity, h, is small, so that R_c is large, the lifetime of the periodic state can be very long indeed.[10] Only by imposing a symmetry that guarantees the equivalence of M_1, M_2, and M_3, or by making special parameter choices which guarantee that flat domain walls move with zero velocity[1] (h=0 in Eq. (1)), can one stabilize periodic states. This is analogous to positioning oneself on the coexistence curve (i.e., low temperature and zero magnetic field) of the equilibrium Ising model, where $R_c = \infty$, and droplets of any initial radius R_0 shrink like[8] $R^2 = R_0^2 - 2\sigma t$.

Though we have discussed only 3-cycles, it seems clear that similar arguments prohibit the existence of stable cycles of any arbitrary length in isotropic, generic, discrete-time systems with short-range interactions. The single exception is the 2-cycle, where the fact that the two states, M_1 and M_2 say, exchange identities at each time step implies that[1] a flat domain

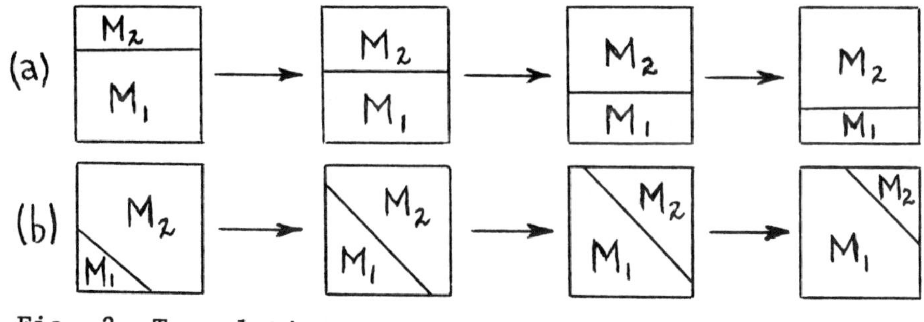

Fig. 2. Translation, in each 3-step cycle, of flat domain walls separating a domain with variable value M_1 from one with value M_2. Domains oriented (a) parallel to, or (b) at 45 ° to, the lattice axes translate to favor (a) M_2 over M_1 , or (b) M_1 over M_2.

wall separating them cannot translate with nonzero velocity. Thus, unlike higher cycles, the 2-cycle has an effective symmetry which ensures the vanishing of h in Eq. (1), and so allows the stabilization of the period-2 state, even under generic conditions.

STABILIZING TEMPORALLY-PERIODIC STATES WITH SPATIAL ANISOTROPY

In writing Eq. (1) and so arguing against stable cycles with periods longer than 2, we implicitly exploited the assumed spatial isotropy of the system. We now show how anisotropic systems (e.g., on a lattice) can exhibit stable states with arbitrary periodicity.[1] Imagine that the 3-cycle system described above is defined on, e.g., a square lattice in two dimensions.

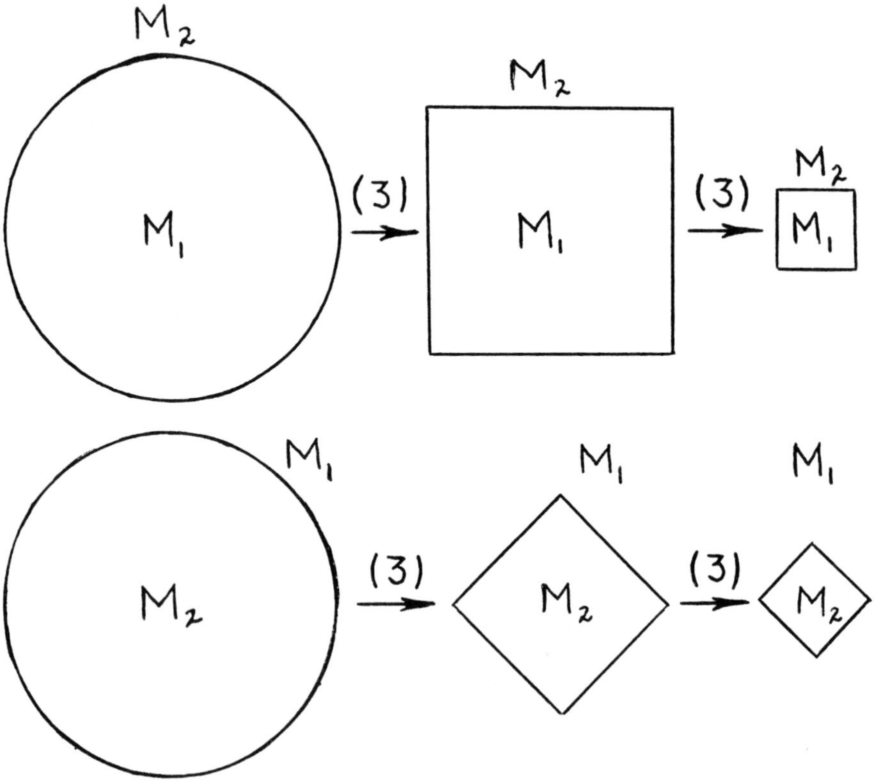

Fig. 3. (a) Droplet of M_1 in a sea of M_2, seen at 3-step intervals. Droplet distorts into square oriented with lattice axes, and then shrinks. (b) Droplet of M_2 in a sea of M_1, seen at 3-step intervals. Droplet distorts into square oriented at 45 ° to the lattice axes, and then shrinks.

While it remains true that flat domain walls separating any two of the states, M_1, M_2, and M_3, must translate under generic conditions, the sense in which they translate can depend on their orientation with respect to the lattice axes. It is easy, e.g., to construct rules (a specific example is given in ref. 1) wherein walls between M_1 and M_2 oriented parallel (at 45° to) the axes translate to favor M_2 over M_1 (M_1 over M_2), as in fig. 2. For finite droplets, this has the effect shown in fig. 3: A droplet of M_1 in a sea of M_2 grows fastest along those parts of its boundary oriented at 45° to the lattice axes. Thus it distorts into a rectangular shape whose boundaries are oriented along the lattice directions. Since domain walls with lattice-axis orientation move to favor M_2, the droplet then shrinks and vanishes! (For large droplets this situation can be described phenomenologically by the equation $dR/dt = -h$, with h positive. Thus droplets shrink <u>linearly</u> with time, a property which, as shown in ref. 1, allows one to distinguish metastable periodic states from stable ones in two dimensions.) Similarly, a droplet of M_2 in a sea of M_1 distorts so that its boundaries orient at 45° to the axes, whereupon it also shrinks and disappears. Thus, systems with appropriate rules can take advantage of spatial anisotropy to ensure that droplets created by thermal fluctuations do not grow and destroy the temporal periodicity. Such systems can therefore exhibit stable 3-cycles, and by obvious generalization, higher cycles as well.

CONTINUOUS TIME, METASTABILITY, AND EXPERIMENTS

The arguments presented here do not apply directly to systems which evolve in continuous time, since domain walls between phases do not retain their identity, but broaden with time, becoming as wide as the entire system. This phenomenon, a consequence of the fact that the time-translation symmetry is <u>continuous</u> in this case, is familiar from ordinary equilibrium systems with continuous symmetry, such as Heisenberg ferromagnets. As in equilibrium situations, we expect that it is at least as difficult to break a continuous symmetry as a discrete one, and hence that the stabilization of periodic states in continuous time can likewise only be accomplished through the exploitation of spatial anisotropy. It is obvious, however, that this argument is rather loose, and that a more substantive one is required before the understanding of the continuous-time case is as firm as for discrete time. Note too that, even in discrete-time systems, the stabilization of periodic states with periods <u>incommensurate</u> with the basic time step is a question equivalent to the continuous-time problem; again, therefore, the conditions under which such states are generically stable are still somewhat speculative. It is important to emphasize, however, that experimental study of the stability of commensurate cycles in the discrete time problem is in principle feasible. Application of an external ac field of frequency ω to a sample changes the continuous-time translation symmetry to a discrete one; i.e., only time-translations by $2\pi/\omega$ leave the system invariant. This makes the time-evolution effectively discrete, allowing the commensurate cycles treated above to be investigated in the laboratory.

Experimentally-realizable nonequilibrium (e.g., fluid mechanical) systems typically occur in the continuum rather than on a lattice. However, the stabilization mechanism described above requires only that there be some special direction in the problem, so that domain walls oriented in different directions need not translate in the same sense. Such anisotropy is a ubiquitous feature of (even fluid mechanical) systems which have periodic states. In Rayleigh-Benard systems, e.g., the occurrence of periodic states occurs only at Rayleigh numbers above the convective instability where the system first develops rolls.[3] The axes of the rolls define a preferred direction which the system chooses spontaneously, thereby breaking its spatial isotropy. The system could, in principle, then make use of anisotropy as described above to produce stable periodic states.

It is important to note that only certain rules can successfully exploit spatial anisotropy to stabilize periodic phases. (The general requirements are made clear in ref. 1.) For example, we now believe that the periodic states we observed numerically in the coupled map lattices of ref. 9 are not genuinely stable, but rather metastable with a lifetime orders of magnitude larger than the longest computer runs currently feasible.[10] Similarly, metastable periodic states may readily be mistaken for stable ones in real experiments. It is interesting to ask, e.g., whether the periodic states observed in Rayleigh-Benard and other systems use the anisotropy mechanism described above to achieve true stability, or whether they are simply metastable with lifetimes long compared to typical measuring times. Comments about distinguishing between these two possibilities can be found in ref. 1.

I am very grateful to my collaborators in this research: Charles Bennett, Yu He, Ciriyam Jayaprakash, and David Mukamel.

REFERENCES

1. C.H. Bennett, G. Grinstein, Y. He, C. Jayaprakash, and D. Mukamel, to appear in Phys. Rev. A.
2. E.g., N.G. van Kampen, Stochastic Processes in Chemistry and Physics (North-Holland, Amsterdam, 1981).
3. E.g., M. Giglio, S. Musazzi, and U. Perini, Phys. Rev. Lett. 47, 243 (1981).
4. E.g., S. Ciliberto and J.P. Gollub, J. Fluid Mech. 158, 381 (1985).
5. A. Brandstater and H.L. Swinney, Phys. Rev. A35 , 2207 (1987).
6. Noise of bounded amplitude may be insufficient to nucleate droplets, but such noise is rather unphysical. In equilibrium statistical mechanics, e.g., finite systems with bounded noise can exhibit broken symmetries, an impossibility with the more realistic unbounded noise.
7. C.H. Bennett and G. Grinstein, Phys. Rev. Lett. 55, 657 (1985), and references therein.
8. See, e.g., J.D. Gunton, M. San Miguel, and P.S. Sahni, in Phase Transitions and Critical Phenomena, edited by C. Domb and J.L. Lebowitz (Academic, London, 1983), Vol. 8, and references therein.
9. T. Bohr, G. Grinstein, Y. He, and C. Jayaprakash, Phys. Rev. Lett. 58, 2155 (1987).
10. We now believe that the periodic states observed numerically in the coupled maps of ref. 9 (on samples of up to 200 x 200, and hundreds of thousands of updatings per spin), were actually metastable rather than stable. By measuring the rate (h in Eq. (1)), at which flat domain walls move in the 4-cycle regime of that reference, we estimate that the critical droplet size for destruction of the periodicity is of O(3000), far beyond one's numerical grasp.

THE RESPONSE FUNCTION OF NON-STATIONARY MEDIA AND BERRY'S WAVE FUNCTION

L. P. Pitaevskii
Institute for Physical Problems, Academy of Sciences of the USSR,
117454 Moscow, Kosygina, 2

When the Hamiltonian of a quantum system contains a slowly varying time dependent parameter $\lambda(t)$ the response function $\alpha(\omega)$ contains an additional term of the form $(i/2)(\partial^2\alpha/\partial\omega\partial t)(d\lambda/dt)$.

Let us consider a quantum system with the Hamiltonian H_0 depending on the parameter λ, so that $H_0 = H_0(\lambda)$. We assume also that the external action on the system can be described by a perturbation operator that can be introduced in the Hamiltonian as an additional term

$$V(t) = -x\,(fe^{-i\omega t} + f^* e^{i\omega t})/2\ , \tag{1}$$

where x is an operator of some physical quantity describing the system. Then the quantum mean value, $<x>$, can be written in the first order to the f in a form:

$$<x> = (1/2)\,[\alpha f e^{-i\omega t} + \alpha^* f^* e^{-i\omega t}] \tag{2}$$

In the absence of the perturbation $<x>=0$. Here α is the response function, which depends on the parameter λ:

$$\alpha = \alpha(\omega, \lambda)\ .$$

Assuming now that the parameter λ (and hence the Hamiltonian H_0) depends on time, we can introduce a time-dependent response function $\alpha(\omega, t)$ which is defined by the same Eq. (2). The function $\alpha(\omega, t)$ will not coincide with the function $\alpha_0(\omega, t) = \alpha(\omega, \lambda(t))$, i.e. with the function of the static value of λ where we will put $\lambda = \lambda(t)$. The fact is that the function $\alpha(\omega, t)$ generally speaking, is not determined by the value of λ at the same moment of time but depends also on the time derivatives of λ. If λ slow changes with time the dependence of α on λ can be limited only by the term $d\lambda/dt$, so we can take into account only the linear terms and rewrite the equation for $\alpha(\omega, t)$ in the form:

$$\alpha(\omega, t) = \alpha_0(\omega, t) + a(\omega, \lambda(t))(d\lambda/dt)\ .$$

The aim of this paper is to calculate the quantity $a(\omega, \lambda)$. We will see that it can be expressed it in the terms of the function α_0.

It was shown in paper [1] that for a "transparent" nondissipative medium, i.e. in the conditions when $\alpha(\omega)$ is real, the following equation is valid

$$\operatorname{Im} a = (1/2)\,\partial^2\alpha_0/\partial\omega\partial\lambda\ . \tag{3}$$

As a proof we have considered a capacitor filled with dielectric which can be characterized by the dielectric susceptibility $\alpha_0(\omega)$. (In this case an electric field E plays a role of a "force" f .) The derivation was based on application of the conservation energy law to a resonant circuit consisting of this capacitor and an inductance. With change in the parameter λ the energy U of the circuit also changes; this change can be expressed in terms of the function a. On the other hand, an adiabatic invariant of the system remains constant during an infinitely slow change in the environment. In any linear oscillating system this invariant is defined as a ratio of the oscillation energy to the circuit frequency ω_0. Thus one can get $\delta(U/\omega_0)=0$, i.e. $\delta U = U(\delta\omega_0/\omega_0)$. Comparing two expressions for the change of energy one can obtain Eq. (3).

It is impossible, however, to generalize this procedure to the case of a dissipative medium.

In the present paper we will show (by the way of direct calculations) that in general case the following expression is valid:

$$a = (i/2)\,\partial^2\alpha_0/(\partial\omega\partial\lambda) \tag{4}$$

i.e. using the definition of a:

$$\alpha(\omega, t) = \alpha_0(\omega, \lambda(t)) + (i/2)\,\partial^2\alpha/\partial\omega\partial\lambda\, d\lambda/dt$$

$$= \alpha_0 + (i/2)\,\partial^2\alpha/\partial\omega\partial t\ . \tag{5}$$

(For a nondissipative medium we recover, of course, the previous result (3)).

The proof consists in a direct quantum mechanical calculation of the mean value $<x>$ in the perturbed system with the perturbation operator (1). Let $\Psi_n(t)$ be the wave functions of unperturbed system which can be described by the Schrödinger equation

$$i(\partial\Psi_n/\partial t) = H_0(\lambda(t))\Psi_n \tag{6}$$

Using the usual time depending perturbation theory one can find the perturbation of the function Ψ_n and calculate the mean value $<x>$. It is easy to see that $<x>$ can be expressed in the form of Eq. (2) with the response function α defined by the Kubo-type equation:

$$\alpha(\omega, t) = i\sum_m \int_0^\infty d\tau \exp[i(\omega + i0)\tau]\{<\Psi_n(t-\tau)|x|\Psi_m(t-\tau)><\Psi_m(t)|x|\Psi_m(t)> - \text{c.c}\}\ . \tag{7}$$

The matrix elements $<\Psi_n(t)|x|\Psi_m(t)>$ are defined with respect to the functions $\Psi_n(t)$

from Eq. (6). In a particular case when the parameter λ does not depend on time the functions $\Psi_n(t)$ can be rewritten in a form

$$\Psi_n(t) = \exp(iE_n t)\Psi_n(\lambda) , \tag{8}$$

where the functions $\Psi_n(\lambda)$ satisfy the equation

$$H_0(\lambda)\Psi_n = E_n(\lambda)\Psi_n . \tag{9}$$

For simplicity, we suppose that the body has no magnetic structure and there is no external magnetic field. Then the wave functions $\Psi_n(\lambda)$ can be taken as real quantities. Substituting Eq. (8) into Eq. (7) we obtain the well known expression for $\alpha_0(\omega)$:

$$\alpha_0(\omega) = i\sum_m \int_0^\infty d\tau \exp(i\omega\tau)[<\Psi_n | x | \Psi_m>]^2 \{\exp(-i\omega_{nm}\tau) - \text{c.c}\} .$$

Calculate now the quantity $\partial^2\alpha_0/\partial\omega\partial\lambda$, which will be important in what follows. Carrying out the differentiation we get

$$\partial^2\alpha_0/(\partial\omega\partial\lambda) = -\sum_m \int_0^\infty d\tau\, \tau \exp(i\omega\tau)$$

$$\times\{d/d\lambda[<\Psi_n | x | \Psi_m>]^2 - i\tau(d\omega_n/d\lambda)\exp(-i\omega_{nm}\tau) + \text{c.c.}\} \tag{10}$$

Let go back now to the general expression (7) to carry out its expansion up to the first term in $d\lambda/dt$. In this procedure one can take into account that the functions $\Psi_n(t)$ must be calculated with the same accuracy. This problem was solved by Berry in his well known paper [2]. It follows from this paper, the wave function $\Psi_n(t)$ can be written as

$$\Psi_n(t) = \exp[-i(\int_0^\infty E_n(t')dt' - \gamma_n)]\{\Psi_n(\lambda(t)) + i\Psi_{1n}\} \tag{11}$$

where

$$\dot{\gamma}_n = <\Psi_n | d\Psi_n/d\lambda> d\lambda/dt, \quad <\Psi_m | \Psi_{1n}> = (1/E_m - E_n)<\Psi_m | d\Psi_n/d\lambda> d\lambda/dt . \tag{12}$$

As was assumed before, the functions Ψ_n are real and therefore the phase γ_n is equal to zero and the function Ψ_{1n} is real.

Using the functions from Eq. (12) we calculate the matrix elements of the quantity x. With the accuracy required we get:

$$<\Psi_n(t-\tau) | x | \Psi_m(t-\tau)><\Psi_m(t) | x | \Psi_n> = \exp(i\int_t^{t-\tau} \omega_{nm}(t')dt')$$

$$\times<\Psi_n(t-\tau) + i\Psi_{1n}(t) | x | \Psi_m(t-\tau) + i\Psi_{1m}(t)><\Psi_m(t) + i\Psi_{1m}(t) | x | \Psi_n(t) + i\Psi_{1n}(t)>$$

$$= \exp(-i\omega_{nm}\tau)\{[<\Psi_n(t)|x|\Psi_m(t)>]^2 - (i/2)(d\omega_{nm}/dt)\tau$$

$$-(1/2)\tau d/dt[<\Psi_n(t)|x|\Psi_m(t)>]^2\} . \qquad (13)$$

(Here we have expanded the slow varying functions in τ. Note that the term containing Ψ_{1n} in first approximation cancels on the right hand side of Eq. (13). Substituting expression (13) into Eq.(7) and comparing with Eq.(10), we get Eqs.(4)-(5).

REFERENCES

1. L. P. Pitaevskii, Soviet Physics JETP 12, 1008 (1961).
2. M. V. Berry, Proc. R. Soc. A 322, 45 (1984).

SOME PROPERTIES OF FAMILIES OF EXCITONIC QUASIPARTICLES

Joseph L. Birman, M. Artoni, and Bing Shen Wang
Department of Physics, City College of CUNY,
Convent Avenue at 138th Street, NY, NY 10031

Three closely related types of excitonic quasiparticles have been recently studied: polaritons, phonoritons and photoritons. We show that the well known polariton displays frequency-tunable intrinsic squeezing. Related to this is a new Quantum Optics Extinction Theorem . The phonoriton is associated with a "reconstruction" of the dispersion equation for the parent-polariton: tests have so far proven elusive. Two new experiments are proposed and analyzed for identifying the phonoriton: Non-Linear Resonance Brillouin Scattering, and Non-Linear Reflectivity. Some remarks are made about the photoriton.

POLARITONS AND SQUEEZING

It is well-known that the correct elementary excitation describing an exciton in a crystal in interaction with the electromagnetic field, is the coupled mode polariton. The Hamiltonian of this interacting system can be obtained in several ways. If we take an extreme "tight binding model" for the exciton, and we include all dipolar interactions, plus the exciton-radiation interaction, we obtain a polariton Hamiltonian with the following generic form

$$H=\sum_k [H_k+H_{-k}]+h.c.$$

with

$$H_k \equiv E_k^{ph}\, a_k^+ a_k + E_k^{ex}\, b_k^+ b_k + B' a_k a_{-k} - C'_k b_k b_{-k}$$

$$+ i A_{1k} a_{-k} b_k + i A_{2k} a_k^+ b_k + h.c. \tag{1}$$

where a_k^+, a_k are photon operators and b_k^+, b_k are exciton operators. This form of exciton-radiation Hamiltonian was discussed by Hopfield [1]. Note the appearance of terms $b_k b_{-k}$ denoting exciton-exciton (dipolar) interactions.It is worthwhile remarking that certain simplifications are commonly made in this Hamiltonian:
(i) Hopfield himself in the initial paper (Ref. [1], eqn. (6)) omitted the terms ($b_k b_{-k}$ +h.c.) and also in effect took $A_{1k}=A_{2k}$. Thus the "traditional" Hopfield model has the form

$$H_k^{Hop} = E_k^{ph}\, a_k^+ a_k + E_k^{ex}\, b_k^+ b_k + B\, a_k a_{-k} + i A_k (a_{-k} b_k + a_k^+ b_k) \tag{2}$$

with coefficients given from an initial assumed Lagrangian for the exciton-photon system. The Hamiltonian in eqn. (2) is diagonalized by introducing the polariton operator

$$\alpha_k = w' a_k + x' b_k + y' a_{-k}^+ + z' b_{-k}^+ \tag{3}$$

and requiring

$$[\alpha_k, H] = \lambda_k' \alpha_k \tag{4}$$

This produces the secular equation for polariton dispersion

$$\lambda_k' = \frac{k^2 c^2}{\omega^2} = \varepsilon(k, \omega) \tag{5}$$

where $\varepsilon(k,\omega)$ is the familiar dielectric function. The solutions of that dispersion equation are written as ω_α: i.e., $\omega_{LP}(k)$ and $\omega_{UP}(k)$ for lower and upper polariton branches respectively with $\hbar\omega(k)=\lambda'$. The operators $\alpha^+_{k,LP}$ and $\alpha^+_{k,UP}$ are polariton creation operators and the state $\alpha^+_{k,\sigma}|0>$ is a one polariton state (σ=UP or LP) on the appropriate branch.

(ii) The polariton Hamiltonian is sometimes presented in a further truncated form [2] by omitting the term $a_k a_{-k}$ (i.e. B_k=0). Polariton operators are introduced as in eqn. (3) and a similar dispersion obtained as in eqn. (5).

In what follows we use the more complete form (eqn. 1). Again defining the new normal modes (polaritons) via the Bogoliubov transformation:

$$\eta_k = w a_k + x b_k + y a^+_{-k} + z b^+_{-k} \tag{6}$$

and $$[\eta_k, H] = \lambda_k \eta_k \tag{7}$$

we obtain a dispersion equation biquadratic in λ_k: we write this (dropping the k index)

$$\lambda^4 - \lambda^2((2E^{ph})^2 + (2E^{ex})^2 - (2B')^2 - (2C')^2 + \frac{1}{4}(A_2^2 - A_1^2)) + f_0 = 0 \tag{8}$$

where f_0 is quadratic in the coefficients. If we now solve for λ_k^2we obtain a polariton dispersion equation. By suitably choosing the coefficients (parameters) of eqn. (1) we can and do duplicate the measured polariton dispersion $\omega(k)$. Thus the "enlarged Hopfield" Hamiltonian in eqn. (1) is a satisfactory starting point for the work: it permits derivation of the physically observed dispersion, and incorporates the relevant interaction (exciton and photon). Mathematically also, eqn (1) (H_k+H_{-k}) is an element in the complete Lie algebra $Sp(8, C)_k$.

We proceed with eqn. (1) in two stages. First define certain intermediate modes $a_\pm$ by

$$a_+ \equiv \alpha_k a_k + \exp(2i\chi)\beta b_k \tag{9}$$

(for a_- replace k by (-k)). The Hamiltonian (1) becomes

$$H_k = \Omega_k(a_+^{+} a_+ + a_-^{+} a_- + 1) + 2i\zeta_k(a_+ a_- + a_-^{+} a_+^{+}) \tag{10}$$

where Ω_k and ζ_k are "c functions" (combinations of the coefficients). The intermediate modes a_+, a_- which are bilinearly coupled can now be subjected to a further linear Bogoliubov transformation as:

$$\eta_k = A_+ a_+ + A_- a_- \tag{11}$$

with η_k diagonalizing H, as in eqn. (7). Of course we obtain the identical eigenvalues λ_k as in (7); but the same eigenvector is now expressed in an alternate form. This latter diagonalizing transform also can be written

$$\eta_k = S^{+}(r, \phi) a_\pm S(r, \phi) \tag{12}$$

Here

$$S(r, \phi) = \exp i(\zeta a_+^{+} a_-^{+} - \zeta^* a_- a_+) \tag{13}$$

and $\zeta = r\exp[i\phi]$ is the complex squeezing factor. The real amplitude r characterizes the squeezing, $\phi = 0 \pmod{2\pi}$. By combining previous results, we also can write

$$\hat{\eta}_k = a_k \cos\theta_k \cosh r + b_k \exp(2i\chi)\sin\theta_k \cosh r$$

$$+ a_{-k}^{+} \exp(2i\phi)\cos\theta_k \sinh r + b_{-k}^{+} \exp(2i[\phi - \chi])\sin\theta_k \sinh r \tag{14}$$

now $\hat{S}(r, \phi)$ as eqn. (13) is the two-mode squeeze operator [3] (i.e. the squeezing transform is on a_+, a_-). From eqn. (14) we note that various angles (θ, r, χ, ϕ) can be determined by using the relevant correspondence to previous equations for (x, y, z, w) and thus determined from the initial Hamiltonian of eqn. (1). The real squeezing factor is

$$r = \tanh^{-1}\left[\frac{2\zeta_k}{\lambda_k + \Omega_k}\right] \tag{15}$$

The quantities λ_k, Ω_k, ζ_k are as before, and their dependence upon the parameters in H is completely specified.

Recapitulating we recognize $H_k(a_+, a_-)$ as a standard two-mode squeezing Hamiltonian, as given in Schumaker [3]. Thus passing from the generalized polariton Hamiltonian (1) of our work, to the diagonal form

$$H_k = \lambda_k \eta_k^{+} \eta_k \tag{16}$$

of excitonic polaritons, via introduction of the intermediate modes a_+, a_-, reveals the squeezed structure of the exciton polariton: the intermediate modes are squeezed [4]. But this in turn produces squeezed electromagnetic and exciton modes.

Now we turn to implement the result by calculating the squeeze amplitude r(ω) as a function of the frequency ω. We have been unable to compute an analytic form for the dependence of r upon ω but, a numerical calculation is straightforward. In Fig. 1 the squeeze amplitude r(ω) is plotted. In Fig. 2 the relative squeezing (ratio of electromagnetic field fluctuation in the squeezed state [polariton] to that in a hypothetical Glauber coherent state) is plotted as function of laser frequency. Several features are of importance: (a) the frequency is tunable with maximum relative squeezing occuring at the "crossing" frequency; (b) the magnitude of squeezing (in one quadrature) is much greater than in previously examined and measured cases of atomic plasma (non-linear atomic polarizability) generated squeezing.

The detection of squeezing of polaritons is of course an extremely important matter. Elsewhere several experimental methods are proposed and analyzed for the detection of the intrinsic polariton squeezing [5]. These include interference and homodyne experiments, in addition to the scattering experiment for detection of squeezing we already discussed [4].

QUANTUM OPTIC EXTINCTION THEOREM

The generalization of the classical extinction theorem [6] to quantum optics is claimed [7]. A proof has not yet been completed but will be outlined here. An incident electromagnetic field in a coherent state is incident upon a bounded dielectric: the dielectric is modeled as an assembly of two-level systems. In the steady state the total resulting field in the medium is the superposition of the incident (coherent state) field plus the sum of all fields reradiated (as "secondary" emission) by the two-level oscillators. Thus

$$\hat{E}^{+}_{TOT}(r,t)=\hat{E}^{+}_{INC}(r,t)+\hat{E}^{+}_{SC}(r,t) \tag{17}$$

these are respectively the total, incident and secondary fields at (r, t). For a monochromatic field (frequency ω_0) the field $\hat{E}_{SC}(r,t)$ has two components: the first simply extinguishes the incident field. The second is the propagating dipole wave with wave number $q=nk_0$ where $k_0=\omega/c$ and n is a refractive index. The resultant total field is

$$\hat{E}^{+}_{TOT}(r,t)\approx \hat{\Pi}_0\,\eta\exp(-i\omega t)\frac{\exp(-iq\cdot r)}{(q^2-k_0^2)} \tag{18}$$

where $\eta \sim (e\omega^2\mathfrak{D}_{12}/4\pi\varepsilon_0 c^2)$, $\mathfrak{D}_{12}$ is the dipole matrix element, and $\hat{\Pi}_0$ is the radiation from a single dipole (two-level) oscillator. As noted by Loudon and Knight [8], such an oscillator can produce squeezed light when interacting with the radiation field. The total field is a single dipole field renormalized to self-consistency by the multiple scattering due to the assembly of dipoles. We claim the Quantum Optic extinction theorem: The incident coherent field is replaced in the medium by the squeezed dipolar field. A reconcilliation of this statement with the results of Carniglia and Mandel [9] on the electromagnetic mode structure in a bounded dielectric will be discussed elsewhere.

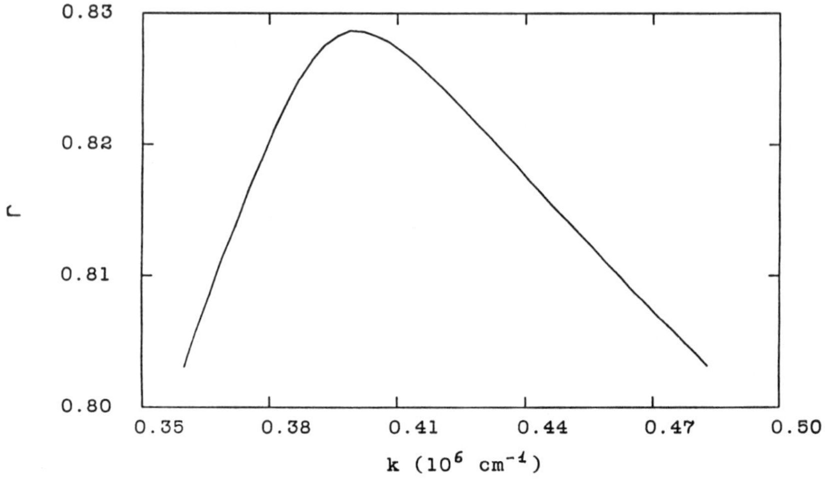

Figure 1. Wavevector dependence of the squeeze factor r for CdS A exciton on the upper polariton branch (UP).

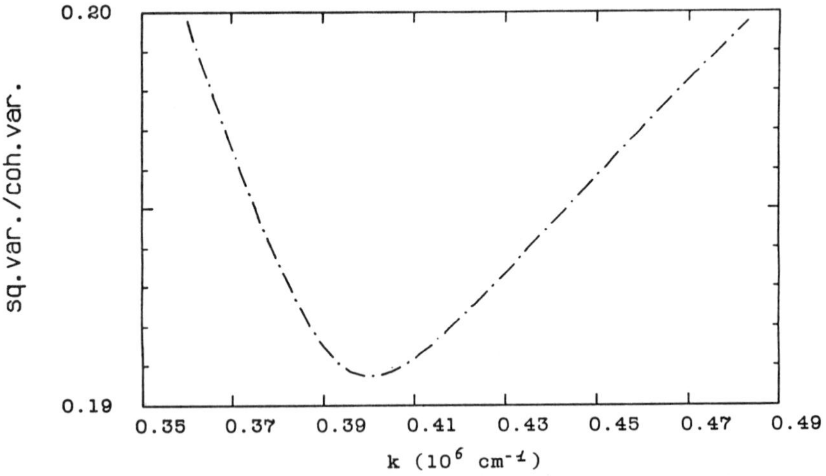

Figure 2. Squeezing ratio of the probability density distribution for a Glauber polariton (UP) standing wave (intrinsically squeezed as in text) to that for a corresponding coherent standing wave at given time (t=0). Maximum squeezing occurs at the crossing value k_0 of bare photon + exciton dispersion curves.

PHONORITONS

To the best of our knowledge, L.V. Keldysh and collaborators were the first to develop a theory [10] encompassing the non-linear generalization of the polariton concept to include the production of phonoritons. We analyzed the physics using a simplified model Hamiltonian including the exciton, photon, phonon and the minimum coupling terms:

$$H = \sum_k H_k \qquad \text{with}$$

$$H_k = E_k^{ex} b_k^+ b_k + E_k^{ph} a_k^+ a_k + \Omega_k^{phon} c_k^+ c_k + \frac{i}{2}\Omega_c (a_k^+ b_k - a_k b_k^+)$$

$$+ \sum_q (iM(k-q) b_k^+ b_q (c_{k-q} + c_{k-q}^+) + c.c.) \tag{19}$$

Here a_k, b_k are photon and exciton operators as before, c_k is a phonon operator, Ω_c is the photon-exciton interaction ($\Omega_c \equiv E_k^{ex}\sqrt{4\pi\beta}$ with β the oscillator strength), iM(k-q) is the exciton-phonon interaction energy. The photon-exciton terms in eqn. (19) are the truncated polariton Hamiltonian and can be simply diagonalized to give polaritons. If we now call the polariton operators B_k (vice $\hat{\eta}_k$ of the earlier part) the Hamiltonian can be written as

$$H_k = \sum_\alpha \lambda_\alpha(k) B_{k\alpha}^+ B_{k\alpha} + \hbar\Omega_k c_k^+ c_k$$

$$+ \sum_q \left[iM'(k-q)_{\alpha\beta} B_{k\alpha}^+ B_{q\beta} (c_{k-q} + c_{-(k-q)}^+) + c.c. \right] \tag{20}$$

with $B_{k\alpha}$ the polariton operator for branch α (vice β), and M' a renormalized polariton-phonon interaction matrix element. For high intensity laser radiation at frequency ω (wave number k_0) we assumed the polariton mode $\omega(k_0)$ is "macroscopically occupied". In the interaction term we make a Bogoliubov ansatz: $B_{k\alpha} \rightarrow \langle B_{k\alpha}\rangle \sim N_0^{1/2}$ i.e.it is as if the state $\omega(k_0)$ is a coherent Glauber (~classical) state, so the operator is replaced by its eigenvalue in the state. The Hamiltonian reduces to

$$H_k = \sum_\alpha (\lambda_\alpha(k) - \lambda_\alpha(k_0)) B_{k\alpha}^+ B_{k\alpha} + \hbar\Omega_k c_k^+ c_k$$

$$+ i\hbar M'(k-k_0)\alpha_{k_0} (B_k c_{k_0-k} - B_k^+ c_{-(k_0-k)}^+) \tag{21}$$

plus a term which is the "Stokes' analogue of the last term ($B_k c_{k-k_0} \rightarrow B_k^+ c_{k_0-k}$ etc.).

Now, introducing the new operator $\hat{\xi} = \mu B_k + \nu c_{k-k_0}$ we diagonalize H_k by solving $[\xi, H_k] = \omega_\sigma \xi$ to find $H_k = \hbar\omega_\sigma \hat{\xi}_\sigma^+ \hat{\xi}_\sigma$ with ($\sigma=\pm$);

$$\omega_\pm = \frac{1}{2}[\omega(k) - \omega(k_0) + \Omega_{k_0-k}] \pm \frac{1}{2}\{[\omega(k) - \omega(k_0) - \Omega_{k_0-k}]^2 + \Psi Q^2\}^{1/2}$$

where $\Psi(k)$ is the weight of the exciton in the polariton of fixed branch α (=LP, UP) and $Q \equiv (VN_0)^{1/2} M'(k_0-k)$. Note that the effective coupling between the macroscopically occupied branch $\omega(k_0)$ and the scattered polariton at k, depends upon the intensity of the radiation via the factor $N_0^{1/2}$. (The polariton lifetime can be included by modifying $\omega(k_0)$, as $\omega(k_0) \rightarrow [\omega(k_0) - i\Gamma]$).

This brief sketch illustrates several important parts. First, at some level of approximation, the phonoriton effects can be understood using an effective two-level model (an analogy to the Inverse Raman Scattering is been explored [11]). Secondly the phonoriton arises via an effective intensity-dependent polariton-phonon coupling which is so strong as to pass beyond the usual weak coupling to give a spectrum reconstruction with the creation of a new quasiparticle: the phonoriton. Thirdly, spectrum reconstruction results in opening a gap in the "parent" polariton continuous dispersion. the magnitude of the gap is intensity-dependent (via the effective coupling M'): in fact we find a threshold intensity for reconstruction [11]. The present experimental situation is not conclusive regarding confirmation of these effects. The luminescence experiments reported give only indirect indication of the presence of a changed spectrum.

We have proposed two new experiments to directly study spectrum reconstruction: details are given elsewhere [12]. A Non-Linear Resonance Brillouin Scattering (NLRBS) experiment involves using a pulsed laser at $\omega(k_0)$ to produce the macropopulation of polaritons and then a weak probe laser for the RBS from the reconstructed dispersion. Significant changes are predicted in the location and shape of the RBS scattering. An illustration is given in Fig. 3 for the Stokes configuration. The dotted continuous curve shows the usual polariton RBS; the shifted peaks show the predicted effect. A non-linear reflectivity experiment (NLR) involves excitations by a highly intensity pump then elastic scattering (from the opposite side of a slab). This is illustrated on Fig. 4 which shows the remarkable change from the smooth reflection profile when phonoritons are present. Further details are given in [11,12], along with indication of how the predicted effects scale with applied intensity.

The phonoriton represents an interesting and novel "quasiparticle" whose properties deserve further study. We have initiated an examination of the squeezing effects associated with phonoriton formation. This would be reported elsewhere, along with studies on the photoriton - another of the new family of composites whose parents are exciton-polaritons.

This work was supported in part by PSC-BHE Faculty Research Award Program, and Naval Air System Command.

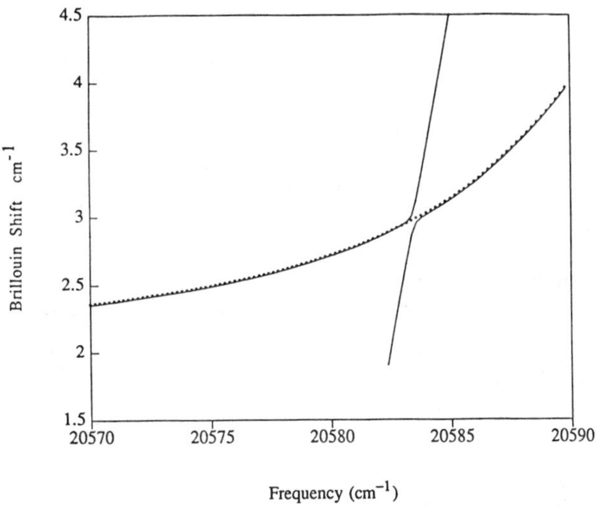

Figure 3. The Non Linear Resonance Brillouin Scattering (NLRBS) Stokes shift in CdS due to phonoriton creation (solid curves). For comparison the usual linear RBS shift is shown as the dotted continuous curve. Brillouin shift is plotted versus incident laser frequency of the probe; the pump intensity is 10 MW/cm^2.

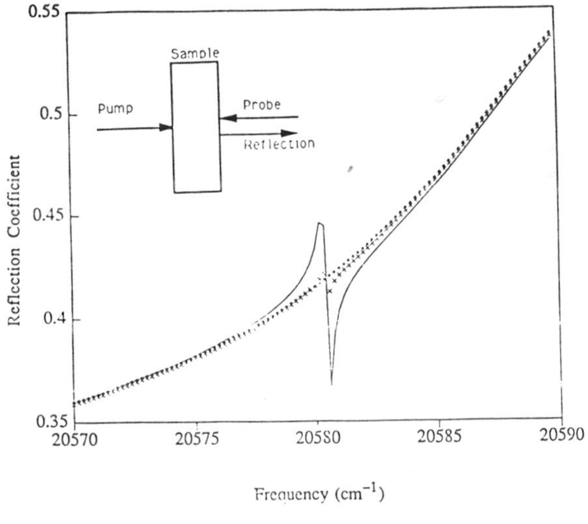

Figure 4. Non-Linear Reflectivity (NLR) in CdS due to phonoriton creation (solid curve, pump power is 100MW/cm^2; crosses, pump power is 10MW/cm^2). The reflection coefficient is plotted versus laser frequency of the probe. The dotted smooth curve is the normal linear reflectivity versus laser frequency. Note that pump and probe are counterpropagating (see insert to Figure).

REFERENCES

[1] J.J. Hopfield, Phys. Rev. **112**, 1535 (1958), see eqns. 17-28, and refs.
[2] C. Kittel, "Quantum Theory of Solids" J. Wiley, NY 2nd edition (1987), eqn. 55, p. 46.
[3] B.L. Schumaker, Phys. Reports **135,** 317 (1985).
[4] M. Artoni, and J. L. Birman, Quantum Opti. **1**, 91-97 (1989).
[5] M. Artoni, and J. L. Birman, (in preparation).
[6] See L. Rosenfeld, "Theory of Electrons" North Holland Publ., (NY) 1951.
[7] J. L. Birman, (in preparation).
[8] R. Loudon, and P.L. Knight, J. of Mod. Optics **34**, 709 (1987).
[9] C.K. Carniglia, and L. Mandel, Phys. Rev. **D3**, 288 (1971); I. Bialynica-Birula, and J.B. Brojan, Phys. Rev. **D5**, 485 (1972); D.M. Fradkin, and W. B. Rolnick, ibid. p. 482.
[10] A.L. Ivanov, and L. V. Keldysh, Sov. Phys. JETP **57**, 2341 (1983); additional references are given in ref. 12.
[11] B.S. Wang, and J. L. Birman, (in preparation).
[12] B.S. Wang, and J. L. Birman, (submitted for publication, Dec. 1989).

Reflections on The Founding of The Landau Institute of Theoretical Physics And Other Matters

I found out what an after-dinner speech was only last year, and I am still not quite sure that my understanding of it is correct since I arrived at it during the performance of two outstanding physicists Vitaly Ginzburg and Phil Anderson. Therefore I shall try to do it my own way, -I hope, a more peaceful way,- and I am determined not to be offensive.

I would like to tell you the history of the Landau Institute, which has never been written, or published. Now in the conditions of "glasnost" I feel it my duty to do so.

Everything began in the spring of 1963, a year after Landau's car accident, when it became clear that he would not come back to science. At the time I was acting as Head of the Theory Department of the Kapitza Institute for Physical Problems. In my memoirs, published in the May issue of "Physics Today," I pointed out what a difficult person Piotr Kapitza had been to deal with. I also mentioned the not quite always correct jokes of Kapitza addressed to theoreticians: For example "Ask a theoretician, and do the contrary" compared to "Ask your wife and do the contrary," which is one of the versions of the well-spread mot. We could take these derogative jokes OK having Landau at our side, but without Landau they were absolutely unbearable for his close associates, namely, Lifshitz, Abrikosov, Gor'kov, Dzyaloshinsky, Pitaevsky and myself.

As long as Landau was one of us, we felt to some degree privileged and honored to be "Landau's boys." But the idea of being "Kapitza's boys" did not make us happy at all. By time most of us had already got invitations to head the theory departments of different scientific centers. Thus there was a threat that the Landau team would fall apart, which would mean the end of the school of Landau with its top-level principles and traditions. This was the situation of the spring of 1963. But the actual trigger for the decisive action was an incident provoked by Kapitza. In the institute under Kapitza it was my responsibility to organize evening seminars, the so-called "kapichniks," initiated by him when he was at Cambridge. That historical day (and in our country each day is historical) in the morning I got in touch with Prof. Panovsky, who was at the time on a visit in Dubna, and agreed with him about his lecture at the next "kapichnik". I saw Kapitza, informed him about it, and left the institute for a short while since the problem of enlisting my youngest daughter into a specialized English school required my personal involvement as well as interference of my influential friends even on the governmental level (this is evidence of the command-administrative system in action).

During my short absence Kapitza wanted to speak to me about some minor staff problem and was displeased to find me out. When I was back I showed up in Kapitza's room with a certain delay. Kapitza was extremely rude to me, so very rude that I was forced to tell him that I could no longer stand such treatment. I was furious.

This was the straw that broke the camel's back, and it occurred to me it was time to set up an Institute for Theoretical Physics with the Landau Team as its core which would carry out the staff policy, independent of Kapitza, and which would preserve the tradition of Landau school. The next day the idea ripened, and when our team got together I initiated them into the conspiracy. Since we had no backing whatsoever, the idea seemed quite crazy but it was crazy enough so that, in conformity with Niels Bohr's principle, it was regarded as serious and worth considering. The idea inspired enthusiasm amongst us and appealed to everyone. But it was clear from the start that Evgeny Lifshitz, who was at the moment over head and ears involved with the course in theoretical physics, would stay behind in Kapitza's institute, where he was so used to working. The only person who sympathized with the idea but did not believe anything would pan out, was Lev Pitaevsky, present here. Thus "the gang of four," Abrikosov, Gor'kov, Dzyaloshinsky and myself, sprang into being. Abrikosov and myself, even in the 1950s, enjoyed a rather "bad" reputation and were called "brothers-highway robbers".

Now it was necessary to work out the strategy and tactics for bringing this crazy project to life. I was entrusted with this task, since I was believed to be no mean chess and checkers player. At that time the President of our Academy, Professor Keldysh, was developing a plan to create a network of scientific centers around Moscow by analogy with Oxford and Cambridge. One of such centers was being set up by the Nobel Prize winner, Professor Semenov, in the village of Chernogolovka where he had formerly had a testing area for his experiments. Therefore it was natural for me to approach Semenov first. Semenov, having listened to me, caught the meaning at once and his backing was guaranteed. Then the most important thing was to get the support of physicist. The next barrier I cleared was Anatoly Alexandrov, with whom I have maintained a very good relationship since the time he had been Director of the Institute of Physical Problems when Kapitza had been in exile in his country dacha. Alexandrov at the time had succeeded Kurchatov in the post of Director of the Institute of Atomic Energy, but he always took a lively interest in what was happening in the Institute of Physical Problems, having put so much energy and effort into it. Let into the conspiracy, Alexandrov was immediately inflamed with the idea, challenged by it because he understood that something revolutionary was fermenting in the institute. He took up the receiver, called up Lev Artsimovich, Head of the Department of General Physics and Astronomy of the Academy, whose backing was indispensable to us. Over the phone, he said the following words for the sake of which I am telling you this story. He said: "Lev, look here, Khalat came to see me. Theoreticians made up their mind to set up their own gipsy encampment. They need your help".

After that everything went very swiftly and on the day of Kennedy's assassination the Presidium of our Academy passed a resolution to establish the Institute for Theoretical Physics, and on the day of Khrushchev's dismissal this resolution was approved by the Government. So, our institute was the first-born of the stagnation period, and the most amazing thing is that in the course of these long 25 years we have, as becomes real gipsies, found no shelter, roaming from Moscow to Chernogolovka and back, like nomads. Now that the doors of our country are wide open, our nomadic way life is acquiring an international character. A small part of our gipsies settled down for a while in Princeton and we hope that other members of our gipsy encampment will find shelter some place in Europe.

But do not, be scared. We are civilized gipsies. Although we have acquired all the customs and traditions of gipsies, we do not steal horses!

Now Christmas is coming. Our seminar precedes this wonderful holiday always accompanied with Christmas presents. So, we could not be left aside from it. But I found myself in a difficult position trying to figure out what would be a good Christmas present for Joe Birman. But an event that took place in our institute just before our departure, somehow helped me. You probably know that our perestroika goes hand in hand with a complete breakdown of our trade and markets. The trade has now been transferred to our offices, and all kind of goods are sold out of there, which, you may guess, does not promote effective work. Our institute was also lucky, and my colleagues got a chance to lay their hands on some absolutely necessary goods like bras and panty-hose, And although these goods, as always were short, I succeeded in wringing something for my friend Joe Birman. Since I was ignorant of the size of the bra he wears, I brought him fashionable Greek stretchable unisize panty-hose.

So, I would like this invaluable present to become a symbol of strengthening ties between Soviet and American physicists!

I. M. KHALATNIKOV

DENSITY OF STATES AND THE METAL - NON-METAL TRANSITION IN THE 2-D ELECTRON GAS

A. L. Efros
Department of Physics, University of California Riverside, CA 92521
and A. F. Ioffe Institute, Leningrad, 194021, USSR

In heterostructures with thick spacer layers the random potential in the channel created by remote impurity ions is smooth and can be treated classically, so that a percolation interpretation of the metal - non-metal transition (MNMT) is possible. The critical electron density, n_c of the MNMT is found as a function of the spacer width and the ion density. The connection between the critical density n_c for the transition without magnetic field and the width of the plateaus of the integer quantum Hall effect (IQHE) in a sufficiently strong magnetic field is established. The width of the plateau as a function of spacer thickness and magnetic field is discussed.

The most important manifestations of the disorder in electronic properties of heterostructures are the metal - non-metal transition (MNMT) in the 2-dimensional electron gas (2DEG) without magnetic field and the existence of the plateaus in the quantum Hall effect (QHE). There are two main sources of the disorder in heterostructures both connected with the charged impurities - the remote ions, which are separated from the 2DEG by the spacer layer, and the residual ions located in the vicinity of the 2DEG. In the structures with thick spacer the remote ions are not important for the mobility of the 2DEG which is probably controlled by the residual ions, but I argue here that the remote ions may be responsible for the MNMT and for the width of the plateaus of the IQHE. In the modulation doped structures one can change the width of the spacer layer, which is an important tool for studying the effect of disorder, because in this way the spatial scale of the random potential is altered. Moreover, in the case of thick spacers the physical description can be essentially simplified, because the fluctuations of the potential are smooth compared with the magnetic length and with the mean distance between electrons. In this case a simple approach to the QHE [1-5] and to the problem of linear and non-linear screening by the 2DEG [6-7] is possible.

Only the disorder connected with the remote ions is considered here. The model consists of the two planes with distance s between them, where s is the width of the spacer layer. Randomly distributed ions with the average density C are in one plane and the other one contains the 2DEG with the density n. The random potential $F(\mathbf{r})$ in the plane of the 2DEG, created by the fluctuations of the ion density $C(\mathbf{r})$ has a form:

$$F(\mathbf{r})=\frac{2\pi e^2}{\kappa}\int C_q \exp(i\mathbf{q}\cdot\mathbf{r}-qs)\frac{d^2q}{q+q_s}\ , \tag{1}$$

where **r** and **q** are the 2D-vectors, κ is the dielectric constant and q_s, is the inverse screening radius of the 2DEG, $q_s = 2\pi e^2 (\kappa)^{-1} dn/dE_F$, E_F is the chemical potential of 2DEG. The random function C_q is given by the relation:

$$<C_q C_{q'}> = C \frac{\delta(\mathbf{q}+\mathbf{q}')}{(2\pi)^2} \tag{2}$$

where < ... > means the averaging over different sets of the random potential. One can see from Eqs. (1) and (2) that without screening the integral over q for $<F^2(r)>$ diverges at small q:

$$\sqrt{<F^2(r)>} = \sqrt{2\pi} \frac{e^2 C^{1/2}}{\kappa} \sqrt{\ln \frac{R}{2s}} \,, \tag{3}$$

where R^{-1} is a lower limit of the integration.

The interpretation is as follows. The random function $C(\mathbf{r})$ contains all spatial harmonics. But the harmonics with $R \ll s$ give an exponentially small potential at a distance s from the plane with randomly distributed ions. The excess number of ions in the square $R \times R$ is $(CR^2)^{1/2}$ and these ions create the potential of the order of $e^2 (cR^2)^{1/2} / \kappa R = e^2 C^{1/2} / \kappa$ which is independent of R. Each spatial harmonic which is larger than s give the same contribution to the random potential, so that $<F^2>$ diverges logarithmically.

With the electron screening one has:

$$\sqrt{<F^2>} = \sqrt{2\pi} \frac{e^2 C^{1/2}}{2\kappa q_s s} \tag{4}$$

for $2q_s s \gg 1$. One can see that the screening strongly decreases the random potential. For $s = 40$ nm the factor $2q_s s$ is about 20 (for GaAs-structures). For the mobility calculations the transport cross-section is necessary. In the case of a smooth potential it contains another small factor $(2q_F s)^{-2}$, where q_F is the wavevector of electrons with the energy E_F. This is why the remote ions do not control the mobility at large s. But the linear screening theory used in Eq. (1) is valid only at high electron density, when the redistribution of the density in the process of the screening is small compared with the average density. It means also that the Fermi energy is larger than $(<F^2>)^{1/2}$ as given by Eq. (4). (See Fig. 1a).

In the framework of the linear screening q_s is independent of the electron density n. At low densities, however, the non-linear screening regime occurs, where the density is strongly non-homogeneous (Fig. 1b). In the regions occupied by the electrons (shaded in Fig. 1b) the Fermi energy referred to the local value of the band bottom is small compared with the amplitude of the random potential, which is now of the order of $e^2 \kappa^{-1} C^{1/2}$. Then following [8] in the first approximation we put $F(\mathbf{r}) = E_F$ within these regions. where $F(\mathbf{r})$ is now the self-consistent random potential created by the ions and the redistributed electrons. Thus we obtain a nonlinear electrostatic problem: divide the plane with the 2DEG into metallic (M) and dielectric (D) regions, so that $F(\mathbf{r}) = E_F$ within M and $F(\mathbf{r}) > E_F$ within D. It was shown [7] that this problem has only one solution at given n or E_F. It is clear that at large enough n there is percolation through M - regions, but the total area of these regions decreases with decreasing

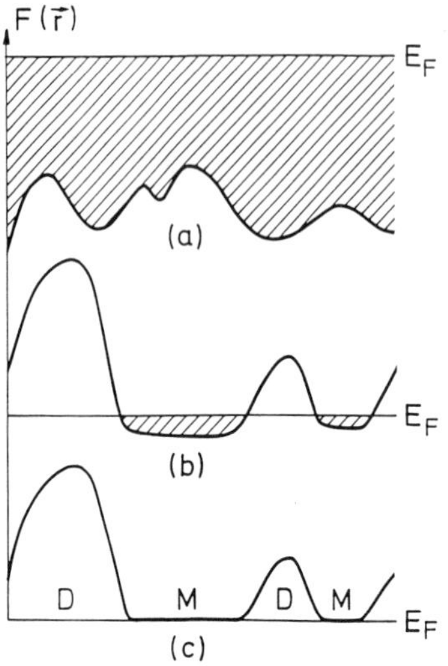

Fig. 1. The bottom of the band of 2DEG bended by the random potential. The occupied states are shaded. (a) Linear screening regime. (b) Non-linear screening regime. (c) Random potential $F(\mathbf{r})$ in the approximation of electrostatic non-linear screening theory. M and D regions are shown.

n, so that the percolation disappears at some critical value n_c, which corresponds to the metal - non-metal transition.

To estimate n_c one notes that an excess density of ions in the square $R \times R$ is of the order of $(CR^2)^{1/2}/R^2$. If $n \gg C^{1/2}/R$, the electrons can screen the R-harmonic of the random potential by a small redistribution of their density. This means the linear screening regime for this harmonic. If $n < C^{1/2}/R$, it is the non-linear screening regime, and the 2DEG density becomes strongly nonhomogeneous on the scale of the order of R. The excess density increases as R decreases. But it was mentioned above that the potential created by the harmonics with $R < s$ is exponentially small. Then with the decreasing 2DEG density n the non-linear screening regime appears for the first time for the spatial harmonic with $R = s$. It occurs when n is of the order of $C^{1/2}/s$. Then

$$n_c = \beta C^{1/2}/s\,. \tag{5}$$

The numerical coefficient β should be found by computer simulation. Rough estimates [9] give $\beta \approx 0.1$.

Now we discuss the behavior of the low-temperature conductivity near the MNMT. In the framework of the percolation theory the conductivity σ has a form [8]:

$$\sigma = \alpha\sigma_M(n_c, s)\left(\frac{n - n_c}{n_c}\right)^t. \tag{6}$$

where α is a numerical coefficient. $\sigma_M(n, s)$ is the metallic conductivity at $n \gg n_c$ and $t \approx 1.2$. Eq. (6) is valid at $(n - n_c) \ll n_c$.

An important point is that the mean free path l calculated in the Born approximation is of the order of s at $n = n_c$, so that $k_F l$ is of the order of $(C^{1/4}s^{1/2})$ which is larger than unity. It follows then that the weak localization of 2DEG is of no importance near the percolation threshold.

The experimental data on the MNMT in the heterostructures with thick spacer layers have been obtained recently by Jiang *et al.* [10]. Their results are shown in Fig. 2 by the full symbols. The sharp dependence of the mobility on the density n at small n should be considered as a manifestation of the transition. One can see that at large n the dependence of the mobility on the spacer width is rather weak, i.e., in this region the mobility is not controlled by the remote ions mainly. This dependence becomes, however, very sharp near the transition. This means that the transition is caused by the smooth potential created by the remote ions. Jiang *et al.* interpret their data in terms of the perturbation theory, which is valid if the electron density is only slightly nonhomogeneous. I think the percolation interpretation must be better in this case. Unfortunately it is impossible to get the value of t from these data, but we can check the dependence $n_c(s)$ given by Eq. (5).

It is important that near the threshold the factor $[(n-n_c)/n_c]^t$ in Eq. (6) gives the main part of the density dependence of the conductivity. It depends on the ratio n/n_c only. Then one can check the theoretical dependence $n_c(s)$ by plotting conductivity (or mobility) against n/n_c rather than against n. In this new scale one should get a more or less universal dependence for all values of s.

Jiang *et al.* [10] have used the samples with different s, but these samples may have also different concentration of charged impurities C. So Eq. (5) gives the dependence $n_c(s)$ only if the function $C(s)$ is known. One can propose two suggestions about this function: (i) C is independent of s. It is true if all donors in the samples are ionized, and the total numbers of donors are approximately the same; (ii) $C=n_0+n_d$, where n_0 is the concentration of donors at zero gate voltage, and n_d is the charge density in the depletion layer. This suggestion has been made by the authors of [10]. It is valid for thick enough layers of doped material.

It follows from Eq. (5) that suggestion (i) leads to the universal mobility dependence for all s if one plots the mobility against the product ns. In Fig. 2a the experimental data are replotted against $\log(ns/75\text{nm})$ (open symbols). One can see that all open symbols and full circles ($s=75$ nm) form now the universal dependence which confirms the above consideration.

The suggestion (ii) is checked in Fig. 2b, where the same experimental data for the mobility are plotted against $\log\{(ns/75\,\text{nm})[1.7\times10^{11}\text{cm}^{-2}/(n_0+n_d)]^{1/2}\}$. The values of n_0+n_d are taken from the paper [10], and $n_0+n_d=1.7\times10^{11}\text{cm}^{-2}$ at $s=75$ nm. The rescaled points are shown in Fig. 2a by the open symbols. In this case the open symbols and full circles do not give such good universal dependence as in the case (i), but one can conclude that Eq. (5) with both suggestions about the $C(n)$ dependence describes the experimental data satisfactorily. An important feature of our model is that the local values of the mobility in the regions occupied by the electrons are high. Therefore the collision times measured not by the dc method, but say by cyclotron resonance should not have the drop corresponding to the MNMT in the dc conductivity. The recent experiment by Liu *et al.* [11] shows that the linewidth of the cyclotron resonance in the structures of such type is very small.

Now I come to the density of states in a strong magnetic field and to the width of the plateau of the IQHE. The problem of the plateau width in the case of a smooth random potential has been discussed previously [6,7]. It has been proposed that the width is determined by the competition between the random potential and the electron-electron interaction. Without interaction the density of electrons of a partly occupied Landau level is strongly non-homogeneous at low temperatures even in a weak

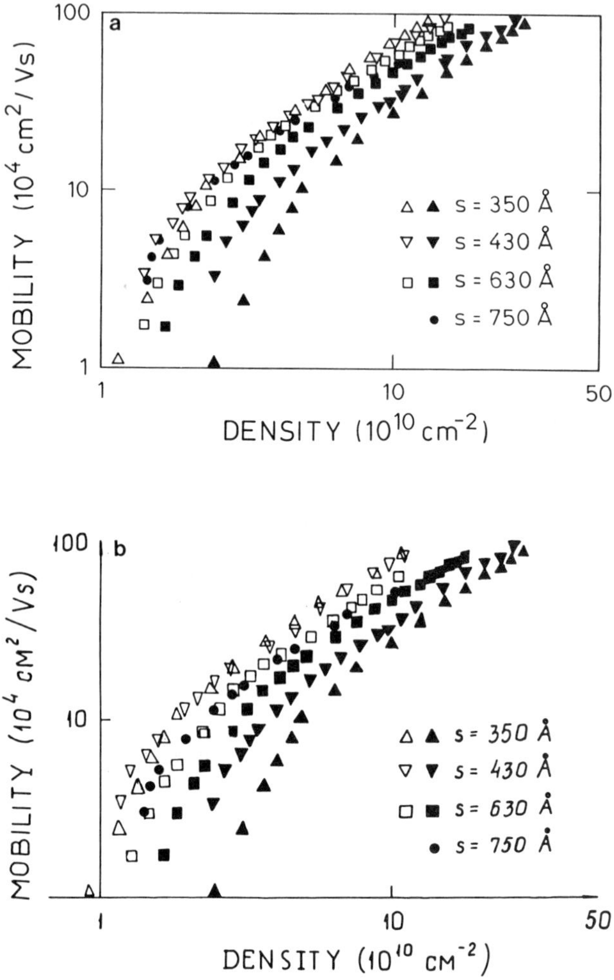

Fig. 2. The experimental data by Jiang *et al.* [10] (full symbols), plotted against log(n) for different spacer widths. The same data are rescaled as a function of log(ns/75 nm) (open symbols in Fig. 2a) and as a function of $\log\{(ns/75\,\text{nm})[1.7\times10^{11}\,\text{cm}^{-2}/(n_0+n_d)]^{1/2}\}$(open symbols in Fig. 2b). For $s = 75$ nm the scales do not change. Other points are moved to the left.

potential (Fig. 3). Only the regions where the energy of the level is lower than E_F are occupied. The delocalized states in this picture are near the percolation level E_p and their energy width is very small [2-4]. In this case the low-temperature Hall-resistance must have very wide plateaus, steep steps, and no fractional QHE because corresponding states are localized.

Electron-electron repulsion destroys this scheme. It tends to keep the electron

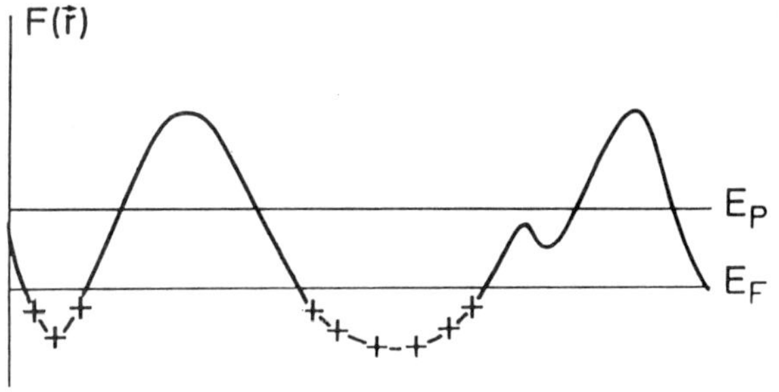

Fig. 3. The Landau level bended by the random potential. The occupied states are shown by ×. E_p is the percolation level.

density homogeneous and to prevent the electrons from condensation in the separate lakes. It is shown in [6] that due to the compressibility the electron gas can screen weak enough smooth random potential by a small redistribution of its density. It can be described as a usual linear screening with $q_s = 2\pi e^2 (\kappa)^{-1} dp/dE_F$, where $E_F = dH/dp$, H is the the energy per area of the interacting 2DEG of a given (not entirely occupied) Landau level, p is the electron density of this level, $p = n - n_0 M$, $n_0 = \lambda^{-2} = eB/hc$ is the electron density of an entirely occupied level, M is the number of occupied levels, λ is the magnetic length. and B is the magnetic field.

In fact the energy density H has many singularities, corresponding to the fractional QHE. However, at a given value of the random potential all weak singularities are smeared out, so one can speak about the energy density H, which is a smooth function in some interval of filling factor. One can get this function in the Hartree-Fock approximation.

The screening radius q_s^{-1} is of the order of the mean distance between electrons at a given Landau level if $p \ll n_0$ or of the order of the mean distance between empty states (which we call holes) if the level is almost occupied. In both limiting cases the screening becomes weak, and the random potential tears the electron liquid into pieces. The amplitude of the potential becomes of the order of $e^2 C^{1/2}/\kappa$. If it is smaller than the distance between the Landau levels Δ, only one level is involved in the screening. In this case the random potential $F(\mathbf{r})$ can be found from the non-linear electrostatic problem formulated above for zero magnetic field. Suppose the density of electrons p with the Landau number N is small ($p \ll n_0$). In this case the plane with 2DEG should be divided into M and D regions, so that electrons with number N occupy only M regions. The density of electrons with the Landau numbers smaller than N is homogeneous. The resulting random potential $F(\mathbf{r})$ is constant within M. This is exactly the same electrostatic problem as without magnetic field because it only uses the condition $F(\mathbf{r}) = E_F$ within M and does not take into account the microscopic properties of the M-phase.

The percolation through M regions disappears when the concentration p becomes smaller than n_c, where n_c is given by Eq. (5). We suppose here that the concentration of the empty states with the Landau number $N-1$ at the percolation threshold is still very small (as compared with n_c). This is true for large enough separations between Landau levels. At smaller total concentration (or at higher magnetic field) the concentration of the empty states becomes larger than n_c. This is the percolation threshold for $N-1$ level. So the width of the plateau in terms of electron density is $2n_c$, where n_c is the critical concentration of the MNMT without magnetic field. There are no singularities corresponding to the fractional QHE within this width.

For $C=10^{11}$ cm^{-2} and $s=40$ nm one obtains n_c of order of 10^{10} cm^{-2} which is much less than n_0 in the interesting cases. It means that the width of the plateau, extrapolated to zero temperature, must be small. If n_c is of the order of n_0, the one-electron picture (Fig. 3) is valid which gives wide plateaus. In the structures with low mobility this is the case [12]. The experimental data for the structures with high mobility show rather narrow plateaus of the IQHE, but their width increases with decreasing temperature. It follows from the above consideration that the limit of the width at $T \to 0$ can be much less than n_0 (in the concentration scale). Unfortunately this question has not yet been studied experimentally.

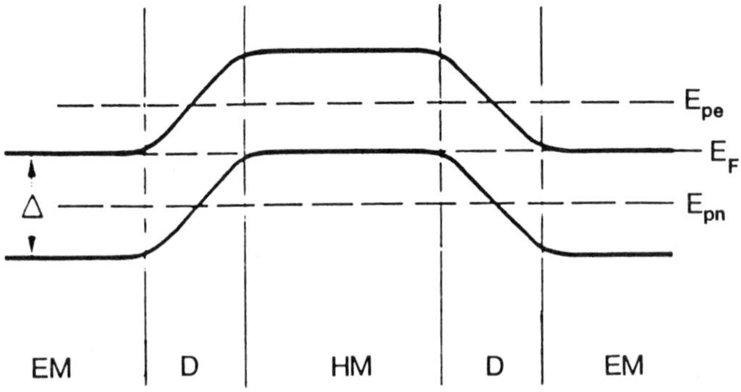

Fig. 4. Two Landau levels, bent by smooth random potential (full line). The straight horizontal lines represent the chemical potential and the percolation levels for electrons and for holes. The vertical straight lines show the boundaries of the EM, the HM, and the dielectric (D) regions.

The width of the plateau equals $2n_c$ and it is independent of the magnetic field only in the limit of the strong field: $e^2C^{1/2}/\kappa \ll \Delta$. Otherwise two Landau levels are involved in the screening simultaneously, and one gets another electrostatic problem with M regions of two kinds: electron metal (EM) and hole metal (HM) [7] (Fig. 4). Then the plateau width decreases with a magnetic field. It is shown in [7] that if $e^2C^{1/2}/\kappa \gg \Delta$, the width of the plateau is of the order of $n_c\Delta\kappa/e^2C^{1/2}$ with an unknown numerical coefficient. In this limiting case of a weak field almost all the plane of 2DEG is occupied by EM and HM regions with the narrow D regions between them. If E_F is exactly in the middle between two Landau levels, the characteristics of EM and EH regions are statistically the same for the Gaussian potential, which is considered here. One can see that there is no percolation at this point either through EM or through HM regions, because otherwise the percolation would occur both through EM and EH regions which is impossible for the two dimensional system. This is why the narrow plateaus exist in this pure classical approximation even in a weak magnetic field. Tunneling, however, can change this result, because in a weak field the size of the D regions becomes small. It is estimated in [7] as $s\Delta\kappa/e^2C^{1/2}$. If $\lambda N^{1/2}$ is comparable with this size the D regions become transparent, and the classical localization fails.

That is why I think that minimum values of ρ_{xx} in the Shubnikov-de-Haas oscillations of magnetoresistance do not necessarily tend to zero as $T \rightarrow 0$. Beginning with some Landau number they must have a non-zero limit. It is a kind of MNMT, which separates the QHE regime from the regime of the Shubnikov-de-Haas oscillations. However, weak localization can change this result, and it seems to be very interesting to check it experimentally.

REFERENCES

1. Y.Ono, J.Phys.Soc.Jpn.51, 2055 (1982).
2. S.V. Iordanskii, Solid State Commun. 43, 1, (1982).
3. R.F.Kazarinov and S.Luryi, Phys. Rev. B25, 7626 (1982).
4. S.Luryi and R.F.Kazarinov, Phys. Rev. B27, 1386 (1983).
5. S.Luryi, in High Magnetic Fields in Semiconductors Physics, v.71 of Springer Seres in Solid State Sciences ed. by G. Landwehr (Springer-Verlag, 1987) p.16.
6. A.L. Efros, Solid State Commun. 65, 1281 (1988).
7. A.L. Efros, ibid 67, 1019 (1988).
8. B.I. Shklovskii, A.L. Efros, Electronic properties of doped semiconductors, Springer-Verlag (1984), chapter 13.
9. A.L. Efros, unpublished.
10. C. Jiang, D.C. Tsui, G. Weimann, Appl. Phys. Lett. 53,1533 (1988).
11. C.T. Liu, P. Mensz, D.C. Tsui, Phys. Rev. B40, 1716 (1989).
12. H.P. Wei, D.C. Tsui, M.A. Paalanen, A.M.M. Pruisken, Phys. Rev. Lett. 61, 1294 (1988).

CORRELATION AND LOCALIZATION OF 2-D ELECTRONS IN A STRONG MAGNETIC FIELD

D.C. Tsui
Department of Electrical Engineering
Princeton University, Princeton, NJ 08544

An overview is given of some recent experiments on the localization-delocalization transition in the integral quantum Hall effect and the new quantum liquid ground states giving rise to the fractional quantum Hall effect.

1. INTRODUCTION

The two-dimensional (2D) electrons (or holes) confined to move along the hetero-interface of two different semiconductors constitute a unique electron system [1] for experimenters to study correlation and localization - two difficult strong interaction problems in condensed matter physics. Their perpendicular motion is quantized by the electric field binding them to the interface. A typical binding energy, for an electron in GaAs confined to the GaAs/$Al_xGa_{1-x}As$ interface, for example, is several tens of meV, and the size of the bound state, which is the effective thickness of the 2D system, is less than 100Å. When a magnetic field B is applied perpendicular to the interface, Landau quantization of the in-plane cyclotron motion further reduces the system to pseudo-zero-dimensional. The energy spectrum of the single particle states, as a result, is a series of discrete Landau levels separated by the cyclotron energy, $\hbar\omega_c$. (The cyclotron frequency is given by $\omega = eB/m^*$ and m^* is the effective 2D electron mass.) Each Landau level is highly degenerate. The orbital degeneracy is independent of any material parameter and is given by $N_\phi = B/\phi_o$, where $\phi_o = h/e = 4.14\times10^{-7}G-cm^2$ is the flux quantum. The Landau level filling factor, describing the population of the system at low T by a 2D electron density n_s, is defined by $\nu = n_s/N_\phi = hn_s/eB$. Thus, in a sample with $n_s = 1\times10^{11}/cm^2$, all electrons will reside in the lowest Landau level for $B \leq 4.1T$.

The extremely high degree of orbital degeneracy of the Landau level ($\sim 2.4\times10^{11}/cm^2$ at B = 10T) is the origin of the strange physics of the integral quantum Hall effect (IQHE) [2] and the fractional quantum Hall effect (FQHE) [3] found in such 2D systems. The

IQHE is understood in terms of the single particle picture, wherein the removal of the Landau level degeneracy by disorder broadens each Landau level into a subband. The quantized Hall resistance plateaus (see Fig. 1) reflect the fact that the electronic states at the Fermi energy E_F are localized and the data can be viewed as a visual display of the localization-delocalization transition in the Landau subbands.

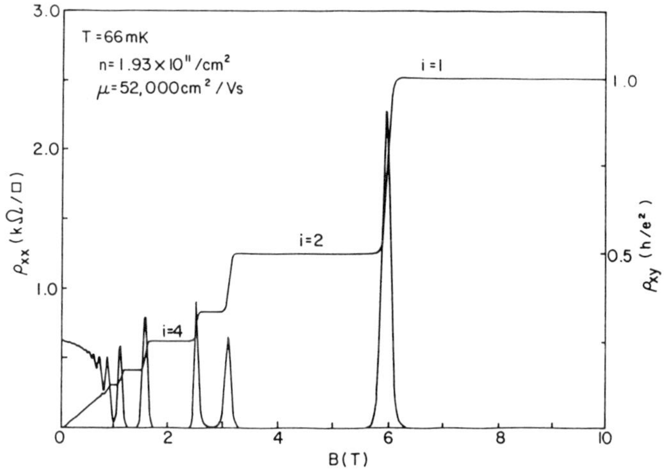

Fig. 1 ρ_{xx} and ρ_{xy} vs. B taken from a lower mobility GaAs/$Al_xGa_{1-x}As$ sample (from H.P. Wei, unpublished).

In Sec. 2, we shall review the results from several recent experiments which show that this transition can be described as a critical phenomenon [4]. Critical exponent, universality, and scaling of the transition have been observed. The FQHE, on the other hand, is only observed in very high mobility samples (Fig. 2) and the phenomenon becomes ever more pronounced as the sample quality improves and the amount of disorder in the 2D system is reduced. The phenomenon is manifestation of a series of new many-body ground states, which are highly correlated quantum liquids [5]. We shall discuss briefly, in Sec. 3, a recent interference experiment measuring the charge carried by a quasi-particle in the $\nu = 1/3$ FQHE state and the problems of the even denominator fractions.

2. SCALING IN THE IQHE

The IQHE data in Fig. 1 are plotted as a function of B. As B is swept down from the high B end, for instance, it decreases the Landau level degeneracy and thus increases the filling factor. As a result, E_F is moved through the single particle energy spectrum of

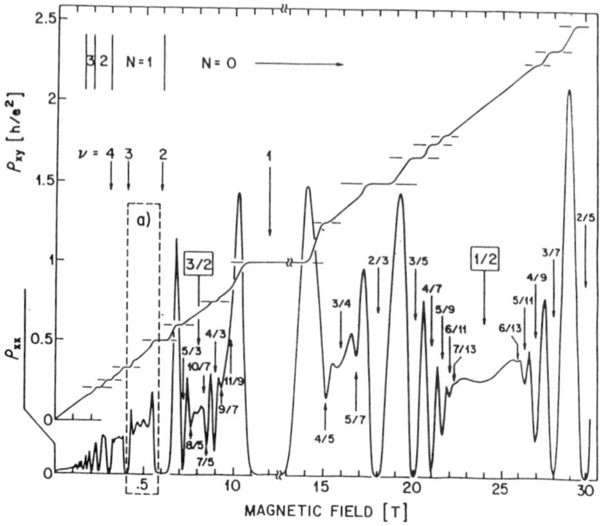

Fig. 2. ρ_{xx} and ρ_{xy} vs. B taken from a high mobility sample. T = 85mK, $n_s = 3\times10^{11}/cm^2$, and $\mu = 1.3\times10^6 cm^2/V_s$. The low field region inside a), taken at T = 25mK, is shown in Fig. 5 (from Ref. 23).

successive Landau subbands. In the ranges of B, where the Hall resistance shows quantized plateaus (ρ_{xy}=h/ie^2 with i = integers) and the diagonal resistivity ρ_{xx} becomes vanishingly small, the electronic states at E_F are localized. The transition regions between two neighboring plateaus, where ρ_{xx} shows peaks of finite widths, indicate that the states at E_F are extended [6]. Wei et al [7-9] have carried out a series of experiments, studying the dependences of ρ_{xx} and ρ_{xy} in these transition regions on T. They have found that there exists a characteristic temperature T_{sc}, above which the diagonal conductivity σ_{xx} (= $\rho_{xx}/(\rho_{xx}^2+\rho_{xy}^2)$) increases with decreasing T and below which σ_{xx} decreases with decreasing T. While the increase of σ_{xx} at higher T is explained by the T dependence of the Fermi-Dirac distribution function, the decrease of σ_{xx} for T < T_{sc} is a manifestation of scaling. Furthermore, T_{sc} is dependent on sample material, as well as sample quality. It can be as low as ~ 0.3K for the 2D electron system in GaAs/$Al_xGa_{1-x}As$, wherein long range potential fluctuations are believed to dominate, to ~ 4.2K for that in the $In_{.53}Ga_{.47}As$/InP alloy system.

In the case of the 2D electrons in $In_{.53}Ga_{.47}As$/InP, they have studied the narrowing of the transition region, as a function of decreasing T by measuring the maximum of $d\rho_{xy}/dB$ and the inverse of the half-width ΔB of ρ_{xx}, defined as the separation between the two extrema of $d\rho_{xx}/dB$. Their data for the transitions between the $i = 1$ and 2, $i = 2$ and 3, and $i = 3$ and 4 plateaus are shown in Fig. 3 and are labeled as the localization-delocalization transitions in the spin split Landau levels $N = 0\downarrow$, $1\uparrow$, and $1\downarrow$ respectively.

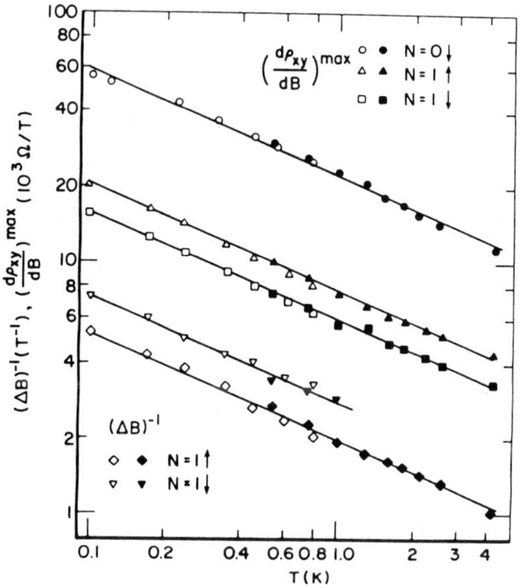

Fig. 3. Data from an $In_xGa_{1-x}As/InP$ sample with $n_s = 3.3\times10^{11}/cm^2$ and $\mu = 3.4\times10^4 cm^2/v_s$. The upper portion shows the T dependence of $(d\rho_{xy}/dB)^{max}$ for Landau levels $N = 0\downarrow$, $1\uparrow$, and $1\downarrow$; the lower portion shows the T dependence of $1/\Delta B$ for the $1\uparrow$ and $1\downarrow$ Landau levels. The slope of the straight lines is 0.42 ± 0.04. (from Ref. 9).

It is clear that, within their experimental error of 10%, the data show that both $(d\rho_{xy}/dB)^{max}$ and $(\Delta B)^{-1} \sim T^{-\kappa}$, with $\kappa = 0.42 \pm 0.04$. More recently, they have extended the measurements to lower T and measured two higher derivatives and found that $d^2\rho_{xx}/dB^2 \sim T^{-2\kappa}$ and $d^3\rho_{xy}/dB^3 \sim T^{-3\kappa}$.

These experimental results are direct confirmation that there is scaling in the localization-delocalization transition in the IQHE [10]. The exponent κ is directly related to the critical exponent ν, describing the divergence of the localization length ξ as E_F approaches the transition through $\xi \sim |E_F - E_c|^{-\nu}$. Here, E_c is expected to be the

Landau level energy and the relation is $\nu = \frac{p}{2\kappa}$, from the fact that the effective sample size for scaling is the dominant inelastic scattering length L_{in} [11]. Given that the temperature exponent for L_{in} is $\sim T^{\frac{-p}{2}}$, it is apparent that $\Delta B^{-1} \sim (E_F - E_c)^{-1} \sim L_{in}^{\frac{1}{\nu}} \sim T^{\frac{-p}{2\nu}}$. Thus, $\kappa = p/2\nu$. If we assume $p = 1$, as is the case for $B = 0$, $\nu = 1.2 \pm 0.1$.

3. FRACTIONAL QUANTUM HALL EFFECT

In the absence of disorder, it is intuitively clear that electrons in the lowest Landau level will take advantage of the extremely high degree of orbital degeneracy and correlate their motion to minimize the Coulomb repulsion energy. In the limit $B \rightarrow \infty$, it has long been anticipated that the correlated ground state is a triangular Wigner crystal [12]. The FQHE, however, is a manifestation of a series of unanticipated new ground states. Our current understanding of the phenomenon is based on Laughlin's theory [5] that these ground states are uniform density, incompressible electron liquids. Their quasiparticle/quasihole excitations are fractionally charged and are separated in energy by finite gaps. In this section, we shall first mention a recent experiment measuring the fractional charge carried by the quasiparticle and then discuss the problem of the even denominator fractions.

3.1 Fractional Charge

The strange notion that the quasiparticle/quasihole excitations of the FQHE states are fractionally charged objects is a most remarkable feature of Laughlin's theory. Several recent experiments have attempted to verify it [13-15]. The one I wish to briefly mention here is based on the Aharonov-Bohm effect that a quantum mechanical object, carrying charge e^*, traversing a loop of area A in a B field will gain a phase factor with a phase angle $\Theta = (e^*/h)BA = 2\pi\phi/(h/e^*)$. In the experiment of Simmons et al. [15], they find quasi-periodic noise structures in ρ_{xx} and ρ_{xy} for B adjacent to the quantized Hall plateaus in narrow samples of high mobility 2D electrons in a $GaAs/Al_xGa_{1-x}As$ heterostructure. Fig. 4 is an example of their data showing that, while the period is $\sim$ 150G for structures adjacent to all the IQHE plateaus (shown in the top two panels of Fig. 4 for $\nu = 1$ and $\nu = 2$, respectively), an approximately three times larger period ($\sim$ 500 G) is observed for structures adjacent to the $\nu = 1/3$ FQHE plateau.

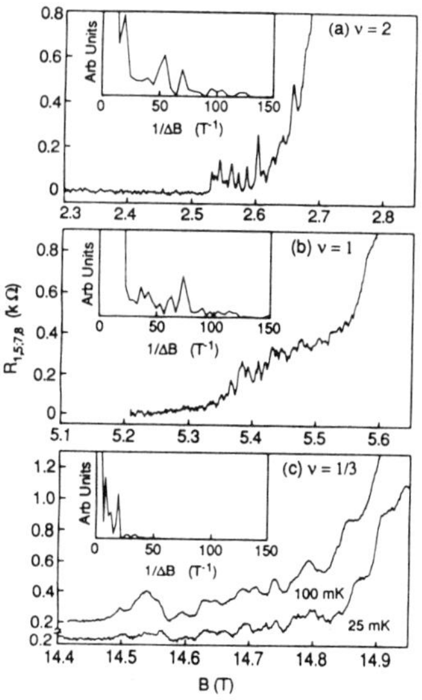

Fig. 4: The resistance along current flow, $R_{1,5;7,8}$, near the high field sides of resistance minima for (a) $\nu = 2$ at 25mK, (b) $\nu = 1$ at 25 mK, and (c) $\nu = 1/3$ at 25 and 100 mK, all plotted with the same field scale. Insets show the Fourier power spectra of the region of fluctuations for each respective ν (from Ref. 19).

They attribute the noise structure to resonant tunneling via states magnetically bound to some submicron size potential hills or valleys [16], and the periodicity of the structure, a manifestation of the phase change in the Aharonov-Bohm effect. The striking difference of the factor 3 in the observed periodicity is evidence that the charge carried by a quasiparticle of the $\nu = 1/3$ FQHE state is $e^* = e/3$ [17].

3.2 Even denominator fraction

For the N=0 Landau level (B>6T in Fig. 2), partial filling is observed at sufficiently large B that removal of the spin degeneracy by the Zeeman effect is assumed to split it into two spin-polarized levels. In the $\nu < 1$ regime, when the lowest energy, N = 0 and spin ↑ Landau level is partially occupied, it is now quite clear that only odd denominator states exist. To date, the observed fractions include: $\nu = 1/3, 2/3; 1/5, 2/5, 3/5, 4/5; \underline{1/7}, 2/7, 3/7, 4/7, 5/7; 2/9, 4/9, 5/9; 2/11, 3/11, \underline{4/11}, \underline{5/11}, \underline{6/11}, \underline{7/11}; \underline{3/13}, \underline{4/13}, \underline{6/13}, \underline{7/13}, \underline{9/13}; \underline{4/15}$.

Except for the underlined fractions, quantized Hall plateaus have been observed and the quantization is better than 1%. A broad ρ_{xx} minimum centered around $\nu = 1/2$, however, was discernible even in the early experiments [18]. It was attributed to the convergence of two sequences of higher order states, emanating from the $\nu = 2/3$ and 1/3 states. In most recent experiments on lower disorder samples [19,20], this broad minimum becomes much stronger than the neighboring higher order states and it persists to temperatures at which the FQHE states are not observable. However, there is no evidence for Hall plateau development at $\nu = 1/2$ and the structure is considered to be other than the indication of a FQHE state [20]. Data on the fractional states in the N=0 and spin ↓ Landau level ($1 < \nu < 2$) is similar, and a similarly strong ρ_{xx} minimum has been observed in samples with lower disorder at ν=3/2.

In case of the N=1 Landau level, an even denominator state at $\nu = 5/2$ is observed [21]. This is shown in Fig. 5, where a Hall plateau quantized to a conductance value of $\frac{5}{2}\frac{e^2}{h}$ and a ρ_{xx} minimum are developing at $\nu = 5/2$.

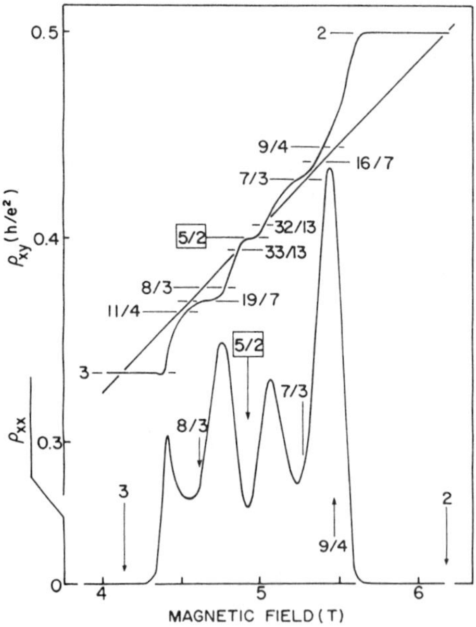

Fig. 5: Blowup of the ρ_{xx} and ρ_{xy} data of the low field section a) of Fig. 2 showing the even fraction state at $\nu = 5/2$ (from Ref. 23).

More recently, Einstein et al [22] have extended the experiment to lower T on a higher quality sample with a lower n_s. They have observed the fully formed ρ_{xy} plateau and an excitation gap of $\approx$

105mK from the T dependence of ρ_{xx} at $\nu = 5/2$. Experimentally, the 5/2 state is observed at considerably lower B. Consequently, the Zeeman energy is sufficiently small that electrons in the partially filled N=1 Landau level cannot be regarded as spin polarized and that exchange may be important. Indeed, it has been demonstrated that the Zeeman energy is important. When it is increased by tilting B away from the surface normal, while keeping the normal component of B constant, the 5/2 state is observed to decrease abruptly in strength [23]. On the theoretical side, Haldane and Razayi [24] are able to show that a spin-singlet liquid state can exist in the N=1 Landau level at $\nu = 5/2$, using their so-called hollow core model interaction. However, the degree to which their model interaction approximates the real Coulomb interaction has been criticized [25] and, moreover, other effects, such as polarization of electrons in the N=0 Landau level and mixing of higher Landau levels, not yet considered in the theory may all be important in the experiment.

ACKNOWLEDGEMENTS

This work is a result of a long collaboration with my colleagues R.L. Willett, H.P. Wei, H.L. Stormer, J.A. Simmons, M. Shayegan, A.M.M. Pruisken, L.N. Pfeiffer, A.C. Gossard, V.J. Goldman, L. Engel, J.P. Eisenstein, A.M. Chang and G.S. Boebinger, to whom I express my gratitude. It is supported by the NSF, ONR, AFOSR and a grant from the NEC Corporation.

REFERENCES

1. For a review, see T. Ando, A.B. Fowler, and F. Stern, Rev. Mod. Phys. **54**, 437 (1983).
2. K. von Klitzing, G. Dorda, and M. Pepper, Phys. Rev. Lett. **45**, 494 (1980).
3. D.C. Tsui, H.L. Stormer, and A.C. Gossard, Phys. Rev. Lett. **48**, 1559 (1982).
4. A.M.M. Pruisken, in *The Quantum Hall Effect*, ed. by R.E. Prange, S.M. Girvin (Springer, Berlin, 1987), p. 233.
5. R.B. Laughlin, Phys. Rev. Lett. **50**, 1395 (1983).
6. M.A. Paalanen, D.C. Tsui, and A.C. Gosssard, Phys. Rev. B **25**, 5566 (1982).
7. H.P. Wei, D.C. Tsui, and A.M.M. Pruisken, Phys. Rev. B **33**, 1488 (1986).

8. H.P. Wei, D.C. Tsui, M.A. Paalanen, and A.M.M. Pruisken, in *High Magnetic Fields in Semiconductor Physics*, edited by G. Landweber (Springer-Verlag, Berlin, 1987) p. 11.
9. H.P. Wei, D.C. Tsui, M.A. Paalanen, and A.M.M. Pruisken, Phys. Rev. Lett. **61**, 1297 (1988).
10. A.M.M. Pruisken, Phys. Rev. B **32**, 2636 (1985).
11. A.M.M. Pruisken, Phys. Rev. Lett. **61**, 1297 (1989).
12. H. Fukuyama, P.M. Platzman, and P.W. Anderson, Phys. Rev. B **19**, 5211 (1979).
13. R.G. Clark, J.R. Mallett, S.R. Haynes, J.J. Harris, and C.T. Foxon, Phys. Rev. Lett. **60**, 1747 (1988).
14. A.M. Chang and J.E. Cunningham (unpublished).
15. J.A. Simmons, H.P. Wei, L.W. Engel, D.C. Tsui, and M. Shayegan, Phys. Rev. Lett. **63**, 1731 (1989).
16. J.K. Jain and S. Kivelson, Phys. Rev. Lett. **60**, 1542 (1988).
17. S.A. Kivelson and V.L. Pokrovsky, Phys. Rev. B **40**, 1373 (1989).
18. D.C. Tsui, H.L. Stormer, J.C.M. Hwang, J.S. Brooks, and M.J. Naughton, Phys. Rev. B **28**, 2274 (1983).
19. M. Shayegan, V.J. Goldman, M. Santos, T. Sajoto, L. Engel, and D.C. Tsui, Appl. Phys. Lett. **53**, 2080 (1988).
20. H.W. Jiang, H.L. Stormer, D.C. Tsui, L.N. Pfeiffer, and K.W. West, Phys. Rev. B, (to be published).
21. R.L. Willett, J.P. Eisenstein, H.L. Stormer, D.C. Tsui, A.C. Gossard, and J.H. English, Phys. Rev. Lett. **59**, 1776 (1987).
22. J.P. Eisenstein, R.L. Willett, H.L. Stormer, L.N. Pfeiffer, and K.W. West, Proc. of the Int. Conf. on Electronic properties of two-dimensional systems, Grenoble, 1989 (to be published in Surf. Sci.).
23. J.P. Eisenstein, R.L. Willett, H.L. Stormer, D.C. Tsui, A.C. Gossard, and J.H. English, Phys. Rev. Lett. **61**, 997 (1988).
24. F.D.M. Haldane and E.H. Razayi, Phys. Rev. Lett. **60**, 956 (1988).
25. A.H. MacDonald, D. Yoshioka, and S.M. Girvin, Phys. Rev. B **39**, 8044 (1989).

PHONON ABSORPTION BY 2 - D ELECTRONS IN STRONG MAGNETIC FIELD

S. V.Iordansky and B. A. Musykantsky
L. D. Landau Institute for Theoretical Physics,
Academy of Science of USSR

Phonon thermocoductivity in layered heterostructures GaAs/AlGaAs has been measured recently [1]. It has been shown that at sufficiently low temperatures, when all charge carriers are concentrated in two-dimensional layers, there are quantum oscillations of thermoconductivity with magnetic field. Phonon absorption by two-dimensional electrons and its effect on phonon thermoconductivity have not yet been considered theoretically. In the present paper we calculate the corresponding life time for phonons.

In the absence of impurities, phonon absorption in two dimensions is possible only with transitions between Landau levels and is exponentially small at low temperatures $T \ll \hbar\omega_c$ where $\omega_c = eH/mc$ is cyclotron frequency, m is the effective mass and H is magnetic field. Thus absorption essentially depends on properties of random potential affecting electrons in the $2d$ layer. At present there is some evidence that the electronic density of states in MOS transistors and heterostructures (see ref. [2] and review [3]) is determined by large scale charge fluctuations far from the layer containing electrons. Therefore we assume that there is smooth random potential described by Gaussian distribution with some homogeneous and isotropic correlation function. $<U(\mathbf{r})U(\mathbf{r}')> = R(\mathbf{r}-\mathbf{r}')$, the scale of which $L \gg l_h$, where l_h is magnetic length $l_h^2 = c\,\hbar/eH$.

1. PHONON LIFETIME

We start from the well-known formulae of perturbation theory for probability of a transition with absorption of one phonon of momentum $\mathbf{k}$, polarization s and frequency $\omega(\mathbf{k}, s)$ (see e.g. ref. [4]):

$$\frac{1}{\tau(\mathbf{k}, s)} = \frac{2\pi}{\hbar}\sum_{i,f}\delta(\varepsilon_f - \varepsilon_i - \hbar\omega(\mathbf{k}, s))\,|<0, \phi_f|H_{\text{e-ph}}|1, \phi_i>|^2(n_i - n_f)\ . \qquad (1.1)$$

Here ε is the energy and μ is the chemical potential of electrons, $n_i = [\exp(\varepsilon_i - \mu)/T) + 1]^{-1}$ is Fermi function, $H_{\text{e-ph}}$ is the Hamiltonian of electron-phonon interaction. We neglect electron-electron interaction and consider only acoustical phonons at low enough temperatures.

The presence of free charge in the layer can result in screening potential. However, we will neglect this screening taking into account strong space dispersion due to a rather short wavelength of thermal phonons(we assume $kl_h \sim 1$.)

The full electron-phonon interaction taking into account deformation potential and piezoelectricity is (see e.g. [5])

$$H_{\text{e-ph}}=\sqrt{\frac{\hbar}{D^3}}\,V_s(\mathbf{k})a_{\mathbf{k},s}\int e^{i\mathbf{k}\cdot\mathbf{r}}\phi^*(\mathbf{r})\phi(\mathbf{r})+c.c. \tag{1.2}$$

Here D is the sample dimension, $a^{\dagger}_{\mathbf{k},s}$, $a_{\mathbf{k},s}$ are creation -annihilation operators for phonons, and ϕ is the electronic wave function. The vertex $V_s(\mathbf{k})$ has the form

$$V_s(\mathbf{k})=\frac{i}{(\rho c_s k)^{1/2}}\frac{k_m d_n(\mathbf{k},s)+k_n d_m(\mathbf{k},s)}{2}(\Xi_{mn}-\frac{4\pi}{\chi k^2}k_j\beta_{jnm})\,, \tag{1.3}$$

where $\mathbf{d}(\mathbf{k},s)$ is the phonon polarization vector, c_s is the sound velocity, ρ is the density, χ is the dielectric constant, Ξ_{nm} is the deformation potential, and β_{jnm} are piezomoduli all for GaAs. The 2d layer is about ten times the lattice spacing, therefore, to sufficient accuracy Ξ_{nm} and β_{jnm} assume their bulk values. The electron wave functions have the form $\phi=f_0(z)\phi_n(\mathbf{r}')$,where $f_0(z)$ corresponds to the wave function of the ground state of dimensional quantization across the layer, and the ϕ_n are wave functions of electron states in the layer. Substituting this expression into eq. (1.2) we obtain the effective vertex $V_s^{\text{eff}}(\mathbf{k})=J(k_z)V_s(\mathbf{k})$ with the extra factor $J(k_z)=\int\exp(ik_z z)|f_0(z)|^2dx$.

It is easy to show that the phonon lifetime after averaging over the random potential takes the form (the sample containing N heterojunctions):

$$\frac{1}{\tau}=\frac{|V_s^{\text{eff}}(\mathbf{k})|^2N}{2\pi\hbar^2D}\int e^{i\mathbf{k}'\cdot\mathbf{r}'}\langle G(\varepsilon_1,0,\mathbf{r})G(\varepsilon_2,\mathbf{r},0)\rangle$$

$$\times(n(\varepsilon_1)-n(\varepsilon_2))\delta(\varepsilon_1-\varepsilon_2+\hbar\omega(\mathbf{k},s))d\varepsilon_1d\varepsilon_2d^2\mathbf{r}\,, \tag{1.4}$$

where the brackets denote average over the random potential, N/D is the number of heterojunctions per unit length, and $\mathbf{k}'$ is the component of wave vector in the plane of the layer.

Since we ignore electron-electron interaction the Green functions entering (1.4) (essentially this is a difference of retarded and advanced Gf) are one electron Gf which can be presented by a Feynman path integral [6]:

$$G(\varepsilon,\mathbf{r},0)=-i\int_{-\infty}^{\infty}dt\int D\mathbf{q}(\tau)\exp\left[\frac{iS(t,\mathbf{r},0)+i\varepsilon t}{\hbar}\right]$$

where $S(t,\mathbf{r},0)$ is the one electron action in a uniform magnetic field and impurity potential, $\mathbf{q}(0)=0$, $\mathbf{q}(t)=\mathbf{r}$. Integration over t_1, t_2 implies as usual the convergence of the integral at $t\to\pm\infty$ due to a small imaginary part of ε. Performing Gaussian average over U we obtain for the polarization operator [7]

$$\Pi(\mathbf{k}',\omega)=\frac{1}{2\pi}\int e^{i\mathbf{k}'\cdot\mathbf{r}+i\varepsilon_1t_1+\varepsilon_2t_2}\delta(\varepsilon_1-\varepsilon_2+\hbar\,\omega)(n(\varepsilon_1)-n(\varepsilon_2))$$

$$\times\int D\mathbf{q}_1 \int D\mathbf{q}_2 \exp\left[\frac{iS_{\text{eff}}(t_1, t_2, \mathbf{r})}{\hbar}\right] dt_1 dt_2 d^2\mathbf{r} d\varepsilon_1 d\varepsilon_2 \,, \tag{1.5}$$

where

$$S_{\text{eff}} = \int_0^{t_1} L_0(\mathbf{q}_1, \dot{\mathbf{q}}_1)d\tau + \int_0^{t_2} L_0(\mathbf{q}_2, \dot{\mathbf{q}}_2)d\tau + \frac{i}{\hbar}\left\{\int_0^{t_1} d\tau_1 \int_0^{t_1} d\tau_2 R(\mathbf{q}_1(\tau_1) - \mathbf{q}_1(\tau_2))\right.$$

$$\left. +2\int_0^{t_1} d\tau_1 \int_0^{t_2} d\tau_2 R(\mathbf{q}_1(\tau_1) - \mathbf{q}_2(\tau_2)) + \int_0^{t_2} d\tau_1 \int_0^{t_2} d\tau_2 R(\mathbf{q}_2(\tau_1) - \mathbf{q}_2(\tau_2))\right\} \tag{1.6}$$

Here L_0 is the Lagrange function of free electron in uniform magnetic field, the integrals over t_1 and t_2 converge at $\pm\infty$ without imaginary additions to ε_1 and ε_2 due to presence of function R. Therefore the integrand is analytic and the contour of integration can be shifted from the real axis to the complex plane.

2.CALCULATION OF THE POLARIZATION OPERATOR.

The calculation of the polarization operator will be performed assuming the phonon wave vector sufficiently large $kl_h \sim 1$ and the correlation length of the random potential large compare to magnetic length. This case corresponds to quasi-classical situation and the functional integral (1.5) can be calculated by saddle-point method.

Varying the effective action (1.7) we obtain electron operator equations of motions

$$m\ddot{\mathbf{q}}_1 = \frac{e}{c}[\dot{\mathbf{q}}_1 \,\mathbf{H}] + \frac{i}{\hbar}\frac{\partial}{\partial \mathbf{q}_1(\tau)}\left[\int_0^{t_1} d\tau_1 R(\mathbf{q}_1(\tau) - \mathbf{q}_1(\tau_1)) + \int_0^{t_2} d\tau_1 R(\mathbf{q}_1(\tau) - \mathbf{q}_2(\tau_1)\right],$$

$$m\ddot{\mathbf{q}}_2 = \frac{e}{c}[\dot{\mathbf{q}}_2 \,\mathbf{H}] + \frac{i}{\hbar}\frac{\partial}{\partial \mathbf{q}_2(\tau)}\left[\int_0^{t_1} d\tau_1 R(\mathbf{q}_2(\tau) - \mathbf{q}_1(\tau_1)) + \int_0^{t_2} d\tau_1 R(\mathbf{q}_2(\tau) - \mathbf{q}_2(\tau_1))\right], \tag{2.1}$$

with boundary conditions $\mathbf{q}_1(0) = \mathbf{q}_2(t_2)$, $\mathbf{q}_1(t_1) = \mathbf{q}_2(0) = \mathbf{r}$. Equations (2.1) in spite of their integral form possess the energy integral for each particle in accordance with elastic scattering by random static potential. Particles exchange their momenta due to inhomogeneity of interaction via function R.

It is very difficult to find solutions of equations (2.1) satisfying the boundary conditions, even if we suppose that R varies slowly. We restrict ourselves to the case where the distances covered by electrons in times t_1 and t_2 (which should be also defined) are small compared to the correlation length L. Then the function R entering Eqs. (2.1) can be expanded to the second order in its argument and a solution can be found analytically. At very large correlation length corrections to free motion of electron are small so that we can restrict ourselves to first order of perturbation theory: i.e., we can calculate the action S_{eff} directly for a free electron trajectory in magnetic field satisfying the boundary conditions set above. The corresponding expression is rather cumbersome. As seen later at low temperatures saddle point values of t_1, t_2 are such that $|t_1+t_2| \ll |t_{1,2}|$, and $\omega_c |t_{1,2}| \gg 1$ where $t_{1,2}=t_1$ or t_2. The expression for the action then can be reduced to

$$S_{\text{eff}} \approx r^2 \left[\frac{m\omega_c}{4} \left\{ \cot\left(\frac{\omega_c t_1}{2}\right) + \cot\left(\frac{\omega_c t_2}{2}\right)\right\}\right.$$

$$\left. + \frac{iR''(0)t_1 t_2}{2\hbar} \left\{ \cot\left(\frac{\omega_c t_1}{2}\right) + \cot\left(\frac{\omega_c t_2}{2}\right)\right\}^2 \right] + \frac{iR(0)}{2\hbar}(t_1+t_2)^2 . \quad (2.2)$$

It is necessary to point out that t_1 and t_2 are imaginary and the singularities related to the poles of (2.2) are not important.

The correction to the extremal action (2.2) should be quadratic in deviations from the extremal path so that to obtain the polarization operator (1.5) Gaussian integration over periodic paths beginning and ending at $\mathbf{q}_1=\mathbf{q}_2=0$ should be performed. Performing this integration we neglect second derivatives of R because they give terms small in parameter $1/\omega_c t$ which can be ignored. The result of this integration can be united with the factor corresponding to free electrons in the extremal action. As a result the product of two Green's functions of free particles becomes

$$\int D\mathbf{q}(\tau)\exp\left(\frac{iS^0(t,\mathbf{r},0)}{\hbar}\right) = \frac{1}{Dl_h}\sum_{n,\kappa}\exp[-i\omega_c t(n+1/2+i\kappa y)]\Phi_n(x/l_h+\kappa l_h)\Phi_n(\kappa l_h) ,$$

where n is the number of Landau level, Φ_n are normalized oscillator functions, and the Landau gauge is chosen (Π is gauge independent.) Substituting this expression into the polarization operator a sum arises in which the terms correspond to transitions between different Landau levels. The integral over t_1 and t_2 corresponding to such a term is easily performed:

$$\frac{\pi\hbar}{k' l_h^2[R(0)|R''(0)|]^{1/2}}\exp\left[-\frac{(E_1+E_2)^2}{8R(0)} + \frac{(E_1-E_2)^2}{2R''(0)q^2 l_h^4}\right],$$

where $E_i=\varepsilon_i-\hbar\omega_c(n_i+1/2)$. This integral is defined by the extremal point of appropriate Gaussian exponent

$$t_1^0 - t_2^0 = -\frac{2i\hbar(E_1-E_2)}{R''(0)k'^2}, \quad t_1^0+t_2^0 = \frac{i\hbar(E_1+E_2)}{2R(0)}, \quad (2.3)$$

(deriving this expression we have assumed $|R''(0)|k'^2 l_h^4 \ll R(0)$.) We see that only transitions with $n_1=n_2$ are important because of the condition $|R''(0)|q_2 l_h^4 \ll \hbar\omega_c^2$ valid due to the large correlation length L of the random potential. Deriving Eqs. (2.2) we have assumed that $|t_1| = \hbar\omega / (|R''(0)| q^2 l_h^4) \gg \omega_c^{-1}$ whence at $k' l_h \sim 1$, $(L^2 \hbar 2\omega_c \omega)/(l_h^2 R(0)) \gg 1$. The latter inequality is valid at sufficiently large L also.

It is easy to show that the aforementioned conditions (large L and small ω) also imply the validity of perturbation theory.

Finally we get the following formula for the polarization operator performing some integrals with oscillator functions connected with Laguerre polynomials L_m (see e.g. [8]) and integrating over ε_1 and ε_2 with Fermi functions

$$\Pi(\omega, k') = \frac{\hbar^3 \omega}{4\pi l_h^2 \sqrt{R(0)|R''(0)|k'^2 l_h^4}} \sum_{m,\sigma} \exp\left[-\frac{(\mu - E_{m,\sigma})^2}{2R(0)} + \frac{\hbar^2\omega^2}{2R''(0) k'^2 l_h^4}\right] H_m(k' l_h)^2 ,$$

$$E_{m,\sigma} = \hbar\omega_c (m + 1/2) + g\mu_B \sigma$$

$$H_m(r) = \exp(-r^2/4) L_m(r^2/2) , \tag{2.4}$$

here μ_B is Bohr magneton, g is g-factor, and summation over spin orientation σ is included.

Expression (2.4) corresponds to the following physical picture: electrons of a given energy ε are localized near the line of level of random potential $U(\mathbf{r}) = \varepsilon$. Due to large correlation length L, the curvature of this line can be ignored so that the wave function can be locally written as $\phi_\varepsilon = \Phi_n(\eta/l_h - k_s l_h)\exp(ik_s s)$, where s is the coordinate along the line of level, η is the normal coordinate, and k_s is such that $U(l_h^2 k_s) = \varepsilon$. Substitution of the wave functions into expression (1.1) after the averaging over the potential U gives the same result as (2.4).

For such local treatment to be valid in the calculation of the matrix element in (1.1) it is necessary that the phonon wavelength should be smaller than the correlation length L. Since in the experiment of ref. [1] the ratio L/l_h is not very large we will assume that Π ceases to grow with decreasing k' at $k' l_h \sim 1$. Further we will substitute k'^2 in the denominator of the factor before the sum sign by $k'^2 + l_h^{-2}$.

3. CORRECTIONS TO THERMAL CONDUCTIVITY

To obtain the phonon lifetime, we substitute (2.4) into (1.4)

$$\frac{1}{\tau} = \frac{\hbar\omega N}{l_h^2 D} \frac{|V_s^{\mathrm{eff}}|^2}{[R(0)|R''(0)|(k'^2 + l_h^{-2})]^{1/2}}$$

$$\times \exp\left[\frac{\hbar\omega^2}{2R''(0) k'^2 l_h^4}\right] \sum_{m,\sigma} H_m^2(k' l_h) \exp\left[-\frac{(\mu - E_{m,\sigma})^2}{2R(0)}\right] . \tag{3.1}$$

We use simple gas kinetic expression for thermal conductivity $\kappa = C_{\mathrm{ph}} c^2 \tau_{\mathrm{ph}}/3$, where C_{ph} is the phonon specific heat, c is the sound velocity, and τ_{ph} is the total phonon lifetime including scattering off the boundaries. As usual we sum inverse lifetimes (independent transitions) $1/\tau_{\mathrm{ph}} = \langle 1/\tau_0 \rangle + 1/\tau$, where the brackets mean average over the thermal phonon distribution. The value of τ_0 is associated with residual scattering independent of 2d electrons and magnetic field. According to ref. [1] the temperature difference at fixed heat flow has been measured, i.e. up to a constant factor the value $\Delta\kappa^{-1} \sim \langle\tau^{-1}\rangle/(C_{ph}c^2)$. The temperature dependence of $\Delta\kappa^{-1}$ according to the experiment is T^3 [1], hence $\langle\tau^{-1}\rangle$ does not depend on temperature. Using Eqs. (3.1) and (1.3) we can account for this assuming that the main term in the vertex corresponds to piezoelectricity. Performing the thermal average it is possible to use the fact that only those phonons are essential which have the momentum in the plane of the 2d layer since $R''(0)$ is small. Furthermore, because we have used a rough estimation for κ from the theory of gas kinetics we may to the same accuracy replace k' by thermal phonon momentum $k_T = T/\hbar c$ which gives

$$\Delta\kappa^{-1} = \frac{N}{2\pi l_h^4 D C_{ph}(T) c^2 \sqrt{R(0)|R''(0)|(k'^2 + l_h^{-2})}} \exp\left[\frac{\hbar c^2}{2R''(0) l_h^4}\right]$$

$$\times \sum_{m,\sigma} H_m^2(k_T l_h) \exp\left[-\frac{(\mu - E_{m,\sigma})^2}{2R(0)}\right] \overline{\left[\frac{4\pi\beta_{jnl} k_n k_j d_l}{\chi k'^2}\right]^2}, \tag{3.2}$$

where the bar means averaging in the plane of the 2d layer. We see that the main temperature dependence is determined by phonon specific heat, other factors are either temperature independent or have a weak dependence at $k' l_h \sim 1$.

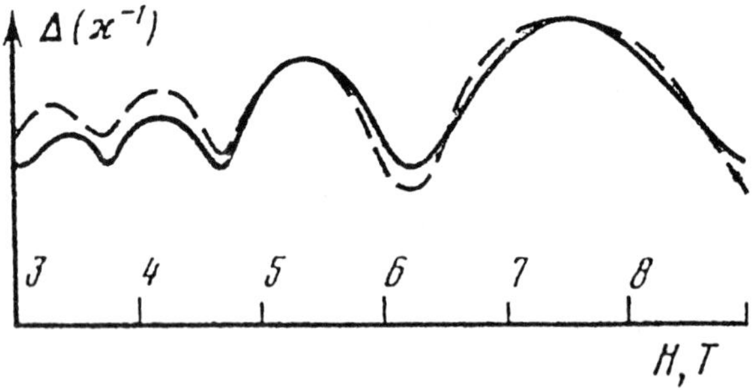

Fig. 1. The value proportional to the correction to inverse thermoconductivity due to phonon scattering by 2D electrons according to Eq.(3.2). The constant factor and a constant additive term not depending on magnetic field chosen to obtain the best correspondence with experimental value given by solid line.

It is known from experiments that the mean square value of random potential oscillates with Landau level occupation number n_s. The least value corresponds to half integer occupation number, when there are free charges and the screening is the strongest. At present a detailed theory verified by experiment does not exist. We use the experimental results of ref. [9] where the thermodynamic density of states has been measured, and it has been shown that the mean square fluctuations of the random potential are described by the empirical expression $R(0)=0.25H$ meV per tesla. Fig. 1 shows $\Delta\kappa^{-1}$ calculated in arbitrary units, according to Eq.(3.2) at 0.4 K, $R''(0)=-3/2\,d^{-2}R(0)$ where $d=300\text{Å}$ the spacer dimension. The value of the chemical potential also was determined according to the experiment [9]. We see rather good agreement with experiment [1].

The authors express their gratitude to I. B. Levinson and E. I. Rashba for helpful discussions.

REFERENCES

1. Eisenstein I.P.,Gossard A.C.,Naryanamurti V., Phys.Rev.Lett.,v.59, **12**, 1341 (1987).
2. Shklovsky B.I.,Efros A.A. Pisma ZhETF 44, 669, (1987).
3. Kukushkin I.V.,Meshkov S.V.,Timofeev V.B., UFN v.155 **2**, 219, (1988).
4. Lifshits E.M.,Pitaevsky L.P.,Physical Kinetics,Moscow,Nauka 1979.
5. Gantmakher V.F.,Levinson I.B.,Rasseyanie nositelei toka v metalakh i poluprovodnikakh, p.58, Moscow,Nauka (1984).
6. Feynman R.,Hibbs A.,Quantum mechanics and path integrals Mc.Grow-Hill Book Company, New York 1965.
7. Abrikosov A.A., Gor'kov L.P., and Dzyaloshinskii J.E., Quantum Field Theoretical Methods in Statistical Physics, Pergamon Press, Oxford (1965).
8. Bychkov Yu.A.,Rashba E.I.,ZhETF v.85,p.1826 ,(1983).
9. Weiss D.,Klitzing K.V.,Mosser V. Springer series in solid state Sciences v.67,Springer 1986.

THE QUANTUM HALL EFFECT IN OPEN CONDUCTORS

M. Büttiker
IBM T. J. Watson Research Center
Yorktown Heights, N.Y. 10598

The quantum Hall effect is discussed in terms of global conductances which describe carrier transport from one contact to another. The theory can treat highly non-uniform samples. Examples are discussed which show a simultaneous quantization of the longitudinal and the Hall resistance as observed in experiments. Contacts which exhibit scattering lead in general to deviations from the quantum Hall effect. In special cases the Hall resistance is quantized at anomalous values.

INTRODUCTION

The discovery [1] of the quantum Hall effect in 1980 was recognized in 1985 by a Nobel Prize [2]. Ever since the discovery this interesting solid state effect has attracted considerable theoretical and experimental attention [3]. One might expect that after a decade of efforts by the scientific community to understand this effect nothing more would be left to do than to fill in a few details which just add to and round off the picture developed in the first few years after the discovery of this effect. Surprisingly, this expectation is not well founded: The past two years have seen exciting new activity in this field which hopefully will bring about a deeper understanding of the quantum Hall effect. Below we stress the important role of contacts used to bring carriers into the sample, remove carriers from the sample and contacts used to make voltage measurements [4,5]. Consideration of the contacts is important not only because they permit a description which is much closer to experimental reality, but also because they permit a sensible definition of a resistance measurement [6].

The accuracy of the quantum Hall effect has led to the belief that this phenomena must have a general explanation independent of the geometry and the microscopic details of the conductor. This belief is at the heart of some of the most widely accepted explanations of the quantum Hall effect and has even led to the characterization of this effect in terms of topological numbers only [7]. Such discussions are concerned with charge transport along a closed path in the interior of the conductor. There is a priori no connection of such charge transport coefficients to the transport coefficients measured in an actual experiment.

The importance of contacts in electrical conduction problems was eventually recognized in a field which since the beginning of this decade has grown tremendously and is concerned with transport in small conductors [8]. Even in this field voltage contacts were initially not explicitly taken into account. A voltmeter ideally exhibits an infinite impedance and, therefore, the net current at a contact used to make a voltage measurement is zero [9]. Nevertheless, such a probe permits carriers to enter and leave the conductor. Since carriers reentering the conductor have a phase and energy which is unrelated to the phase and energy of the carriers leaving the contact, a voltage contact can be dissipative and can be a source of randomization of the quantum mechanical phase of the carriers [10]. A voltage contact can thus be viewed as a simple model to bring phase randomizing and dissipative events into an electrical conduction process.

An important step forward was the experimental [11] and theoretical work [6] which clarified the symmetry of the magneto-resistance of tiny metallic loops. This work led to the recognition, that in general, all the probes connected to a conductor must be considered. Only if all probes are considered and treated on an equivalent footing could the observed magneto-resistance symmetries be explained [6,11].

II. FOUR PROBE RESISTANCE MEASUREMENTS

Measurement of a Hall resistance in the simplest case requires the consideration of a four probe set-up. The conductor in Fig. 1a has four contacts connected to electron reservoirs at chemical potentials μ_i. Two of the contacts are used to make a voltage measurement and two contacts are used as a current source and a current drain. If the chemical potentials deviate only by a small amount from their equilibrium value, the current at probe i is linearly related to the voltages $V_i = \mu_i/e$ of the terminals and given by

$$I_i = \sum_{j \neq i} G_{ij} V_j. \quad (1)$$

Here the transport coefficients are conductances which according to Casimir [12] must obey the symmetry relation $G_{ij}(B) = G_{ji}(-B)$. The theoretical task is to calculate the conductances G_{ij}. In Ref. 6, we obtained Eq. (1) using a point of view long advocated by Landauer [13]. The sample is viewed as a target at which carriers are reflected and which permits transmission of carriers from one contact to another. Inelastic scattering occurs only in the reservoirs. The sample scatters carriers only elastically. These assumptions, which are rather conceptual in nature, permit a simple solution. Ref. 6 finds for the current at probe i,

$$I_i = \frac{e}{h} [(M_i - R_{ii})\mu_i - \sum_{j \neq i} T_{ij}\mu_j]. \quad (2)$$

Here M_i is the number of quantum channels of reservoir i, R_{ii} is the total reflection coefficient of carriers incident from probe i to be scattered back into probe i and T_{ij} is the total transmission probability from j to probe i. Consider now the matrix of transport coefficients of Eq. (2) determined by the four equations for i = 1,2,3, and 4. Current conservation requires that the sum of all transport coefficients for each row and each column of this matrix is zero. Using current conservation, Eq. (2), can be reduced to Eq. (1) and gives, $G_{ij} \equiv (e^2/h)T_{ij}$. Because of the microreversibility of the scattering matrix which describes

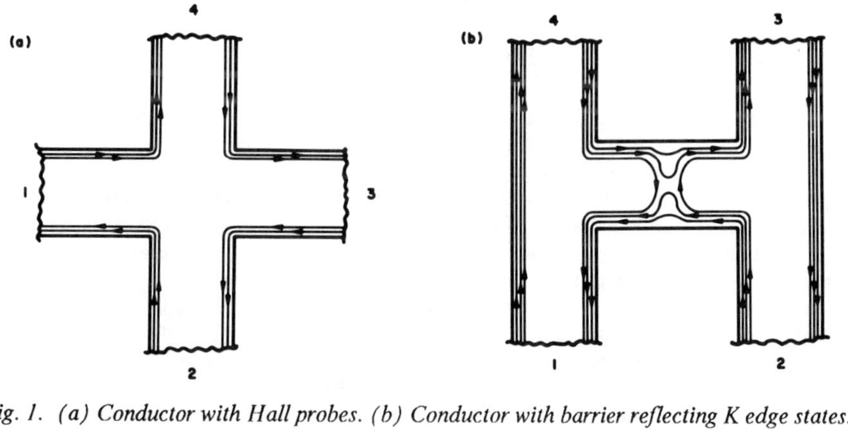

Fig. 1. (a) Conductor with Hall probes. (b) Conductor with barrier reflecting K edge states.

the sample, we also have $T_{ij}(B) = T_{ji}(-B)$ and $R_{ii}(B) = R_{ii}(-B)$. The conductances G_{ij} are not what is directly measured in a four-probe resistance measurement. But Eq. (1) or Eq. (2) can be used to find the resistance. Since current is conserved we must have $I = I_m = -I_n$, where m and n denote the carrier source and sink contact. Further, if probes k and l are used to make the voltage measurement we have, in addition, $I_k = I_l = 0$. The solution of Eq. (2) under these conditions yields [6],

$$R_{mn,kl} = (h/e^2)(T_{km}T_{ln} - T_{kn}T_{lm})/D. \tag{3}$$

D is a subdeterminant of rank three of the matrix formed by the coefficients in Eq. (2), which multiply the chemical potentials. All subdeterminants of rank three of this matrix are equal and independent of the indices m,n,k, and l. Making use of the microreversibility of the transmission probability gives rise to the reciprocity of the four-terminal resistances [6],

$$R_{kl,mn}(B) = R_{mn,kl}(-B). \tag{4}$$

Eq.(4) states that the resistance measured in a configuration of current and voltage sources in the presence of a field B is the same if the exchange of the current and voltage sources is accompanied by a reversal of the magnetic field. This symmetry was indeed found in the experiments of Ref. 11. A very striking illustration of this symmetry is found in electron focusing experiments by van Houten et al. [14]. For additional references we refer to Ref. 15. Ref. 15 discusses conductors with more than four terminals and shows that a mapping exists such that a four-terminal measurement on a conductor with a current source and current sink and an arbitrary number of voltage contacts can be described by a generalization of Eq. (3). The success of this approach in describing many interesting phenomena in small conductors has led to a number of efforts to derive Eqs. (1-4) from formal linear response theory [15], but a discussion which leads to the same expression of the conductance coefficients as obtained in Ref. 6 was only given recently [16].

Eqs. (1-3) describe a conductor not in terms of local conductivities $\sigma_{\alpha\beta}$ but in terms of global conductance coefficients. All conductance coefficients are treated on the same footing and thus give rise to an equivalent treatment of longitudinal resistances and Hall resistances. It should be noted that Eq. (3) predicts resistances which are not necessarily positive [15]. Even in high magnetic fields absolutely negative resistances can be observed as shown in an experiment by Chang et al. [18]. A simple model which exhibits negative longitudinal resistances in high fields is discussed in Ref. 19.

III. THE QUANTUM HALL EFFECT IN OPEN CONDUCTORS

Let us next consider how the quantum Hall effect is established [4]. According to Eq. (2) we have to study transmission of carriers at the Fermi energy from one contact to the other. For the scope of this paper, it will be sufficient to assume that electron motion occurs in a potential eV(x,y) which varies slowly compared to the cyclotron radius. In such a weakly fluctuating potential electron motion can be described in terms of the trajectories of the guiding center for cyclotron motion. The energy of the guiding center is $E_G = E_F - \hbar\omega_c(n + 1/2)$ and the trajectory is determined by $E_G = V(x,y)$. Thus electron motion is along equipotential lines of the potential. Note, that the higher the Landau index the lower is the available guiding center energy. A similar approach to discuss motion in high magnetic fields has been used in the past in the percolation theory [20]. The essential difference to this earlier discussion is the following: The potential V(x,y) is very small in the interior of the conductor where it fluctuates due to impurities, but the potential is very strong near the boundary of the sample. Away from the center of a Landau level the trajectories of $E_G = V(x,y)$ consist mostly of closed loops (localized states) in the interior of the sample but most importantly there are extended trajectories along the edge of the sample. It is these "edge states" [21] which originate at a metallic contact and terminate at a metallic contact which provide paths for electrical conduction. If there are N Landau levels in the bulk of the sample below the Fermi energy then along the edge of the sample there are N edge states. Let us now assume that there are very few localized states at the Fermi energy. Then electrons incident on a set of edge states have a vanishingly small probability of traversing the interior of the conductor (either through quantum tunneling at zero temperature or via Mott hopping at elevated temperatures). Electrons could only be scattered back against their direction of incidence if they can traverse the interior of the sample and reach the edge at the opposite sample boundary. Due to the suppression of backscattering at high magnetic fields [4] the edge states become one way streets for electrons. Electrons traverse the sample from one contact to another along an edge with a transmission probability equal to 1. Note that even if the surface is rough or if residual impurities cause transitions from one edge state to another, the overall transmission is still quantized. If each edge state is full, i.e., has an incident current of unit amplitude, then regardless of scattering at the sample edge, all the outgoing edge states must also be full. Therefore, in the absence of backscattering, N edge states at the Fermi surface give rise to total transmission probabilities from one contact to another equal to N. In the conductor of Fig. 1a the edge states are indicated by faint lines. Thus in the conductor of Fig. 1a the transmission probabilities are $T_{41} = N$, $T_{34} = N$, $T_{23} = N$, and $T_{12} = N$. All other T_{ij} are zero [4]. We have assumed that a carrier when it reaches a contact can leave the sample with a probability of 1. In this case, the total reflection probabilities are $R_{ii} = M_i - N$. The Hall resistance $R_{13,42}$ is determined by $T_{41}T_{23} - T_{43}T_{21}$, which is equal to N^2. Evaluation of the subdeterminant yields $D = N^3$. All Hall resistances of the conductor of Fig. 1a are quantized and yield $\pm h/e^2N$. The "longitudinal" resistances (for example $R_{12,43}$) are zero.

IV. QUANTIZED BACKSCATTERING

In the conductor of Fig. 1b the cyclical motion of carriers along the edge of the sample has been changed by diverting a number of edge states into a "wrong" contact. This can be achieved with the help of a gate over the conductor. Application of a gate voltage has the consequence that the electron density beneath the gate is reduced and a barrier for electron motion is produced. Suppose the barrier can be characterized by a height V_0. Edge states with guiding center energy $E_G > V_0$ will surmount the barrier and edge states with a guiding center energy $E_G < V_0$ will be reflected. Suppose now that out of the N edge states which exist far away from the gate K edge states are reflected. Application of Eq. (3) predicts Hall resistances [4,22]

$$R_{13,42} = (\frac{h}{e^2})\frac{1}{(N-K)}, \tag{5}$$

$$R_{42,13}(B) = R_{13,42}(-B) = -(\frac{h}{e^2})\frac{N-2K}{N(N-K)}, \tag{6}$$

and *quantized* longitudinal resistances which are symmetric in the field [4,22]

$$R_{12,43}(B) = R_{12,43}(-B) = R_{43,12}(B) = (\frac{h}{e^2})\frac{K}{N(N-K)}. \tag{7}$$

All other four-terminal resistance measurements on the conductor of Fig. 1b are zero. The plateaus predicted by Eqs. (5-7) have been observed in strikingly clear experiments by Washburn et al. [23] and Haug et al. [24]. A recent experiment by Snell et al. [25] uses a point contact geometry to measure all the resistances predicted by Eqs. (5-7).

Note that due to the transmission and reflection of edge states at the barrier, the edge states bringing carriers away from the gate are no longer populated equally. In the derivation of Eqs. (5-7) it is assumed that equilibration occurs at the contacts only.

V. CONTACTS WITH INTERNAL REFLECTION

So far we have assumed that carriers which reach a contact from the interior of the sample can escape into the reservoir with probability 1. This is called a contact without internal reflection [4]. Correspondingly, if carriers approaching a contact have a probability less than 1 to escape into the reservoir, we have a contact with *internal* reflection [4]. A current source contact with internal reflection populates edge states in a non-equilibrium fashion, similar to the barrier discussed above. If contacts with no internal reflection and contacts with internal reflection *alternate* along the perimeter of the sample, all Hall resistances are still quantized (proportional to $1/N$) and all longitudinal resistances are zero. But if two contacts with internal reflection are adjacent, there is at least one Hall measurement which depends on the detailed scattering properties of the contacts [4]. A clear demonstration of this has come with the work of van Wees et al. [26]. In the conductor of Fig. 2a there are two adjacent contacts with internal reflection. Suppose contact 1 by itself is characterized by a transmission probability T_1 and contact 2 by itself is characterized by a transmission probability T_2. The overall transmission probability from contact 2 to contact 1 is denoted by T_{12}. (In general if phase coherence plays a role T_{12} is not a simple function of T_1 and T_2.) Contact 3 and 4 are taken to be perfect. The determinant D is in this case, $D = NT_1T_2$. The Hall resistance $R_{13,42}$ is determined by the number N of bulk edge states, since carrier flow is from contact 1 to contact 4 and contact 4 provides equilibration. But if carrier flow is from contact 2 to contact 4 the Hall resistance is

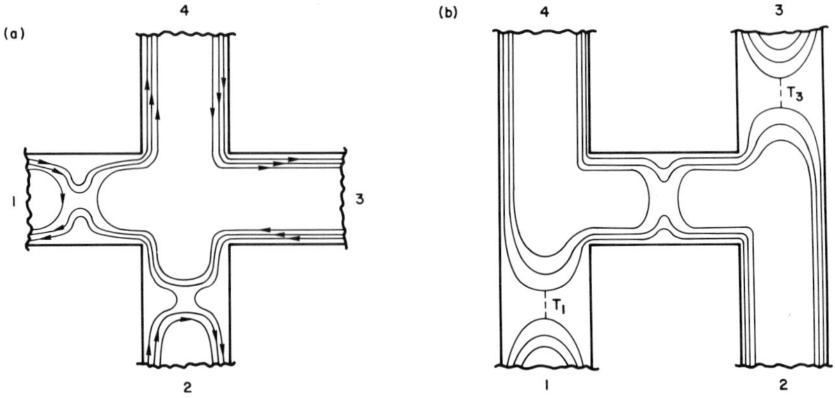

Fig. 2. (a) Conductor with two contacts with internal reflection. (b) Conductor with barrier and two weakly coupled contacts.

$$R_{24,13} = (\frac{h}{e^2})\frac{T_{12}}{T_1 T_2} . \tag{8}$$

Thus for two adjacent contacts with internal reflection and without intervening inelastic scattering the Hall resistance is not quantized as pointed out in Ref. 4. Fig. 2a shows a special case which can be achieved experimentally [26]. Contact 1 allows for the transmission of N_1 edge states and contact 2 for the transmission of N_2 edge states. Thus in this special case we have $T_1 = N_1$ and $T_2 = N_2$. In the van Wees et al. [26] experiment the contacts are only a distance of a micron apart. We are justified in neglecting scattering from one edge state to another. Hence in this case the overall transmission probability is simply determined by the less transparent contact, .i.e., $T_{12} = \min(T_1, T_2) = \min(N_1, N_2)$. In such a situation the Hall resistance is again quantized but at an *anomalous* value,

$$R_{24,13} = (\frac{h}{e^2})\frac{1}{(\max(N_1,N_2))} . \tag{9}$$

Ref. 26 tests Eq. (7) not only at the quantized values but over a wide range of non-quantized transmission probabilities. This is achieved by two split gate microconstrictions whose width can by varied by adjusting gate voltages. The case discussed here is of course very special: We have assumed that transmission through each contact is quantized. The more general case has been studied by Komiyama and Hirai [5].

Next, we would like to emphasize that the quantized backscattering investigated in Fig. 1a leads to a different result if the contacts are not ideal. To show this, consider the conductor [27] in Fig. 2b where two of the contacts are separated by barriers from the main conducting channel. If this barrier forms a smooth saddle, the contacts interact only with the outermost edge state. Let the probability for transmission from this contact to the outermost edge be $T_1 < 1$ at contact 1 and $T_3 < 1$ at contact 3. (We assume here that the number of edge states N is larger than 1). Instead of the Hall resistances found for the conductor of Fig. 1b which were not anti-symmetric, the Hall resistances of the conductor of Fig. 2b is anti-symmetric in the field and is given by $(h/e^2)1/(N - K)$ independent of T_1 and T_2. More dramatically, for the conductor of Fig. 2b, *all longitudinal resistances are zero (!)*. That is a consequence of contacts coupling only to the outer-most edge state and the outer-most edge state is transmitted at the barrier with probability 1. These simple examples show the significance of the properties of contacts. Note that the properties of the contacts in the examples discussed above are important only because we have assumed that there is no inelastic relaxation in the sample. If at a contact differing states are populated differently and the voltage contact is much further away than an equilibration length, then the contact effects discussed above would not be seen.

The discovery which raises the phenomena discussed above out of the domain of transport phenomena in small conductors and elevates them to a genuine property of the quantum Hall effect in macroscopic samples was made by Komiyama et al. [28]. The Komiyama experiments are performed on macroscopic conductors with dimensions larger than 400 μm. The conductor has a structure which is, in principle, similar to that shown in Fig. 2b. The voltage contacts are more than 50μm away from the gate. But in contrast to Washburn et al. [23] and Haug et al. [24] Komiyama et al. [28] find that their results are not described by Eqs. (5-7). Measurement of the contact resistances in their experiment shows indeed that these resistances are not small. The contacts in the Komiyama experiment exhibit (in our terminology) internal reflection. The crucial point of these experiments is that more than 50μm away from the gated region the edge states are not equilibrated. To explain their experiments Komiyama and Hirai [5] have expanded the theory of contacts with internal reflection.

A demonstration of the lack of equilibration of the edge states over large distances was independently also given by van Wees et al. [29]. In this experiment they demonstrate that the Shubnikov-de Haas oscillations depend on the way a contact couples to the two-dimensional electron gas. If contact is made only to the outermost edge state, then as explained in connection with Fig. 2b, backscattering is not observed, i. e., the Shubnikov-de Haas peaks almost vanish. This experiment [29] provided quite a different demonstration that edge states do not interact over macroscopic distances.

A third, very illustrative experiment, was undertaken by Alphenhaar et al. [30]. This experiment, as in the conductor of Fig. 2a, considers two point contacts but now at a distance of more than 80μm. The current source contact is adjusted such that it permits transmission only into the uppermost edge state. Thus $T_2 = 1$ in Eq. (8). Now suppose that due to scattering (elastic and inelastic) a portion of the carriers is transmitted into the other edge states. Denote the current of the k-th edge state arriving at contact 1 by α_k. If we assume that no current is transferred to the bulk we have $\sum \alpha_k = 1$ since there is a unit current incident. Now the contact 1 can be adjusted to transmit one, two or N_1 edge states. Using Eq. (8), the Hall resistance is,

$$R_{24,13} = \left(\frac{h}{e^2}\right)\frac{1}{N_1}\sum_{k=1}^{k=N_1} \alpha_k. \tag{10}$$

Therefore, this experiment allows the determination of the currents α_k in all the edge states and allows us to study to what extent carriers are scattered from one edge state to another. At T = 0.45K Alphenhaar et al. [30] find that for two contacts at a distance of 80μm, in the presence of three edge states, the two outermost edge states almost equilibrate, $\alpha_1 = 0.48$ and $\alpha_2 = 0.44$, but the innermost edge state remains empty even over this large distance, $\alpha_3 = 0.08$. Additional tests confirm this picture: The N-1-outermost edge states over large distances equilibrate but the innermost edge state is nearly completely decoupled. A complete understanding of this behavior has not yet been achieved. It seems that due to the fact that the edge of the sample provides a very soft potential the states at the Fermi energy can be rather far apart. The distance between the innermost edge state and the remaining edge states is especially large since the electric field generated by the confining potential is very week for this state and is possibly dominated by the fluctuations of the impurity potential. The latter fluctuations cause this state to wander into the bulk of the sample.

The approach outlined permits us not only to study exterior contacts but permits us to study interior contacts. An important conclusion, which immediately follows, is that very minor changes in the current flow pattern at the Fermi energy can cause large fluctuations of chemical potentials measured at interior contacts. There is, therefore, no direct and simple relationship between the measured chemical potentials at the interior contact and the current flow pattern [27]. Interior contacts near the sample edge allow preferential coupling to the innermost edge state. Faist et al. [31] contact the interior of Hall bars and find, among other interesting phenomena negative longitudinal resistances despite the fact that the contacts are more than 250μm apart.

The picture which we have developed here also has implications for the scaling of the Shubnikov-de Haas peaks with sample length and sample width. Since even away from a plateau transport through the bulk of the sample competes with transport along edge states, and since in such a situation the contacts exhibit necessarily internal reflection due to backscattering, a simple scaling cannot be expected. A recent experiment by Haug and von Klitzing [32] explicitly points to non-equilibrium effects as an explanation for the non-classical scaling of the Shubnikov-de Haas oscillations.

The emphasis on global conductances has led to a description of the Hall effect which is much closer to experimental reality. This made it possible to understand and to interpret a number of very interesting recent experiments. We are hopeful, therefore, that the approach out-lined here will eventually enable us to understand the quantum Hall effect on a deeper level. Our discussion focused entirely on the integer quantum Hall effect but there are already experimental [33] and theoretical [34] works which indicate that a similar discussion of the fractional quantum Hall effect is needed.

REFERENCES

1. K. von Klitzing, G. Dorda and M. Pepper, Phys. Rev. Lett. **45**, 494 (1980).
2. K. von Klitzing, Rev. Mod. Phys. **58**, 519 (1986).
3. *The Quantum Hall Effect* , edited by R. E. Prange, and S. M. Girvin, Springer Verlag, New York, 1987. page 56.
4. M. Büttiker, Phys. Rev. **B38**, 9375 (1988).
5. S. Komiyama and H. Hirai, Phys. Rev. **B40**, 7767 (1989).
6. M. Büttiker, Phys. Rev. Lett. **57**, 1761 (1986).
7. J. E. Avron, A. Raveh, and B. Zur, Rev. Mod. Phys. **60**, 873 (1988).
8. A. G. Aronov and Yu. V. Sharvin, Rev. Mod. Phys. **59**, 755, (1987); Y. Imry in *Directions of Condensed Matter Physics*, G. Grinstein and G. Mazenko, eds., World Scientific, Singapore, 1986. Vol. 1. page 101; R. A. Webb and S. Washburn, Physics Today **41**, Dec., 46 (1988); B. Alt′shuler and P. A. Lee, ibid. 41 (1988).
9. H. L. Engquist and P. W. Anderson, Phys. Rev. **B24**, 1151 (1981). This calculation, taking the possibility of phase coherent voltage measurements into account, has been re-done by M. Büttiker, Phys. Rev. **B40**, 3409 (1989). An extended account of this work is given in *Analogies in Optics and Micorelectronics*, W. van Haeringen and D. Lenstra, eds., Kluwer Academic Publishers, Dordrecht. Similar calculations and conclusions are reported by I. B. Levinson, Sov. Phys. JETP **68**, 1257 (1989).
10. M. Büttiker, Phys. Rev. **B32**, 1846 (1985).
11. A. D. Benoit, S. Washburn, C. P. Umbach, R. B. Laibowitz, and R. A. Webb, Phys. Rev. Lett. **57** , 1765 (1986).
12. H. B. G. Casimir, Rev. Mod. Phys. **17**, 343 (1945).
13. R. Landauer, IBM J. Res. Dev. **1** , 223 (1957); Z. Phys. **B68**, 217 (1987).
14. H. van Houten, C. W. J. Beenakker, J. G. Williamson, M. E. I. Broekaart, P. H. M. van Loosdrecht, B. J. van Wees, J. E. Moij, C. T. Foxon and J. J. Harris, Phys. Rev. **B39**, 8556 (1989).
15. M. Büttiker, IBM J. Res. Develop. **32**, 317 (1988).
16. A. D. Stone and A. Szafer, IBM J. Res. Develop. **32**, 384 (1988); O. Viehweger, Z. Phys. **B77**, 135 (1989).
17. H. U. Baranger, A. D. Stone, Phys. Rev. **B40**, 8169 (1989).
18. A. M. Chang, G. Timp, J. E. Cunningham, P. M. Mankiewich, R. E. Behringer, and R. E. Howard, Solid State Commun. **76**, 769 (1988).
19. M. Büttiker, Phys. Rev. **B38**, 12724 (1988).

20. S. L. Luryi and R. F. Kazarinov, Phys. Rev. **B27**, 1386 (1983).
21. R. E. Prange and T. Nee, Phys. Rev. **168**, 779 (1968); M. Heuser and J. Hajdu, Z. Phys. **270**, 289 (1974).
22. M. Büttiker, Phys. Rev. Lett. **62**, 229 (1989).
23. S. Washburn, A. B. Fowler, H. Schmid, and D. Kern, Phys. Rev. Lett. **61**, 2801 (1988).
24. R. J. Haug, A. H. MacDonald, P. Streda, and K. von Klitzing, Phys. Rev. Lett. **61**, 2797 (1988); R. J. Haug, J. Kucera, P. Streda and K. von Klitzing, Phys. Rev. **B39**, 10892 (1989).
25. B. R. Snell, P. H. Beton, P. C. Main, A. Neves, J. R. Owers-Bradely, L. Eaves, M. Henini, O. H. Hughes, S. P. Beaumont, and C. D. W. Wilkinson, J. Phys. **C1**, 7499 (1989); P. C. Main, P. H. Beton, B. R. Snell, A. J. M. Neves, J. R. Owers-Bradley, L. Eaves, S. P. Beaumont and C. D. W. Wilkinson, Phys. Rev. **B40**, 10003 (1989).
26. B. J. van Wees, E. M. M. Willems , C. J. P. M. Harmans, C. W. J. Bennakker, H. van Houten and J. G. Williamson, C. T. Foxon and J. J. Harris, Phys. Rev. Lett. **62**, 1181 (1989).
27. M. Büttiker, in *Nanostructure Physics and Fabrication,* edited by M. A. Reed and W. P. Kirk, (Academic Press, Boston, 1989). page 319.
28. S. Komiyama, H. Hirai, S. Sasa, and S. Hiyamizu, Phys. Rev. **B40**, 12566 (1989).
29. B. J. van Wees, E. M. M. Willems, L. P. Kouwenhoven, C. J. P. M. Harmans, H. G. Williamson, C. T. Foxon, and J. J. Harris, Phys. Rev. **B39**, 8066 (1989).
30. B. W. Alphenhaar, P. L. McEuen, R. G. Wheeler, and R. N. Sacks, (unpublished).
31. J. Faist, H. P. Meier and P. Gueret, "Interior Contacts for Probing the Equilibrium between Magnetic Edge Channels in the Quantum Hall Effect", (unpublished).
32. R. J. Haug and K. von Klitzing, Europhysics Lett. **10**, 489 (1989).
33. A. M. Chang and J. E. Cunningham, Solid State Communic. **72**, 651 (1989); L. P. Kouwenhoven, B. J. van Wees, N. C. van der Vaart, C. J. P. M. Harmans, C. E. Timmering, and C. T. Foxon, (unpublished, 1989); J. A. Simmons, H. P. Wei, L. W. Engel, D. C. Tsui and M. Shayegan, Phys. Rev. Lett. **63**, 1731 (1989).
34. C. W. J. Beenakker, (unpublished, 1989); A. MacDonald, (unpublished, 1989); S. A. Kivelson and V. L. Prokovski, Phys. Rev. **B40**, 1373 (1989).

TUNNELING OF ELECTRONS FROM THE TWO-DIMENSIONAL CHANNEL INTO THE BULK

S. V. Meshkov
Institute of Solid State Physics USSR, Academy of Sciences
Chernogolovka, Moscow district, 142432 USSR

It is shown that the interaction of charge carriers quantized in a 2D channel, both with each other and with inhomogeneities, has a fundamental influence on their probability of tunneling into the bulk, and as a result there is a slowing down in the decrease of electron density with distance for carriers with a nonzero kinetic energy of motion along the surface. In the multi-valley case an additional increase in the tunneling probability is obtained, caused by the decrease of the mass in the tunneling direction due to virtual intervalley scattering. Somewhat analogous behavior appears in the presence of a magnetic field parallel to the surface.

Let a planar 2D channel be situated on the surface of a semiconductor crystal. We shall assume that the channel is formed by a one-dimensional potential well $V(z)$ with a single (for simplicity) energy level ε_0, that is smooth enough for the effective-mass approximation to be applicable. It is known that for a nondegenerate quadratic carrier-dispersion law, in the absence of electron-electron interaction and inhomogeneities, the free motion along the channel can be separated completely from the quantized transverse motion. For this reason, the wave function Ψ of the charge carriers situated in the channel, irrespective of their energy $\varepsilon = \varepsilon_0 + \mathbf{p}_\parallel^2/2m$, decays with distance from the surface at the same rate:

$$\Psi \propto \exp\{-\hbar^{-1} \int^z [2m(V(z) - \varepsilon_0)]^{1/2} dz\} \ . \tag{1}$$

The carriers tunnel into the bulk just as in a one-dimensional potential $V(z)$, the energy being not the total energy ε but the energy ε_0 corresponding to the bottom of the surface band. The kinetic energy of the motion along the channel remains irrelevant owing to the conservation of momentum.

The present paper is intended to describe the breakdown of the above idealized picture in real 2D channels - MOS structures and heterojunctions. There are a few different scattering processes due to which the wave functions of the surface state are not as simple as (1), but the important thing is that the kinetic energy along the surface changes the situation qualitatively because the electron density distribution gains a component decaying with distance as if wave functions of surface states are (1) with ε_0 replaced by ε. Note that, however weak the scattering is, at sufficiently large distance z from the channel this component will dominate. In addition, if the dispersion law of carriers is complicated (multi-valley or degenerate), they tunnel into the bulk with that mass which provides the slowest decay of tunneling exponential.

The statement formulated above is fairly obvious and can probably be proved in an extremely general form. This exponential effect must be principally observable in some tunnel experiments and, more actually, by photoluminescence with participation of the holes located far from the surface. The real problem is to estimate the distances at which the exponential effects under discussion become appreciable. To do this it is necessary to know the preexponential coefficients depending on the precise scattering mechanism. As only one experiment [1] that seems to demonstrate the effect considered here is known to the author, I shall restrict myself to a theoretical discussion.

The direct way to treat the problem theoretically is to consider the one-particle equilibrium density matrix $\rho_\varepsilon(\mathbf{r}, \mathbf{r}')$. In the absence of inhomogeneities and many-particle effects this value is a sum

$$\rho(\mathbf{r}, \mathbf{r}' \mid \varepsilon) = \sum_n \Psi_n(\mathbf{r}) \Psi_n^*(\mathbf{r}')\, \delta(\varepsilon - \varepsilon_n) \tag{2}$$

over all occupied one-particle states, but for a general case it is determined through the imaginary part of the causal Green function

$$\rho(\mathbf{r}, \mathbf{r}' \mid \varepsilon) = \pi^{-1} \Theta(\varepsilon_F - \varepsilon)\, \mathrm{Im}\, G(\mathbf{r}, \mathbf{r}' \mid \varepsilon)\ . \tag{3}$$

In turn, the Green function can be calculated by use of the Feynman diagram technique with the different scattering processes being taken into account. It should be noted that the diagonal component of the density matrix, the electron density $\nu(\mathbf{r} \mid \varepsilon) = \rho(\mathbf{r}, \mathbf{r}' \mid \varepsilon)$, is the most important function and is slightly easier to calculate. In order to permit the evaluation of the transition intensities under the control of different selection rules it is expedient to consider the full density matrix.

Below, for illustration, a very simple model of the channel is discussed in which the exponential dependences are trivial. That is, the channel is assumed to be formed by a very narrow and deep potential well on the surface of a semi-infinite crystal ($z > 0$). The bottom of the conduction band is assumed to be absolutely flat and taken as reference point of energy. The opposing wall of the well is assumed to be infinitely high, so that the channel can be described entirely by the effective boundary condition

$$\left.\frac{\partial \Psi(\mathbf{r})}{\partial z}\right|_{z \Rightarrow +0} = -\kappa_0 \Psi(\mathbf{r}), \quad \kappa_0 = (-2m\varepsilon_0/\hbar^2)^{1/2}\ , \tag{4}$$

where the binding energy of a single (for simplicity) bound state $\varepsilon_0 < 0$ enters as the only characteristic of the channel. In this model the bound surface states are described by wave functions

$$\Psi_{\mathbf{k}}(\mathbf{r}) = (\kappa_0/2\pi) \exp[-\kappa_0 z + i\mathbf{k}\boldsymbol{\rho}]\ , \tag{5}$$

where $\boldsymbol{\rho}$ and $\mathbf{k}$ are the two-dimensional position vector and wave vector in the xy plane. The surface band $\varepsilon(\mathbf{k})$ is assumed to be filled up to the Fermi level

$$\varepsilon(\mathbf{k}) = \varepsilon_0 + \hbar^2 \mathbf{k}^2/2m\,, \quad \varepsilon_F = \varepsilon_0 + \hbar^2 \mathbf{k}_F/2m < 0\,, \tag{6}$$

which results in zeroth-approximation density matrix

$$\rho^0(\mathbf{r}, \mathbf{r}' \mid \varepsilon) = \Theta(\varepsilon_F - \varepsilon_0)\Theta(\varepsilon - \varepsilon_0)(m/\pi\hbar^2)\kappa_0 J_0(k \mid \boldsymbol{\rho} - \boldsymbol{\rho}' \mid)\exp[-\kappa_0 \mid z + z' \mid] \,. \tag{7}$$

($J_0(x)$ is a Bessel function.) This matrix falls off exponentially with the distances $\mathbf{r}$ and $\mathbf{r}'$ from the channel, and does not depend on ε in the interval $\varepsilon_0 \leq \varepsilon \leq \varepsilon_F$.

THE GREEN FUNCTION

First of all we must construct the unperturbed Green function, which is defined usually as the sum

$$G^0(\varepsilon) = \sum_n \frac{\Psi_n^*(\mathbf{r}')\Psi_n(\mathbf{r})}{\varepsilon - \varepsilon_n + i\,0\cdot\text{sign}(\varepsilon_n - \varepsilon_F)} \tag{8}$$

over the complete set of the stationary states n with appropriately normalized wave functions Ψ_n. This set include, besides the bound states (5), the delocalized states

$$\Psi_{\mathbf{k}, k_z}(\mathbf{r}) = \exp(i\mathbf{k}\cdot\boldsymbol{\rho})\cos[k_z z + \text{arctg}(\kappa_0 / k_z)] \tag{9}$$

with energies corresponding to the bulk dispersion law

$$\varepsilon(\mathbf{k}, k_z) = \hbar^2(\mathbf{k}^2 + k_z^2)/2m \,. \tag{10}$$

The most convenient form of Green function for developing the perturbation expansion is the Fourier representation along the surface combined with the z coordinate. The required coordinate representation must be obtained by inverse 2D Fourier transformation at the end of calculations. To obtain the zeroth approximation it is simpler not to use the summation (8), but to solve the Schrödinger equation

$$\left[\varepsilon - \frac{\hbar^2}{2m}\left(\mathbf{k}^2 - \frac{\partial^2}{\partial z^2}\right)\right] G^0(\mathbf{k}, z, z' \mid \varepsilon) = \delta(z - z') \tag{11}$$

with the boundary condition (4). The result is

$$G^0(\mathbf{k}, z, z' \mid \varepsilon) = (\kappa^2 + k^2)^{-1/2}\left\{-\left(\frac{m}{\hbar^2}\right)\exp[-(\kappa^2 + k^2)^{1/2} \mid z - z' \mid] + \right.$$

$$\left.\frac{1}{2}\,\frac{[(\kappa^2 + k^2)^{1/2} + \kappa_0]^2}{\varepsilon - \varepsilon_0 - \hbar^2\mathbf{k}^2/2m + i\,0\cdot\text{sign}(k - k_F)}\exp[-(\kappa^2 + k^2)^{1/2}(z + z')]\right\} \,, \tag{12}$$

where $\kappa = (-2m\varepsilon/\hbar^2)^{1/2}$. The physical meaning of the two terms in the curly brackets is clear: the first term, which does not contain ε_0, describes the purely bulk tunneling, while the second corresponds to tunneling with reflection from the surface.

PERTURBATION THEORY

The perturbation-theory series for the one-particle Green function are constructed from unperturbed function (12) according to usual rules (see e.g. [2,3]). However, the problem is slightly unusual, because we are interested in the asymptotic form of the density matrix at large distances from the surface. As one can see from the expression (12), only the real part of Green function has a contribution which falls off with distance more slowly than zeroth-approximation density matrix (7). Thus we must take into account those diagrams whose imaginary part does not vanish when, in the external G^0 -lines, only the real parts are retained. The main diagrams of required type shown in Fig. 1 are of the second order in perturbation. We shall not describe the procedure of the calculation in detail and concern ourselves only the main points simplifying the evaluation of diagrams.

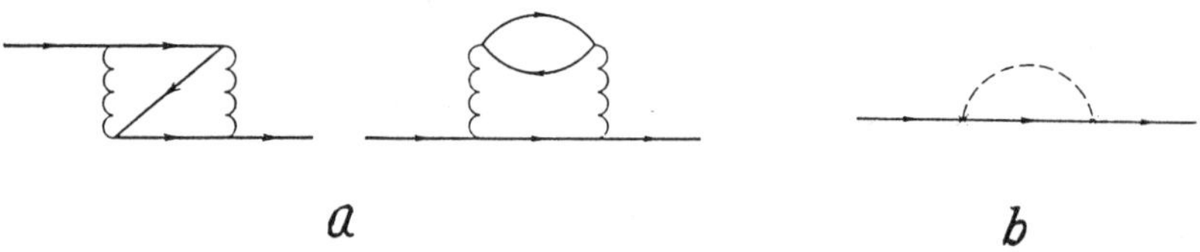

Fig. 1. Principal diagrams describing the main contribution of electron-electron interaction (a) and impurity scattering (b) to the electron Green function. The solid lines are the unperturbed Green functions, the wavy lines are matrix elements of the interaction, and the crosses are matrix elements of impurity potential (the dashed line indicates that the two crosses correspond to the same impurity).

At first we note that because only the imaginary part of the G^0 -function is important in all the internal lines we can replace it by the purely two-dimensional function defined by the sum (8) with only the surface states (5) taken into account

$$G^0(\mathbf{k}, z, z' \,|\, \varepsilon) = \frac{2\kappa_0}{\varepsilon - \varepsilon_0 - \hbar^2\mathbf{k}^2/2m + i\,0\cdot\mathrm{sign}(k - k_F)} \exp[-\kappa_0(z+z')] \, . \tag{13}$$

The next convenience comes from the fact that exponential dependence of external G^0 -functions on momentum $\mathbf{k}$ becomes dominant for large distances from the surface making it possible to integrate over $\mathbf{k}$ by the steepest descent method. When doing this we may take $\mathbf{k}=0$ anywhere except in the exponents. After which the steepest descent integration $\int d^2k/(2\pi)^2$ of the product of two external line exponentials gives the factor

$$\frac{\kappa}{\pi(z+z')} \exp\{-\kappa[(z+z') - (\rho - \rho')^2/2(z+z')]\} \approx \frac{\kappa}{\pi R} \exp(-\kappa R) \tag{14}$$

the latter expression including the shortest distance $R = [(\rho - \rho')^2 + (z+z')^2]^{1/2}$ between

r and **r′** points with reflection from the surface. The remaining part of each of external G^0 function may be taken in the form

$$G^0(\mathbf{k}, z, z' \mid \varepsilon) = -(m/\hbar^2\kappa)\{\exp(\kappa z) + [(\kappa_0+\kappa)/(\kappa_0-\kappa)]\exp(-\kappa z)\}\ , \qquad (15)$$

where only one coordinate z (or z') enters corresponding to the vertex-side end of external lines.

As one can find when analyzing the above simplifications, the problem is reduced to calculation of the imaginary part of the second order contribution to a self-energy (mass) operator $\sum(\varepsilon, \mathbf{k})$ described by diagrams Fig. 1 without external lines. The 2D vertices are to be obtained by integration of the corresponding 3D ones over z with z -dependent exponentials of G^0 -functions (13) or (15). The value of momentum $\mathbf{k}$ can be taken to equal zero after or when doing the evaluation of $\sum(\varepsilon, \mathbf{k})$.

ELECTRON-ELECTRON INTERACTION

For simplicity we consider not a realistic Coulomb-like interaction but that described by a short-range potential $U(\mathbf{r})$. In this case the resulting expression for the density matrix is

$$\rho(\mathbf{r}, \mathbf{r}' \mid \varepsilon) = [32\kappa_0^3\kappa(\kappa_0-\kappa)^{-2}(9\varepsilon_0-\varepsilon)^{-2}\exp(-\kappa R)/R\,]\cdot\mathrm{Im}\sum(\varepsilon, 0)\,, \qquad (16)$$

$$\mathrm{Im}\sum(\varepsilon, 0) = (1/4\pi^3)U_0^2\,\kappa_0(m/\hbar^2)^2(\varepsilon_F-\varepsilon_0)\Phi[(\varepsilon-\varepsilon_0)/(\varepsilon_F-\varepsilon_0)]\ ,$$

$$\Phi(\nu) = \int_{\nu^{1/2}}^{1}\{[1-(\nu/2\eta)^2]^{1/2} - [1-(\eta-\nu/2\eta)^2]^{1/2}\}d\eta/\eta\ ,\quad 0<\nu<1\,,$$

where $U = \int U(\mathbf{r})d^3r$ is the Born matrix element and $\Phi(\nu)$ is a function of order of unity, vanishing at $\nu \to 1$.

SCATTERING BY INHOMOGENEITIES

In this case we use the diagram Fig. 1b describing the double scattering on the same center. Characterizing the bulk scatterers by their concentration n and the Born amplitude $U_0 = \int U(\mathbf{r})d^3r$ of their short-range potential we obtain the density matrix far from the surface

$$\rho(\mathbf{r}, \mathbf{r}' \mid \varepsilon) = (n/2\pi)(m/\hbar^2)^3U_0^2\,\kappa(\kappa_0-\kappa)^{-2}\exp(-\kappa R)/R\ . \qquad (17)$$

The variant of this case is the scattering by the inhomogeneities of the 2D channel itself. The small fluctuations of the channel depth leads to a dependence of the energy ε_0 of the bottom of the surface band on the coordinate $\boldsymbol{\rho}$ in the plane:

$$\varepsilon_0(\rho) = \bar{\varepsilon}_0 + \tilde{\varepsilon}_0(\rho)\ ,\quad \langle\tilde{\varepsilon}_0\rangle = 0\ . \qquad (18)$$

These fluctuations are convenient to represent as being produced by the 3D potential $\tilde{U}(\mathbf{r}) = \tilde{\varepsilon}_0(\boldsymbol{\rho})\delta(z+0)/2\kappa$, after which a result of the same form as (17) is obtained

$$\rho(\mathbf{r},\mathbf{r}'|\varepsilon) = (1/2\pi)(m/\hbar^2)^3 Q[(\kappa_0^2-\kappa^2)^{1/2}](\kappa_0-\kappa)^{-2}\exp(-\kappa R)/R \ , \qquad (19)$$

$$Q(\mathbf{k}) = \int <\tilde{\varepsilon}_0(\rho)\tilde{\varepsilon}_0(\rho')>\exp[-i\mathbf{k}(\rho-\rho')]d^2\rho \ ,$$

where for generality a Fourier component of the correlator of the binding energy, $Q(\mathbf{k})$, is used.

INTERVALLEY SCATTERING

Considering a semiconductor whose conduction band has several identical valleys with an anisotropic effective mass we shall restrict ourselves to a qualitative discussion. It should be noted that anything fundamentally new arises only if the intervalley transitions are allowed and the valley orientations are not equivalent (otherwise a change of variables reduces the problem to that considered above). The deepest surface band is formed from the states of the valleys with the largest mass perpendicular to the surface, whereas for tunneling of an electron it is favorable to be in the valleys with the smallest mass in this direction. Therefore, the virtual transitions from the first valleys into the second should exponentially enhance tunneling. We note that in contrast to the one-valley case this effect occurs even for electrons from the bottom of the surface band.

One fundamental complication appears when considering the electron-electron interaction which conserves the total momentum of the electron system. For real values of the surface Fermi wave vector k_F, an intervalley transition is possible only as a result of many-particle processes in which an electron, crossing over into a favorable valley, gathers momentum from several electrons remaining in the surface band.

As the only quantitative point in this section we give a coordinate dependence of the density matrix for (001) silicon, the six valleys of which in **k**-space are oriented along the axes of a cubic lattice and have large longitudinal and small transverse effective masses $m_\parallel$ and $m_\perp$. Each of valleys parallel to the surface gives the contribution

$$\rho(\mathbf{r},\mathbf{r}'|\varepsilon) \propto \exp[-(-2m_\perp\varepsilon/\hbar^2)R]/R \ , \qquad (20)$$

$$R = [(m_\parallel/m_\perp)[(x-x')^2+(y-y')^2+(z-z')^2]^{1/2} \ ,$$

(for a valley directed along x.)

MAGNETIC FIELD

In conclusion one more factor involving the longitudinal kinetic energy in tunneling processes should be mentioned. This factor is the magnetic field parallel to the surface, e.g. in the y direction. The first order effect is that the electron velocity in the x direction causes a force directed perpendicular from or to the surface, which can be formulated by including the term $(e\hbar/mc)k_xH_yz$ in the Hamiltonian. This term acts as if the barrier shape changes depending on the electron momentum k_x, enhancing tunneling for one sign of k_x and suppressing it for another. Owing to the exponential nature of this effect the enhancement taking place for some k_x is greater than the suppression for $-k_x$, so a net enhancement of tunneling with the electron energy ε must appear.

It should be emphasized that the effect is so simple only at distances from the surface at which the additional term is less than the kinetic energy along the surface. At larger distances the term of the second order in magnetic field and in z, independent of $\mathbf{k}$, becomes dominant. This term suppresses the tunneling for all $\mathbf{k}$, if the bulk impurity scattering [4] is not taken into account.

REFERENCES

1. V.D.Kulakovskii, B.N.Shepel', A.A.Denisov, A.P.Senichkin, Sov.Phys.Semicond., 21, (1987).
2. E.M.Lifshitz, L.P.Pitaevskii, Statistical Physics, Part 2, Pergamon Press, Oxford (1965).
3. A.A.Abrikosov, L.P.Gor'kov, J.E.Dzyaloshinskii, Quantum Field Theoretical Methods in Statistical Physics, Pergamon Press, Oxford (1965).
4. B.I.Shklovskii, A.L.Efros, Sov.Phys.JETP 57,470(1983).

ADIABATIC BALLISTIC TRANSPORT IN MICRO-CONSTRICTIONS

D. E. Khmelnitskii
Institute of Solid State Physics
Chernogolovka, Moscow Distr. 142 432 USSR

Quantization of the resistance of micro-constrictions in a 2D-electron gas, observed experimentally by B. J. van Wees et al [1] and D. A. Wharam et al [2], is explained as a result of the reflectionless propagation of electron waves. The value of the conductance is equal to an integer number of fundamental units $e^2/\pi\hbar$, and the relative width of the step between the plateaus is defined by the shape of the constriction and is equal to $\pi^{-2}(w/2R)^{1/2}$, where w is the width of the constriction and R is its radius of curvature. Applications of the theory to transport in a magnetic field, to non-linearity and to excess noise are discussed.

Progress of modern microelectronics gives the possibility of obtaining a 2D gas of electrons with a very long mean free path on the order of 10 microns. Modern electron beam lithography can form split gates with characteristic dimension of order 0.1 micron, and, as a result, can form micro-constrictions in the 2D layer with an electric field controlled width of submicron size. Recent experiments on micro-constrictions formed in the 2D electron gas have revealed an abrupt change in the conductance G as a function of the width of the constriction w, which is adjusted by varying the gate voltage [1,2]. The values of the corresponding steps have turned out to be equal to the fundamental quantum $e^2/\pi\hbar$. According to Landauer [3], this means that we have reflectionless propagation of electron waves through the constriction, and that the number of transverse channels increases when the constriction width grows. The experimentalists have explained the observed pattern by utilizing the electrical conductivity in a long channel under conditions of transverse quantization of electrons. Within the framework of this theoretical picture they expected to have both the increase in number of transverse channels versus the width and reflectionless transport. The applicability of this picture is doubtful even for a long and narrow channel because of back-scattering at its ends (i.e. in the accommodation regions). It is not applicable at all for the shape of constriction of Ref. [1], where the shape was definitely not approximately one-dimensional. There is accordingly the question of the particular conditions under which one can observe a clear pattern of quantization of resistance.

We show that smoothness of variation in the transverse dimension of the constriction plays a decisive role in determining whether or not the effect can be observed. The existence of universal steps in $G(k_F w)$ does not require a sharply bounded region with well-defined transverse-quantization levels. The condition that the constriction be smooth, like the requirement on the length of the channel, turns out not to be very stringent, because of numerical factors.

A smooth variation $w(x)$ makes possible adiabatic separation of variables in the Schroedinger equation. If we ignore the curvature of the bottom of the quantum well, we can write the wave function $\Psi(x, y)$, which is the solution of the boundary-value problem

$$-\frac{\hbar^2}{2m}\Delta\Psi = E\,\Psi \qquad \Psi[y=\pm w/2]=0 \,, \tag{1}$$

in the form $\Psi=\psi(x)\,\phi_x(y)$, where

$$\phi_x(y)=\left[\frac{2}{w(x)}\right]^{1/2} \sin\left\{\frac{\pi n[2y+w(x)]}{w(x)}\right\} , \tag{2}$$

$$-\frac{\hbar^2}{2m}\frac{d^2\psi}{dx^2}+\varepsilon_n(x)\psi=E\psi \,, \tag{3}$$

$$\varepsilon_n(x)=\frac{\pi^2 n^2 \hbar^2}{2mw^2(x)} \,. \tag{4}$$

Under the condition that the variation $w(x)$ is smooth on the scale of K_F^{-2}, the value n is a well-defined quantum number, and the adiabatic potential (4) in Eq. (3) is semiclassical. For all values of n which satisfy the condition $n<n_{\max}(kw)$, electron reflection efforts are therefore inconsequential, while at $n>n_{\max}(kw)$ there exists a classically forbidden region, and the transmission coefficient is exponentially small. The value $n_{\max}$ is found from the condition that the semiclassical momentum $p_n(x)=\{2m[E-\varepsilon_n(x)]\}^{1/2}$ be real at the narrowest point $x=0$. The result is

$$n_{\max}(kw)=\left[\frac{kw}{\pi}\right], \tag{5}$$

where $[x]$ means the greatest integer in x. Thus, in this approximation the transmission coefficients t_{nl} equal either zero or unity:

$$t_{nl}=\delta_{nl}\,\theta(n-n_{\max}) \,. \tag{6}$$

According to Landauer, the conductance is

$$G(k_F w)=\frac{e^2}{\pi\hbar}\,n_{\max}(k_F w) \,. \tag{7}$$

A sharp, stepped change in $G(k_F w)$ at $k_F w=\pi n$ is a consequence of the semiclassical approximation, which is valid for solution of Eq. (3). Incorporation tunneling and above-barrier reflection leads to spreading of the sharp edge of the steps. The shape of the nth conductance step $\delta G[(k_F w)-n]$ is defined by dependence of ε_n on x near the top of the barrier, and this dependence is defined by the curvature of the constriction wall at the narrowest place $\delta^2 w/\delta x^2=2/R$.

$$\delta G(z) = \frac{e^2}{\pi\hbar}\{1+\exp[-z\pi^2(2R/w)^{1/2}]\}^{-1}, \tag{8}$$

where $z=(k_F w/\pi)-n$. It can be seen from the Eq. (8) that the step width weakly depends on the index n. We wish to call attention to the numerical factor $2^{1/2}\pi^2$ in the exponential function of Eq. (8). This factor makes the step sharp even at $R=d$. The expression (8) gives us the shape of the step at low temperatures $T<\Delta=n\hbar^2/m(2Rw^3)^{1/2}$. At $T>\Delta$, the factor $\pi^2(2R/w)^{1/2}$ in the exponential function in Eq. (8) should be replaced by $\hbar^2\pi^2 n/(mw^2 k_B T)$.

If the constriction length, L, exceeds $(Rw)^{1/2}$ by a factor of several units, the channel width will change over the distance by a value sufficient to allow the voltage drop to concentrate in the adiabatic region of the electron motion. In this case ($L \gg (Rw)^{1/2}$), the accommodation region make no substantial contribution to the overall resistance.

This theory, first discussed in Ref. [4], was extended to describe ballistic magnetoresistense, photogalvanic effect, nonlinearity etc. The most interesting application of the same idea was carried out by G. B. Lesovik [5], who studied spectral intensity of excess noise $S_{ex}(\omega)$ in a ballistic channel. He derived the following Landauer-like formula for spectral intensity of excess noise induced by the current I at $T=0$:

$$S_{ex}(\omega=0) \approx eI\sum t_n(1-t_n). \tag{9}$$

This means that quantum noise is maximal at $t_n=1/2$ when the uncertainty in the reflection process (to reflect or not) is maximal. As a result, the intensity of excess noise versus gate voltage (width of constriction) has narrow spikes at the gate voltage corresponding to step in the ballistic conductance.

I would like to thank Lyonya Glazman, Gordey Lesovik, Volodya Fal'ko and Robert Shechter for their collaboration and many helpful discussions.

REFERENCES

1. B. J. van Wees, H. van Houten, C. W. Beenacker et al, Phys. Rev. Lett. **60,** 848 (1988)
2. D. A. Wharam, T. J. Thornton, M. Pepper et al J. Phys. **C21,** L209, (1988)
3. R. Landauer IBM Jour. REs. Dev **1,** 223, (1957) R. Landauer, M. Buettiker Phys. Rev. Lett **54,** 2049, (1985) D. C. Langreth, E. Abrahams Phys. Rev. **B24,** 2978, (1981)
4. L. I. Glazman, G. B. Lesovik, D. E. Khmelnitskii and R. I. Shechter Pis'ma Zh. Exp. Theor. Phys. **48,** 218 (1988)
5. G. B. Lesovik Pis'ma Zh. Exp. Theor. Phys. **49,** 145 (1989)

COUPLING BETWEEN 2-D ELECTRONS IN QUANTUM WELLS AND 3-D PHONONS

M. Lax
Department of Physics, City College of New York, New York, N.Y. 10031
and AT&T Bell Laboratories, Murray Hill, New Jersey 07974.

W. Cai, P. Hu T. F. Zheng B. Yudanin
Department of Physics, City College of New York, New York, New York 10031

M.C. Marchetti
Physics Department, Syracuse University, Syracuse NY 13244

The proper treatment of the electron-phonon coupling between electrons confined to two dimensions (2-D) by phonons traveling freely in three dimensions (3-D) requires special care because phonon heating produces a bottleneck in the rate of transfer of energy from the carriers to the phonons. Because the carriers interact with phonons primarily when the latter are close to the quantum well, the latter should be described, not by plane waves, but by packets adapted to the shape of the carrier confinement. A quasi-equilibrium technique that retains off-diagonal elements in the phonon wave-vector permits an unrestricted treatment of the density operator equation. That in turn leads to a choice of wave packet that comes from solving the integrodifferential equations rather than by imposition. Moreover, if the carrier distribution is assumed in quasi-equilibrium with a given drift and temperature, the coupleed partial differential equations are reduced to coupled ordinary differential equations that can be solved with modest computer power. Comparison with experimental results for steady flow of energy from carriers to phonons, and for time-dependent relaxation yields quantitative agreement.

1. THE PROBLEM

Almost 4 years ago, I perceived a need for understanding electron (and hole) transport in quantum wells and heterostructures and hired two research associates, W. Cai and M. C. Marchetti, to work in this area. Cai was already an expert in semiconductor physics and Marchetti an expert on transport in liquids. They and the graduate students who joined us, T. F. Zheng and B. Yudanin performed all the calculations reported here. My role was advice and criticism.

The production of microelectronic devices by molecular beam epitaxy, and the high mobility of carriers particularly in modulated heterostructures created the importance of this area of study. Moreover, for transport in small high mobility systems, moderate voltages can lead to strong fields and non-linear effects. There already was experimental evidence [1-4] that the rate of energy transfer from electrons to phonons was an order of magnitude less than perturbation theory would yield [5-7] An excellent review of two-dimensional transport, with an extensive list of references has been given by Ando, Fowler, and Stern [8]

It was clearly perceived by the experimentalists, and a qualitative theory was developed by Price [9]. that the inability of the longitudinal optical (LO) phonon system to dissipate heat fast enough was creating a bottleneck. Price's theory is qualitative, because he was forced to introduce an ad hoc parameter - the number of phonons that interact with an electron: Price recognized the need, and called for a more rigorous treatment.

We concluded [10,11] that the problem was a general one: how should electrons, whose transport is confined to two dimensions, interact with phonons that can propagate freely in three dimensions. More specifically, since the electrons only interact with phonons when they are in the vicinity of the quantum well, a plane wave description for the phonons is inconvenient. One possibility is to retain the phone wave description, but quantize the phonons (in the z direction) over a thickness L comparable to the well width. But such a treatment would be equivalent to Price's with L as the arbitrary parameter.

Cai proposed a resolution of this problem by using a basis set for the z direction (normal to the well walls) that consists in a Gaussian times a set of Hermite polynomials. But it was not known how many terms were needed. The use of a single term, the Gaussian, has as an arbitrary parameter, the width of the Gaussian.

2. THE RESOLUTION

a. Quasi-Equilibrium

After reviewing the work, I suggested that the shape of the phonon "packet" should come out of the problem, not be imposed. Marchetti then suggested the use of a quasi-equilibrium procedure of the sort introduced by Bogolyubov in dealing with classical liquid transport and by Zwanzig [12,13] in a variety of problems. See also Zubarev [14] and Peletminskii and Yatsenko [15] The basic idea is the quasi-equilibrium assumption that he variables of a problem can be divided into slow variables and fast ones. The fast variables assumed to be in equilibrium with the current values of the slow variables. When the fast variables are inserted into the equations for the slow ones, we get an effective set of equations for the slow variables.

b. The Choice of Slow Variables

The success of such a procedure clearly depends on the appropriate choice of slow variables. The proposal made by Marchetti and used by electronic variables $a^{\dagger}_{n\mathbf{k}}a_{n\mathbf{k}}$ that describe the occupancy of a state of transverse momentum $\mathbf{k}$ in the nth subband (on the z motion). The average of this set of variables

$$f_n(\mathbf{k}, t) = \langle a^{\dagger}_{n\mathbf{k}}a_{n\mathbf{k}} \rangle \tag{2.1}$$

is the familiar distribution function for these carriers.

The phonons are described by the three-dimensional wave vector

$$\mathbf{Q} = (\mathbf{q}, q_z) \tag{2.2}$$

associated with a place wave representation $\exp(i\mathbf{Q}\cdot\mathbf{R})$. The variables to be used in this case are $b^{\dagger}_{\mathbf{q},q_z}b_{\mathbf{q}q'_z}$ with average value.

$$n_{\mathbf{q}}(q_z, q'_z) = \langle b^{\dagger}_{\mathbf{q},q_z}b_{\mathbf{q}q'_z} \rangle \tag{2.3}$$

Although the work starts in the plane-wave representation, by allowing off-diagonal elements with respect to q_z, we have prepared the way for an eventual transformation to packets whose shape is as yet unknown.

3. DESCRIPTION OF THE HAMILTONIAN

The Hamiltonian $\hat{H}$ consists in an electron part $\hat{H}_e$, a phonon part $\hat{H}_p$, an electron-phonon interaction $\hat{V}_{ep}$ and a phonon-phonon interaction $\hat{V}_{pp}$ with

$$\hat{H}_e = \sum_{n,\mathbf{k}} E_{n\mathbf{k}}\, \hat{a}^{\dagger}_{n\mathbf{k}}\, \hat{a}_{n\mathbf{k}} \tag{3.1}$$

where $E_{n\mathbf{k}}$ is the energy associated with the state

$$\Psi_{n,\mathbf{k}}(\mathbf{r}, z) = A^{-1/2}\zeta_n(z)\exp(i\mathbf{k}\cdot\mathbf{r}) , \tag{3.2}$$

associated with the transverse wave-vector $\mathbf{k}$, and quantum well state n. Here $\zeta_n(z)$, is the nth quantum state in the well. The unperturbed phonon part is

$$\hat{H}_p = \sum_{\mathbf{Q}} \hbar\omega_L\, \hat{b}^{\dagger}_{\mathbf{Q}} \hat{b}_{\mathbf{Q}} \tag{3.3}$$

$\mathbf{Q} = (\mathbf{q}, q_z)$

The electron-phonon interaction is given by the Frohlich interaction

$$\hat{V}_{ep} = \frac{1}{4\pi} \int d\mathbf{R} \int d\mathbf{R}'\, e\, \hat{n}(\mathbf{R})\, \frac{1}{|\mathbf{R} - \mathbf{R}'|} [-4\pi\nabla'\cdot\hat{\mathbf{P}}(\mathbf{R}')] \tag{3.4}$$

where $\mathbf{P}$ is the phonon induced polarization, so that

$$\hat{V}_{ep} = \frac{i\alpha}{\sqrt{A}} \sum_{n,\,\mathbf{k}} \sum_{n',\,\mathbf{k}'} \sum_{\mathbf{Q}} \left\{ \hat{b}_{\mathbf{Q}}\, \delta_{\mathbf{k}',\,\mathbf{k}+\mathbf{q}}\, G_{n'n}(\mathbf{q}, q_z) - \hat{b}^{\dagger}_{\mathbf{Q}}\, \delta_{\mathbf{k}',\,\mathbf{k}-\mathbf{q}}\, G^{*}_{n'n}(\mathbf{q}, q_z) \right\} \hat{a}^{\dagger}_{n'\mathbf{k}'} \hat{a}_{n\mathbf{k}} , \tag{3.5}$$

when the matrix element G takes the form

$$G_{n'n}(\mathbf{q}, q_z) = \frac{1}{Q\sqrt{L}} \int_{-\infty}^{+\infty} dz\, \zeta^{*}_{n'}(z) \exp(\, i q_z z)\, \zeta_n(z). \tag{3.6}$$

In Eq. (3.5), α is the Frohlich interaction constant

$$\alpha = [2\pi e^2 \hbar\omega_L (1/\varepsilon_\infty - 1/\varepsilon_0\,]^{1/2} \tag{3.7}$$

4. FORM OF THE EQUATIONS OF MOTION

We treat the electron-phonon interaction, $\hat{V}_{ep}$, to second order. But neither the electron nor the phonon system is assumed to be even close to equilibrium. By retaining $a^\dagger_{\mathbf{k}} a_{\mathbf{k}}$ and $b^\dagger_{\mathbf{q},q_z} b_{\mathbf{q},q_z'}$ as the slow variables we obtain coupled integrodifferential equations of the form

$$\frac{\partial f_{n\mathbf{k}}}{\partial t} = \sum_{q}\sum_{q_z}\sum_{q_z'}\sum_{n'\mathbf{k}'} M[f_n\mathbf{k}, f_{n'\mathbf{k}'}, n_{\mathbf{q}}(q_z, q_z', t)] \tag{4.1}$$

$$\left[\frac{\partial n_{\mathbf{q}}(q_z, q_z', t)}{\partial t}\right]_{ep} = \sum\sum\sum_{nkn'\mathbf{k}'q_z''} f_{n\mathbf{k}}\,[1-f_{n'\mathbf{k}'}] \times [\text{ Four terms }]\,. \tag{4.2}$$

where a typical term is given by

$$\delta_{\mathbf{k}',\,\mathbf{k-q}}\,\frac{G_{n'n}(\mathbf{q},\, q_z)\, G^*_{n'n}(\mathbf{q},\, q_z'')}{\varepsilon - i\,[E_{n\mathbf{k}} - E_{n'\mathbf{k}'} - \hbar\omega_L]}[\delta_{q_z'',\, q_z'} + n_{\mathbf{q}}(q_z'',\, q_z',\, t)] \tag{4.3}$$

The detailed equations have been presented in [10,11]. Here we emphasize their form. In particular, let us regard $\mathbf{q}$ and $\mathbf{q}'_z$ as parameters. Then Eq. 2 with Eq. 3 has the form (for the term shown in Eq. 3)

$$\frac{\partial n(q_z, t)}{\partial t} = \phi(q_z, t) + K(q_z)M(t) \tag{4.4}$$

where

$$M(t) = \int H(q_z'')n(q_z'', t)dq_z'' \tag{4.5}$$

after all parameters such as $\mathbf{q}$, q_z' are suppressed. Multiplication of Eq. (4.4) by $H(q_z)$ and integration leads to the reduced *ordinary differential* equation

$$\frac{dM(t)}{dt} = I(t) + AM(t) \tag{4.6}$$

where

$$I(t) = \int H(q_z)\phi(q_z, t)dq_z \quad ; \quad A = \int H(q_z)K(q_z)dq_z \tag{4.7}$$

5. THE SHAPE OF THE PACKET

Not only is the remaining computational task greatly simplified, Eq. (4.5) already informs us that (aside from a choice of normalization) the phonon wave-packet operators are

$$\hat{c}^{\dagger}_{n'n} \sim \sum_{q_z} G^{*}_{n'n}(\mathbf{q}, q_z)\hat{b}^{\dagger}_{\mathbf{q},q_z} \tag{5.1}$$

If there are S subbands of importance, there is one packet operator for each choice of (n', n) or $S(S+1)/2$ differently shaped packets. For the important case in which only the $n = 0$ states participate there is one such packet (for each $\mathbf{q}$):

$$G^{*}_{00}(\mathbf{q}, q_z) = \frac{1}{\sqrt{L}} \int_{-\infty}^{\infty} |\zeta_0(z')|^2 dz' \frac{\exp(-iq_z z')}{\sqrt{q^2 + q_z^2}} \tag{5.2}$$

The shape of this packet in ordinary share may be obtained by multiplication by $\exp(iq_z z)$ and integrating over q_z:

$$\frac{1}{\sqrt{L}} \int R(z - z')dz' |\zeta_0(z')|^2 \tag{5.3}$$

where

$$R(z - z') = \int \frac{\exp[iq_z(z - z')]}{\sqrt{q^2 + q_z^2}} dq_z = 2K_0(q|z - z'|) \tag{5.4}$$

has the form of a modified Bessel function. This form arises from the Coulomb nature of the interaction. If we had used a point interaction $(q^2 + q_z^2)^{-1/2}$ would have been replaced by unity and $R(z - z')$ by $\delta(z - z')$ so that the packet shape would simply be $|\zeta_0(z)|^2$.

6. FURTHER SIMPLIFICATIONS

We have not written the explicit form of the phonon-phonon interaction $\hat{V}_{\rm pp}$ by means of which the relevant longitudinal optical phonons decay into acoustic phonons because we have replaced that process by a relaxation process of the form

$$\left[\frac{\partial n_{\mathbf{q}}(q_z, q_z', t)}{\partial t}\right]_{\rm pp} = -\frac{n_{\mathbf{q}}(q_z, q_z', t) - \delta_{q_z, q_z'} n_L}{\tau_{\rm op}}, \tag{6.1}$$

where the decay time for optical phonons $\tau_{\rm op}$ has been estimated in the experimental papers to be 7 psec.

For times larger than a picosecond it has been found by Monte Carlo calculations [16-19] and by our own quasi-analytical procedure [20] that the electron distribution has equilibrated relative to two macroscopic parameters: an electron temperature $T_e(t)$ and a drift velocity $v_e(t)$, both of which may be time dependent. By introducing these parameters, the equation for the electron distribution is replaced by ordinary differential equations for these parameters.

Finally, since we have been dealing with variables such as $n_q(q_z, q'_z)$ it was convenient to use a wave-packet construction on both left and right wave-vectors. Thus our reduced phonon variables are

$$N_{n'n,\, m'm}(q,\, t) = \frac{\sum\limits_{q_z q'_z}\sum G^*_{n'n}(\mathbf{q},\, q_z)\, n_{\mathbf{q}}(q_z,\, q'_z,\, t)\, G_{m'm}(\mathbf{q},\, q_z')}{F_{n'n,\, m'm}(q)/2q}, \tag{6.2}$$

where

$$F_{m'm,\, n'n}(q) = \int dz_1 \int dz_2\ \zeta^*_{m'}(z_1)\zeta_m(z_1)\zeta^*_{n'}(z_2)\zeta_n(z_2)e^{-q|z_1-z_2|}\ . \tag{6.3}$$

Thus, in general, we will get coupled ordinary differential equations for $T_e(t)$, $v_e(t)$ and $N_{n'n,\, m'm}(t)$.

7. THE REDUCED EQUATIONS

For simplicity, we shall write here only the special case in which only the lowest subband contributes. Thus we shall set

$$N_0(\mathbf{q}, t) = N_{00,\,00}(\mathbf{q}, t) \tag{7.1}$$

The phonon equation can then be written

$$\frac{\partial N_0(q,\, t)}{\partial t} = \frac{N_0(q,\, t)+1}{\tau_e(q,\, T_e(t))} - \frac{N_0(q,\, t)}{\tau_a(q,\, T_e(t))} - \frac{N_0(q,\, t) - N_L(T_L)}{\tau_{\mathrm{op}}} \tag{7.2}$$

where the rate of phonon emission $1/\tau_e$ and the rate of phonon absorption $1/\tau_a$ are given by

$$\frac{1}{\tau_e(q,\, T_e)} = \frac{\pi\alpha^2}{\hbar}\ \frac{1}{q}\ \frac{F_{00,\,00}(q)\, I^{(+)}_{00}(q)}{|\varepsilon_{00,00}(q, \omega_L)|^2}, \tag{7.3}$$

$$\frac{1}{\tau_a(q,\, T_e)} = \exp\,[\ \hbar\omega_L/k_B T_e(t)]\ \frac{1}{\tau_e(q,\, T_e(t))}. \tag{7.4}$$

where $F_{00,\,00}(\mathbf{q})$ was defined in Eq. (6.2), and the matrix element $I^{(+)}$ is given by

$$I^{(+)}_{00}(q,\, T_e(t)) = \frac{m}{\pi^2\hbar^2 q} \int\limits_{\frac{q}{2}+\frac{m\omega}{\hbar q}}^{\infty} kdk\ \frac{1}{[k^2 - (\frac{q}{2} + \frac{m\omega}{\hbar q})^2]^{1/2}}$$

$$\times f\,(E_k,\, T_e(t))[1 - f\,(E_k - \hbar\omega_L,\, T_e(t))]\ , \tag{7.5}$$

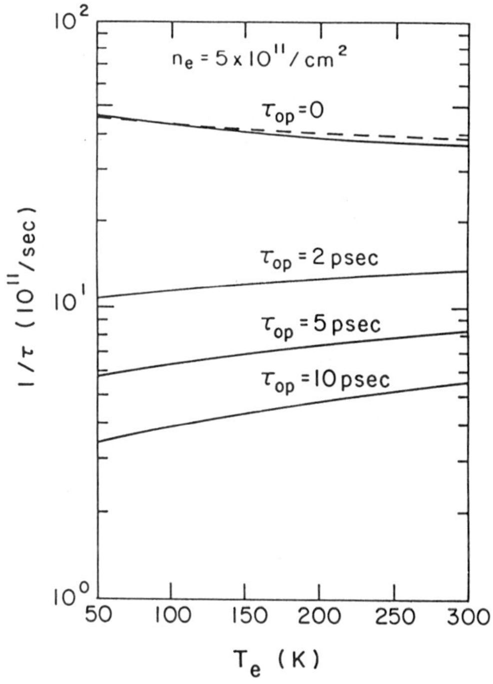

Fig. 1. $1/\tau$ as a function of the electron temperature T_e, for $\tau_{op} = 0, 2, 5,$ and 10 psec and $(n = 0)$ lowest subband occupation. Dynamic screening is included (solid curves). The dashed curve shows for comparison a calculation for $\tau_{op} = 0$ when static screening is used. [11]

Eq. (7.4) shows that the ratio of rates of absorption and emission is governed by the instantaneous electron temperature.

8. ENERGY LOSS RATE IN STEADY STATE CASE

The first explicit calculation made is for the steady state case. In that case, the electron temperature $T_e(t) = T_e$ will be independent of time and assumed given. The problem is to calculate the rate of energy transfer from the electron gas into the LO phonons, and to express the result in the form used by the experimentalists. The latter fit their experimental data with an expression of the form:

$$P_e(t) = \frac{\hbar\omega_L}{\tau} \exp\left(-\frac{\hbar\omega_L}{k_B T_e}\right) \tag{8.1}$$

Then $1/\tau$ is plotted as a function of electron temperature.

We calculate $1/\tau$ by using the expression:

$$P_e(t) = \frac{\hbar\omega_L}{N_e} \sum_{\mathbf{q}} \frac{\partial n_{\mathbf{q}}(q_z, q_z, t)}{\partial t} \Big|_{ep} \tag{8.2}$$

and then comparing with the previous equation. A plot of $1/\tau$ obtained in this manner is shown in Fig. 1.

We note that in the case in which $\tau_{op} = 0$, equilibration of the LO phonon

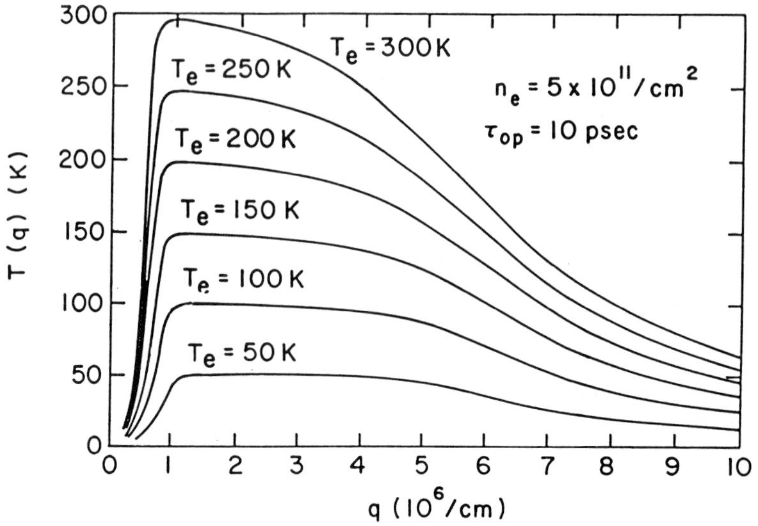

Fig. 2. "Phonon temperature" $T(q)$ as a function of q for 50 K $\leq T_e \leq$ 300 K and lowest subband occupation. [11].

modes takes place instantaneously. Thus no bottleneck effect will occur. But the effective energy transfer rate is reduced by an order of magnitude if one takes $\tau = 10$ psec. The dashed curve demonstrates that neglect of screening would have little effect on the results.

If we apply the steady state condition:

$$\frac{dN_0(\mathbf{q}, t)}{dt} = 0 \tag{8.3}$$

we can solve for $N_0(\mathbf{q})$, the non-equilibrium phonon occupancy associated with the transverse phonon wave-vector $\mathbf{q}$.

Using the Planck formula,

$$N_0(\mathbf{q}) = 1/\{\exp[\hbar\omega_L / k_B T(\mathbf{q})] - 1\} \tag{8.4}$$

the results can be expressed in the form of a temperature $T(\mathbf{q})$ for the phonons of a given transverse wave-vector $\mathbf{q}$. The results displayed in Fig. 2 show that the predominant heating occurs for small wave-vector phonons.

We have also made calculations when electrons and holes are simultaneously present, and when more than one sub-band is occupied. A comparison between experiment and theory for this case is shown in Fig. 3.

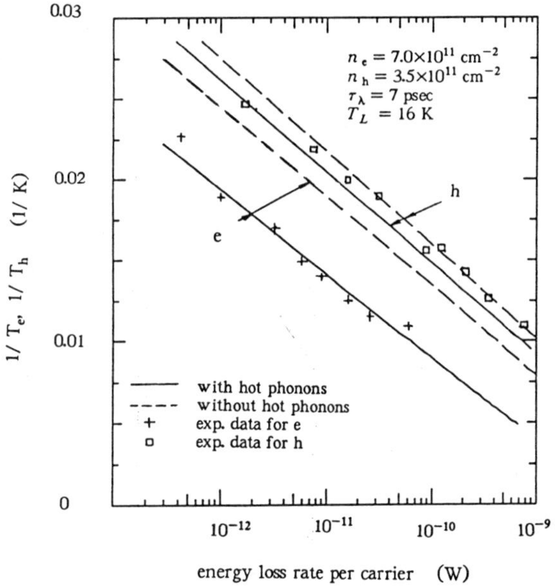

Fig. 3. The carrier temperature is plotted against the energy loss rate per carrier for the electron and hole case. For the electron case, the effect of hot phonons is found to be an order of magnitude, whereas for holes it significant but much less than an order of magnitude. The points are experimental data Shah et al. [21]

9. TIME-DEPENDENT RELAXATION

Our starting equations are valid for the time-dependent case. We simply do not assume time derivatives vanish. The electron energy can be written as a sum over the transverse **k** vector:

$$<E_e(t)> = \frac{2}{A} \sum_{\mathbf{k}} E_{\mathbf{k}} \, f(E_{\mathbf{k}}, T_e(t)) \tag{9.1}$$

For simplicity, we are again considering here only one sub-band. The electron temperature change rate is then given by:

$$\frac{\partial T_e}{\partial t} = \frac{1}{C_\upsilon(T_e(t))} \frac{\partial <E(t)>}{\partial t} \tag{9.2}$$

where C_υ is the specific heat:

$$C_\upsilon = \frac{\partial <E(t)>}{\partial T_e(t)} \tag{9.3}$$

Since the phonon equations depend on the instantaneous temperature $T_e(t)$, we now have coupled ordinary differential equations connecting the phonon occupancies $N_0(\mathbf{q})$ and the electron temperature. It is assumed, of course, that $T_e(t = 0)$ is given. A comparison is given in Fig. 4 of our theoretical results with experiment [4].

As the electron gas relaxes, its temperature decreases as shown by the solid curve in Fig. 5. The associated rise in the phonon temperature is shown by the dashed curve which merges with the solid curve as the combined system relaxes to the lattice temperature. When τ_{op} is set equal to zero, the phonon bottleneck effect disappears,

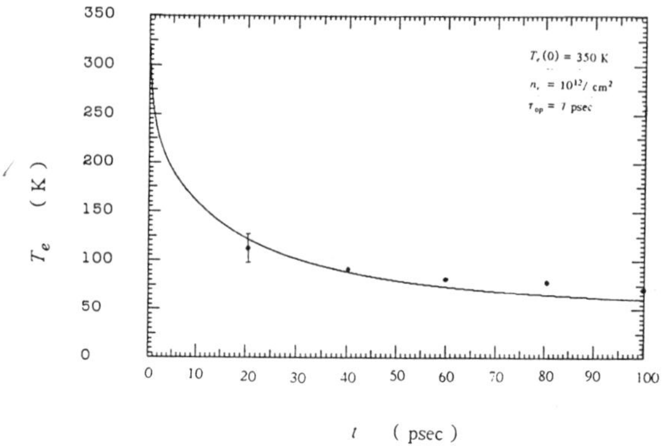

Fig. 4. Time dependent relaxation of an electron gas of density $n_e = 10^{12}/\mathrm{cm}^2$ starting at an initial temperature of $T_e(0) = 350\mathrm{K}$. The optical phonon decay rate has been given the accepted value of $\tau_{op} = 7\mathrm{psec}$. The experimental data are from Ryan *et al.* [4] corresponding to a 3D electron density $n = 5\times 10^{18}\mathrm{cm}^{-3}$ and to a maximum power absorbed by the sample of ~ 50 mW.

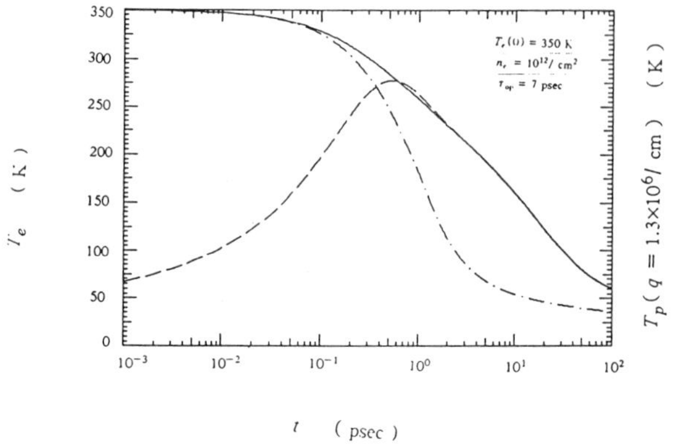

Fig. 5. A plot of electron temperature T_e against time (solid curve for $\tau_{op} = 7$ psec and dot-dash curve for $\tau_{op} = 0$) and of the phonon temperature $T_p(q)$ at $q = 1.3\times 10^6\mathrm{cm}^{-1}$.

and the electron temperature falls more quickly as shown by the dot-dash curve.

10. HOT ELECTRON TRANSPORT

When a strong dc field is applied, the electron gas acquires an elevated temperature as well as a drift velocity. Moreover, the drift mobility is reduced more in the presence of a phonon bottleneck than in its absence.

The methods described earlier involving the introduction of a phonon packet remain valid. It is convenient, in addition to separate the center of mass motion of the electron gas from its relative motion. The density of electrons will again be assumed sufficiently high that the electron gas can be assumed in quasi-equilibrium with a given drift velocity and a given temperature. The intracollisional effect [22,23], namely the effect of the electric field during the course of a collision can be neglected for the fields considered in our work [24].

With the above approximations, the problem is again reduced to coupled ordinary nonlinear differential equations. [24] Results for the mobility reduction and temperature increase with applied field are shown in Fig. 6.

Fig. 6. The normalized mobility of electrons, $\mu(\mathbf{E})/\mu(\mathbf{E} \to \mathbf{0})$ (solid curves), and the electron temperature, T_e, (dashed curves) as functions of the external electric field $|\mathbf{E}|$ at $\tau_{op} = 0$ and $\tau_{op} = 7$ psec. The crosses represent the experimental data from Fig. 2(c) of Ref. [25] at $T_L = 77$ K.

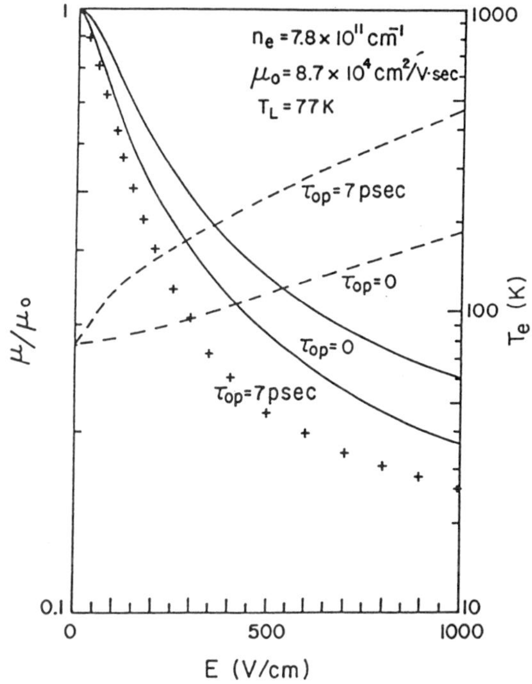

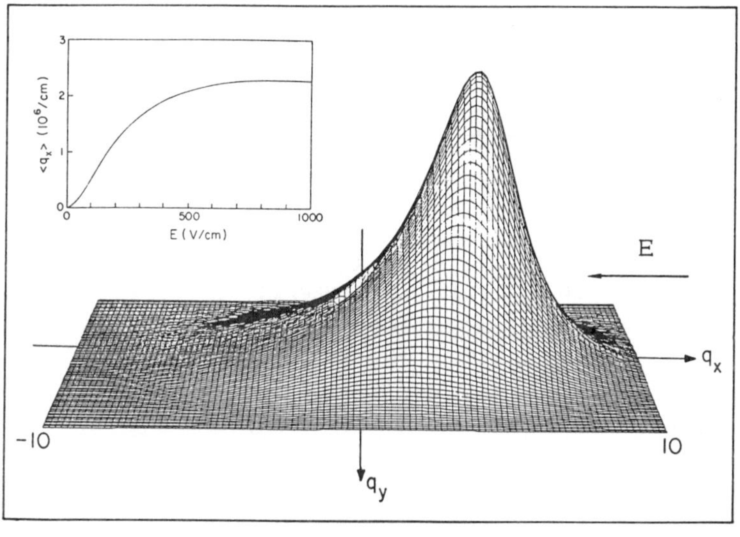

Fig. 7. "Optical phonon temperature", $T_{op}(\mathbf{q})$, as a function of $\mathbf{q}$ (unit: 10^6 cm^{-1}) at $|\mathbf{E}| = 500$ V/cm. The q_x-q_y plane represents $T_{op} = 77$ K, the peak value of T_{op} is 696 K. Inset: The average wave vector of nonequilibrium LO phonons, $< q_x >$, as a function of the electric field, $|\mathbf{E}|$.

The phonon heating produced by the strong electric field is not isotropic, but is more effective for phonons whose propagation direction direction is in the direction of motion of the electrons. The phonons also acquire a mean "momentum" $<q_x>$. A contour plot of the phonon temperature rise, and an inset of the mean momentum are shown in Fig. 7.

ACKNOWLEDGMENT

The work at the City college of City University of New York was partly supported by the U. S. Army Research Office and the U. S. Department of Energy.

REFERENCES

1. J. Shah, A. Pinczuk, A. C. Gossard, and W. Wiegmann, Phys. Rev. Lett. 54, 2045 (1985)
2. C. H. Yang, J. M. Carlson-Swindle, S. A. Lyon, and J. A. Worlock, Phys. Rev. Lett 55, 2359 (1985)
3. J. Shah, A. Pinczuk, H. L. Störmer, A. C. Gossard, and W. Wiegmann, Appl. Phys. Lett. 44, 322 (1984)
4. J. F. Ryan, R. A. Taylor, A. J. Turberfeld, A. Maciel, J. M. Worlock, A. C. Gossard and W. Wiegmann, Phys. Rev. Lett. 53, 1841, (1984).
5. D. K. Ferry, Surf. Sci. 75, 86 (1978)
6. K. Hess, Appl. Phys. Lett. 35, 484 (1979)
7. B. K. Ridley, J. Phys. C 15, 5899 (1982); 16, 6971 (1983)
8. T. Ando, A. B. Fowler, and F. Stern, Rev. Mod. Phys. 54, 437 (1982)
9. P. Price, Ann. Phys. (NY) 313, 217, (1981), Surf. Sci. 31B, 199, (1985)
10. W. Cai, M. C. Marchetti, and M. Lax, Phys. Rev. B 334, 8573 (1986, Dec. 15)
11. W. Cai, M. C. Marchetti, and M. Lax, Phys. Rev. B 35, 1369 (1987, Jan. 15)
12. R. Zwanzig, J. Chem. Phys. 33, 1338 (1960)
13. R. Zwanzig, Phys. Rev. 124, 983 (1961)
14. D. N. Zubarev, Nonequilibrium Statistical Thermodynamics (Consultant's Bureau, NY, 1974)
15. S. Peletminskii and A. Yatsenko, Zh. Eksp. Teor. Fiz. 53, 1327 (1967) Sov. Phys. JETP 26, 773 (1968)
16. P. Lugli and S. M. Goodnick, Phys. Rev. Lett. 59, 716 (1987)
17. M. A. Osman and D. K. Ferry, Phys. Rev. B 36, 6018 (1987)
18. D. W. Bailey, M. A. Artaki, C. J. Stanton, and K. Hess, Appl. Phys. Lett. 62, 4638 (1987)
19. S. M. Goodnick and P. Lugli, Phys. Rev. B 38, 10135 (1988)
20. T. F. Zheng, W. Cai, P. Hu, and M. Lax, Phys. Rev. B 40, 1271 (1989)
21. J. Shah, A. Pinczuk, A. C. Gossard, W. Weigmann, and K. Kash, Surf. Sci. 174, 363 (1986)
22. J. R. Barker, J. Physics C 6, 1663 (1973)
23. S.K. Sarker, Phys. Rev. B 33, 7263 (1986)
24. W. Cai, M. C. Marchetti and M. Lax, Phys. Rev. B 37, 2636 (1988)
25. M. Keever, W. Kopp, T. J. Drummond, H. Morko, and K. Hess, Jpn. J. Appl. Phys. 21, 1489 (1982)

LOCALIZATION EFFECTS IN HIGH-TEMPERATURE SUPERCONDUCTORS

M. V. Sadovskii
Institute for Electrophysics, USSR Academy of Sciences,
Ural Branch, Sverdlovsk, 620219, USSR

Basic facts on the interplay of Anderson localization and superconductivity in high-T_c oxides are presented. The "minimal metallic conductivity" for the quasi-two-dimensional case is enhanced owing to a small overlap of electronic states on the nearest neighbor conducting planes. This leads to a much stronger influence of localization effects than in ordinary (three- dimensional) superconductors. From this point of view high-temperature oxides are very close to the Anderson transition even for rather weak disorder. Anomalies of the upper critical field are also analyzed as well as degradation of T_c under disordering, due to the enhanced Coulomb effects caused by the disorder-induced decrease of localization length.

INTRODUCTION

The concept of Anderson localization [1] is basic for the modern theory of electrons in disordered systems [2-6]. According to this concept the introduction of sufficiently large disorder transforms the initial metallic system to an insulator, because of the transition from extended to spatially localized electronic states at the Fermi level.

At the same time, since the classic BCS-paper [7], it is well known that even the slightest attraction of electrons near the Fermi level leads to superconductivity, which is relatively insensitive to disorder which conserves time-invariance (normal, nonmagnetic impurities, etc.) [8,9].

Thus, an interesting problem arises of the possible interplay of the localization transition and superconductivity in a strongly disordered metal. This problem is important from both the theoretical point of view, because it is a question of the interplay of apparently opposite kinds of ground states (insulator versus superconductor), as well as the experimental point of view, because in many cases the superconducting transition is observed close to the metal-insulator transition in highly disordered systems. It is specially important for high-T_c oxides, which are close to the metal-insulator transition from the very beginning.

The general picture of the interplay of Anderson localization and superconductivity was analyzed in several papers in recent years [10-17]. Here we present the basic results applied especially to high-T_c oxides, where many experimental results were also obtained, mainly for the case of radiation-disordering by fast neutrons [18-21].

High-T_c oxides are quite appropriate for studying the interplay of localization and superconductivity. First of all, the high values of T_c are important in order to overcome rather strong mechanisms of the degradation of T_c after disordering [11].

Secondly, the smallness of Cooper pairs in these materials is also very important, because of the basic criteria of the possible coexistence of localization and superconductivity - the Cooper pairs must be smaller than the localization length in the insulating phase [10, 11]. These strict criteria are difficult to satisfy in ordinary superconductors. And finally the high-T_c oxides are close to a metal-insulator transition probably of the Mott-Hubbard type. The parent compounds such as La_2CuO_4 and $YBa_2Cu_3O_6$ are antiferromagnetic insulators, but disorder effects are also quite important owing to inherent disorder present in all real samples. These effects are manifested among other things by the variable range hopping behavior of conductivity in these insulating phases [22,23] as well as the remnants of it in disordered superconducting phases [19-21].

Fast neutron irradiation is probably the purest method for investigating the effects of disordering on physical properties of high-T_c superconductors because of the absence of any chemical effects in case of low temperature irradiation. The appropriate growth of structural disorder leads to rather drastic changes in the behavior of both single-crystalline and ceramic samples [18-21]:

(a) continuous metal-insulator transition at very slight disordering;

(b) rapid degradation of T_c;

(c) apparent coexistence of hopping conductivity and superconductivity at intermediate disorder and anomalous (exponential) growth of resistivity with defect concentration;

(d) approximate independence of the derivative of the upper critical field H_{c2}' on the degree of disorder.

These anomalies were interpreted using the ideas of possible coexistence of Anderson localization and superconductivity and in the following we present mainly theoretical aspects of this problem for strongly anisotropic (quasi-two-dimensional) high-T_c systems. For the extensive discussion of experimental data we refer to Refs. 18-21.

1. LOCALIZATION AND SUPERCONDUCTIVITY IN QUASI-TWO-DIMENSIONAL SYSTEMS.

All the known high-T_c systems (with $T_c > 30K$) are strongly anisotropic, or even quasi-two-dimensional conductors. For such systems it is natural to expect the strong enhancement of localization effects due to the known role of spatial dimensionality of $d = 2$: in the purely two-dimensional case localization appears for infinitely small disorder [3-6].

Exact electronic states $\phi_\nu(r)$ in a disordered system are defined by the solution of the appropriate Schroedinger equation in a random field. These states may be both extended or localized. Cooper pairing can be realized between the time-reversed partners $\phi_\nu(r)$ and $\phi_\nu^*(r)$. For the case of the self-averaging superconducting order parameter this problem was solved by Anderson [9] , who showed that for the given value of pairing interaction, T_c is essentially independent of the nature of these states: either extended or localized. The only limitation for the latter case is due to the known effects of the discrete level repulsion in the localization region [2,11]. It is clear that Cooper pairing can be realized in the localized phase only for the electrons with the centers of localization within the volume of the characteristic size determined by R_{loc} - the localization length, because only such electrons have overlapping wave functions and can interact with each other. However, these states are split in energy on the scale of the order of $[N(E_F)R_{loc}^3]^{-1}$, where $N(E_F)$ is the density of states at the Fermi level.

Obviously, to observe superconductivity we must demand that this splitting be smaller than the value of the superconducting energy gap at T=0:

$$\Delta, T_c \gg [N(E_F)R_{loc}^3]^{-1} , \tag{1}$$

or for strongly anisotropic high-T_c systems:

$$\Delta, T_c \gg [N(E_F)R_{loc}^a R_{loc}^b R_{loc}^c]^{-1} , \tag{2}$$

where we introduced the appropriate values of localization lengths along the axes of an orthorhombic lattice. This inequality is equivalent to the condition of rather large localization lengths [10,11], e.g. for an isotropic case:

$$R_{loc} \gg [N(E_F)\Delta]^{-1/3} \approx (\xi_0 \hbar^2 / p_F^2)^{1/3} \approx (\xi_0 a^2)^{1/3} , \tag{3}$$

where $\xi_0 \approx \hbar V_F / T_c$ is the coherence length of the BCS theory and $p_F \approx \hbar / a$ is the Fermi momentum (a is the lattice spacing, V_F is the Fermi velocity). For high-T_c oxides with rather large values of Δ and small ξ_0 this condition can can be satisfied rather easily. Actually the physical meaning of of it is very simple: R_{loc} must be much larger than the characteristic size of the Cooper pair in the strongly disordered system [10,11].

The main properties of a superconductor can be analyzed via the Ginzburg-Landau theory, and to do this we must derive the GL-expansion coefficients for the strongly disordered quasi-two-dimensional system near the localization transition [18]. As a one-electron model of the Anderson transition we use a self-consistent theory of localization [3,5] for the quasi-two-dimensional case [24]. The electron motion in such a system is determined by the two-particle Green's function with a characteristic diffusion form:

$$\phi(\mathbf{q}\omega) = -N(E_F) / \{\omega + iD_{\parallel}(\omega)q_{\parallel}^2 + iD_{\perp}(\omega)[1 - \phi(q_{\perp})]\} \tag{4}$$

where $D_{\parallel,\perp}(\omega)$ are the longitudinal and transverse generalized diffusion coefficients (with respect to conducting planes), $\phi(q_{\perp}) = \cos(q_{\perp} a_{\perp})$, and $q_{\parallel,\perp}$ are longitudinal and transverse components of $\mathbf{q}$, $a_{\perp}$ is the interplane distance in the quasi-two-dimensional lattice. For simplicity we assume isotropic motion of electrons in the conducting plane.

The generalized diffusion coefficient can be determined from the self-consistency equation [24]:

$$D_{\parallel,\perp}(\omega) = D_{\parallel,\perp}^0(\omega) - \frac{1}{\pi N(E_F)} \int \frac{d^3q}{(2\pi)^3} \frac{D_{\parallel,\perp}(\omega)}{-i\omega + D_{\parallel}(\omega)q_{\parallel}^2 + D_{\perp}(\omega)(1 - \phi(q_{\perp}))} , \tag{5}$$

where $D_{\parallel}^0 = V_F^2 \tau / 2$, $D_{\perp}^0 = (wa_{\perp})^2 \tau$ are Drude-like diffusion coefficients, w is the interplane transfer integral and τ is the mean-free time in the conducting plane.

The mobility edge position on the energy axis is determined by:

$$E_c = \hbar / \pi\tau \ln(2^{1/2} \hbar / w\tau) , \tag{6}$$

so that $E_c \to \infty$ for $w \to 0$ reflecting the complete localization for $d=2$ [3,6]. For $E = E_c$

the Drude conductivity in the plane is equal to the so called "minimal metallic conductivity" [2]:

$$\sigma_c^{\parallel} = 2e^2 D_{\parallel}^0 (E_F = E_c) = \frac{1}{\pi^2}\frac{e^2}{\hbar a_\perp}\ln(2^{1/2}\hbar/w\tau) . \tag{7}$$

From that expression we can see that σ_c for the quasi-two-dimensional system is significantly enhanced in comparison to the Mott estimates for the isotropic case, due to a logarithmic factor which grows with diminishing overlap of electronic states on the nearest-neighbor planes. Thus for the strongly anisotropic (or quasi-two-dimensional) systems, such as high-T_c oxides, the value of "minimal metallic conductivity" may be larger than the usual estimates of $(3–5)\times10^2\ \mathrm{ohm}^{-1}\mathrm{cm}^{-1}$, and actually can exceed $10^3\ \mathrm{ohm}^{-1}\mathrm{cm}^{-1}$ for typical values of $\sigma_\perp/\sigma_\parallel \propto 10^2$ and $E_F\tau \propto 1$. While there is no rigorous definition of "minimal metallic conductivity" now, owing to the continuous nature of the Anderson transition [3-6], it actually defines the scale of conductivity near the metal-insulator transition caused by disorder. From these estimates it is clear that most of the real samples of high-T_c superconductors are quite close to the Anderson transition and even very slight disordering is sufficient [18-21] to transform them to the Anderson insulators.

Using the expressions for $D_{\parallel,\perp}(\omega)$ from Ref. 24 we can derive the microscopic expression for the coefficient of gradient term in the GL-expansion similar to Refs. 10-13:

$$C_{\parallel,\perp} = N(E_F)\xi_{\parallel,\perp}^2 , \tag{8}$$

where for the coherence lengths $\xi_{\parallel,\perp}$ we have slightly different expressions depending on the values of the dimensionless parameter $w^2\tau/2\pi T_c\hbar$, determining the crossover from anisotropic to quasi-two-dimensional superconductor. For $w^2\tau/2\pi T_c\hbar \gg 1$ we have:

$$\xi_{\parallel,\perp}^2 = (\pi/8T_c)D_{\parallel,\perp}^0(E_F - E_c)/E_c = \xi_{\parallel,\perp}^0 l_{\parallel,\perp}(E_F - E_c)/E_c , \tag{9}$$

where $\xi_\parallel^0 \approx \hbar V_F/T_c$, $\xi_\perp^0 \approx \hbar w a_\perp/T_c$, $l_\parallel = V_F\tau$, $l_\perp = w a_\perp\tau$. Eq.(7) is valid for $\sigma_\parallel > \sigma^*$, where:

$$\sigma^* \approx \sigma_c^\parallel \xi_\parallel^0/l_\parallel (T_c^2/E_F w)^{2/3} , \tag{10}$$

and $w^2\tau/2\pi T_c\hbar \gg 1$ is equivalent to $\xi_\parallel \approx \sqrt{\xi_\perp^0 l_\perp} \gg a_\perp$, i.e. the size of a Cooper pair is larger than $a_\perp$, and we have just the anisotropic superconductor.

In the vicinity of the Anderson transition for $\sigma_\parallel < \sigma^*$:

$$\xi_{\parallel,\perp}^2 \approx D_{\parallel,\perp}^0/[(E_F T_c w)^{2/3}\tau] \approx (\xi_{\parallel,\perp}^0)^2 (T_c^2/E_F w)^{2/3} . \tag{11}$$

For the isotropic case, for $w \approx E_F$ these expressions transform into that of Refs. 10,11, where close to the Anderson transition we obtained (Cf. (3)):

$$\xi \approx (\xi_0 l^2)^{1/3} \approx (\xi_0 a^2)^{1/3} , \tag{12}$$

which is valid for $\sigma<\sigma^*$ with σ^* given by:

$$\sigma^* \approx \sigma_c (p_F \xi_0/\hbar)^{-1/3} \approx \sigma_c (T_c/E_F)^{1/3} , \tag{13}$$

where $\sigma_c = e^2 p_F/\pi^3\hbar^2$ is the Mott estimate for "minimal metallic conductivity". For the case of $w^2\tau/2\pi T_c<1$ corresponding to the quasi-two-dimensional superconductor we get:

$$\xi^2_{\parallel,\perp} \approx \frac{D^0_{\parallel,\perp}}{\pi T_c}\left\{\frac{(E_F-E_c)}{E_c}+\left(\frac{\pi^2}{8}-1\right)\left[1-\frac{1}{2\pi E_F\tau}\ln\left(\frac{1}{2\pi T_c\tau}\right)\right]\right\} , \tag{14}$$

for $\sigma>\sigma^*$, where σ^* is again determined by (10), and for $\sigma<\sigma^*$ we obtain the same expression as in (14) but with first term replaced by (11). For high-T_c oxides $\xi^0_\parallel \approx l_\parallel$, $T_c \approx w$, $T_c \approx 0.1E_F$ and actually $\sigma^* \approx \sigma^c_\parallel$, so that these systems are close to the Anderson transition also in their superconducting behavior. Also, it is clear that for most of these systems we have apparently $w^2\tau/2\pi T_c\hbar \approx 1$, i.e. $\xi_\perp \approx \sqrt{\xi^0_\perp l} \approx a_\perp$, so that they are on the edge of quasi-two-dimensional behavior.

For the derivative of the upper critical field in the isotropic case we have [10,11]:

$$-\frac{\sigma}{N(E_F)}\left(\frac{dH_{c2}}{dT}\right)_{T_c} \approx \begin{cases} 8e^2/\pi\hbar\phi_0 & \sigma>\sigma^* \\ \phi_0/2\pi\sigma/[N(E_F)T_c]^{1/3} & \sigma<\sigma^* \end{cases} \tag{15}$$

where $\phi_0 = \pi c\hbar/e$ is the magnetic flux quantum. The classic Gorkov relation [25] given in the upper expression in (15) is invalid near the Anderson transition, where ($\sigma<\sigma^*$) the value of H_{c2}' becomes independent of conductivity and only slightly depends on disorder via the appropriate behavior of $N(E_F)$ and T_c diminished by the cubic root in the lower expression in (15).

For an anisotropic (quasi-two-dimensional) case we have:

$$(dH^\perp_{c2}/dT)_{T_c} = -\phi_0/(2\pi\xi^2_\parallel T_c)$$

$$(dH^\parallel_{c2}/dT)_{T_c} = -\phi_0/(2\pi\xi_\parallel\xi_\perp T_c) , \tag{16}$$

with $\xi_{\parallel,\perp}$ given above in (9)-(14) and the behavior is similar to that in (15). The most important relation is given by:

$$(H^\parallel_{c2})'/(H^\perp_{c2})' = \xi_\parallel/\xi_\perp = V_F/wa_\perp , \tag{17}$$

and we see that the anisotropy of H_{c2}' is actually determined by the anisotropy of the Fermi velocity irrespective of the regime of superconductivity: from the "pure" limit, through the usual "dirty" case, up to the vicinity of the Anderson transition.

2. SUMMARY OF EXPERIMENTS ON RADIATION DISORDERING IN SINGLE-CRYSTALS OF HIGH-T_c SUPERCONDUCTORS.

The experiments were performed [20,21] on a series of $YBa_2Cu_3O_{7-\delta}$ single-crystals with the sizes of the order of $1\times1\times0.03\,mm^3$ grown from the melt. Initial values of T_c were between 91 and 92K. Anisotropy of electrical resistivity ρ_c/ρ_{ab} at T=300K varied between 40 and 150, and ρ_c demonstrated semiconductor-like temperature behavior. Both ρ_{ab} and ρ_c were measured by the Montgomery method during irradiation by fast neutrons in the core of nuclear reactor at liquid nitrogen temperature. ρ_c increases exponentially with fluence ϕ (i.e. defect concentration) starting from the smallest doses, while ρ_c grows slower, only for $\phi > 10^{19}\,cm^{-2}$ both ρ_c and ρ_{ab} grow with the same rate. T_c rapidly drops with ϕ and there is no superconductivity at $T > 1.5$K for $\phi > 6\times10^{18}\,cm^{-2}$. Anisotropy ρ_c/ρ_{ab} at 80K drops rapidly to the value of order of ≈ 30 for $\phi = 10^{19}\,cm^{-2}$ and then practically does not change. Structural neutronography has shown that lattice changes under such doses of neutron irradiation are rather small [18-21].

From the comparison of these results with earlier data obtained on ceramic samples we may conclude that the exponential growth of electrical resistivity, which was interpreted as a manifestation of hopping-like conduction [18,19] due to localization, is an inherent property of high-T_c superconductors.

The upper critical fields of $YBa_2Cu_3O_{7-\delta}$ single-crystals were measured before and after irradiation in the fields up to 5T. Temperature dependence of H_{c2} in the disordered samples is essentially nonlinear in these fields, especially for samples with low T_c. $(H^{\perp}_{c2})'$ as determined from the high-field region increases with disorder. To obtain unambiguous results it is necessary to perform the measurements in high fields. $(H^{\parallel}_{c2})'$ drops in the beginning and then does not change or increases very slightly. However, the anisotropy of H_{c2} decreases for any field as disorder grows and in the samples with $T_c \approx 10$ K the ratio of $(H^{\parallel}_{c2})'/(H^{\perp}_{c2})'$ is close to unity. These data showing the absence of direct correlation of resistivity and H_{c2} behavior characteristic of "dirty" superconductors (such as the Gorkov relation) can be interpreted as due to closeness to the Anderson transition, or even as a consequence of superconductivity in the localized (insulating) phase [18-21]. From the point of view expressed by (17) the experimentally observed isotropization of slopes of $H^{\parallel}_{c2}$ and $H^{\perp}_{c2}$ under disordering is the manifestation of the isotropization of Cooper pairs. The remanent anisotropy of resistivities may be due to the anisotropy in the scattering mechanism, e.g. due to interplane defects. This behavior shows that just before the destruction of superconductivity the disordered high-T_c oxides become essentially isotropic (from the point of view of their superconducting properties) and we return to three-dimensional H_{c2} behavior. It will be of much interest also to try to observe the predicted $H_{c2}' \propto T^{-1/3}$ behavior (Cf.(15)) as superconductivity vanishes. Note however, that all this analysis assumes more or less smooth disorder dependence of the density of states.

Hall effect data [20,21] obtained on irradiated ceramic samples of $YBa_2Cu_3O_{7-\delta}$ show that the temperature dependence of Hall concentration n_H weakens, remaining linear in the high temperature region as in initial samples. At low T< 100K n_H practically does not change under disordering in sharp contrast with its behavior in oxygen deficient samples. This constancy of n_H for low T under disordering agrees well with the assumption that Anderson localization the main reason for metal-insulator transition under disordering [26,27]. However we must stress that it is difficult to explain the observed temperature dependence of n_H. There is also no unambiguous correlation between n_H and T_c.

3. LOCALIZATION AND DEGRADATION OF T_c.

In the absence of any accepted microscopic theory for T_c in high-T_c oxides it is rather difficult to discuss mechanisms for its degradation under disordering. Among general reasons for a drop of T_c apparently important for any BCS-like model of high-temperature superconductivity we can mention:

(a) growth of Coulomb repulsion within Cooper pairs [11,28];

(b) growth of spin-scattering effects due to the appearance of disorder induced local moments [18];

(c) incipient inhomogeneities due to "statistical fluctuations" near the Anderson transition [17].

Assuming that the experimentally observed exponential growth of resistivity is directly connected with the Mott law for hopping conduction we have analyzed [18-20] the fluence dependence of R_{loc} for radiationally disordered high-T_c oxides. Using these data for R_{loc} and estimating $N(E_F) \approx 5\times10^{33}$ $(\text{erg cm})^{-1}$ (for one electron per lattice cell in the free-electron model) and $\Delta \approx 5T_c$ (as for extremely strong coupling regime of superconductivity) we deduced that the inequality (1) determining the critical value of R_{loc} for the observation of superconductivity becomes invalid for $\phi > (5\text{–}7)\times10^{18}$ cm^{-2}. This is in surprisingly good accord with experimental data on superconductivity destruction by fast neutron irradiation.

To analyze the R_{loc} dependence of T_c in localized phase we used the exact analysis of Ref. 11 for T_c suppression due to the growth of Hubbard-like repulsion in a single quantum state, which becomes important in Anderson insulator [2,29,5]. Owing to the observed isotropization of superconducting properties under disordering we consider only three-dimensional isotropic case. According to Ref. 11 T_c in a disordered superconductor can be determined from the linearized gap equation:

$$\Delta(\omega) = \lambda\Theta(<\omega> - \omega)\int_0^{<\omega>} \frac{d\omega'}{\omega'}\Delta(\omega')\text{th}\left[\frac{\omega'}{2T_c}\right] -$$

$$\Theta(E_F - \omega)\int_0^{E_F} \frac{d\omega'}{\omega'}K_c(\omega-\omega')\Delta(\omega')\text{th}\left[\frac{\omega'}{2T_c}\right], \tag{18}$$

where λ is the pairing interaction constant, $<\omega>$ the characteristic frequency range, where the pairing interaction is important,

$$K_c(\omega) = \int d\mathbf{r}\int d\mathbf{r}'\, V(\mathbf{r}-\mathbf{r}') << \rho_E(\mathbf{r})\rho_{E+\omega}(\mathbf{r}') >>_{E=E_F} \tag{19}$$

is the generalized Coulomb kernel, where:

$$<<\rho(\mathbf{r})\rho(\mathbf{r}')>> = \frac{1}{N(E_F)} < \sum_{\mu\nu}\phi_\nu^*(\mathbf{r})\phi_\mu(\mathbf{r})\phi_\mu^*(\mathbf{r}')\phi_\nu(\mathbf{r}')\delta(E-\varepsilon_\nu)\delta(E+\omega-\varepsilon_\mu)> \tag{20}$$

is the spectral density (averaged over disorder) defined by Berezinskii and Gorkov [30] and $V(\mathbf{r}-\mathbf{r}') = V_0\delta(\mathbf{r}-\mathbf{r}')$ is the point-like interelectron interaction. The appearance of

Hubbard-like repulsion in a single (localized) quantum state is directly connected [5] with the characteristic $\delta(\omega)$ -contribution to (18) within the localization region [30], where we get:

$$K_c(\omega) = V_0 A(E_F)\delta(\omega)+\ldots;\quad A(E_F)\approx R_{\rm loc}^{-3}(E_F)\ . \tag{21}$$

This singular contribution in (18) can be treated exactly and we obtain the following equation for T_c [11]:

$$1 = \lambda\int_0^{<\omega>} d\omega \,\mathrm{th}(\omega/2T_c)[\omega+\mu A(E_F)/2N(E_F)\,\mathrm{th}(\omega/2T_c)]^{-1}\ , \tag{22}$$

where $\mu=N(E_F)V_0$. Approximately (22) reduces to:

$$\ln(T_{co}/T_c) \approx \Psi(1/2+\mu/[4T_cN(E_F)R_{\rm loc}^3]) - \Psi(1/2)\ , \tag{23}$$

where T_{co} is the initial value of T_c before disordering, $\Psi(x)$ is the digamma-function. Assuming the hopping conduction described by the Mott law [2]:

$$\rho(T)=\rho_0\exp(Q/T)^{1/4};\quad Q\approx 2.1\,[N(E_F)R_{\rm loc}^3]^{-1}\ . \tag{24}$$

We may directly express T_c via resistivity $\rho(T_{ex})$ in localized phase of strongly disordered superconductor [18,19] at temperature $T_{ex}>T_c$:

$$\ln(T_{co}/T_c) \approx \Psi\{1/2+0.013\mu T_{ex}[\ln(\rho(T_{ex})/\rho_0)]^4\}-\Psi(1/2)\ . \tag{25}$$

This expression gives a rather satisfactory fit to experimental data on $T_c(\rho)$ dependence in high-T_c superconductors disordered by radiation, assuming $\mu\approx 0.3$–1.0 [18,19]. Though speculative in nature this explanation of T_c degradation due to the growth of Coulomb effects may be of some interest.

Among other important data on high-T_c oxides disordered by radiation, we must also keep in mind the disorder - induced appearance of the Curie-Weiss contribution to magnetic susceptibility [18]. According to Mott [2,29] we may also try to connect this contribution with the appearance of single-occupied states in a narrow energy region below the Fermi level, once again induced by the Hubbard-like Coulomb repulsion in a single (localized) quantum state. We may estimate the value p (in Bohr magnetons) of the localized moment due to this mechanism as [5]:

$$p^2 = \mu R_{\rm loc}^{-3}\Omega_0\ , \tag{26}$$

where again $\mu=N(E_F)V_0$ is the dimensionless Coulomb potential, Ω_o is the volume of a unit cell. For large disorder ($\phi=2\times10^{19}\ \mathrm{cm}^{-2}$) we estimate [18] $R_{\rm loc}\approx 8$ Å and for $\mu\approx 1$ we get $p^2=0.66$ for $YBa_2Cu_3O_{7-\delta}$, which is in precise agreement with experimental value of 0.661 determined from Curie constant. However for smaller degrees of disorder (fluences) p^2 estimated from (26) is considerably smaller than experimental data. Here we must note that (26) is valid only for rather small values of $R_{\rm loc}$ (deep in the localized region), when we can neglect overlap of localized states, and that Curie constants are determined in weakly disordered samples with considerably less accuracy than in case of strong disorder owing to smaller values of the Curie-Weiss contribution and the small interval of T where this contribution is observed (because of high-T_c). If we estimate T_c via (23) using R_{loc} determined from (26) and experimental data for Curie constant, then even at $\phi = 2\times10^{18}\,cm^{-2}$ we obtain $T_c \approx 15$ K. This discrepancy may be not so impressive taking into account the crude nature of our analysis as well as other important contributions to Coulomb suppression of T_c [11,28]. The question remains however, if there is also additional suppression of T_c by magnetic moments themselves, or why do they have so little influence on superconductivity?

CONCLUSION

The extreme sensitivity of high-T_c oxides to disordering may have several explanations, some among them based upon the idea of exotic types of pairing. However, here we presented another point of view: that this instability of T_c can be explained as due to Anderson localization. The quasi-two-dimensional nature of high-T_c oxides (with $T_c > 30$K) leads to significant enhancement of localization effects at relatively weak disorder. This may help to realize rather exotic situation of superconducting transition in the system of localized states (Anderson insulator). High-T_c oxides are especially promising in this respect owing to the small size of the Cooper pairs, so that there may be a wide region where the localization length is larger than the Cooper pair. There is some serious evidence that such a situation is actually realized in high-T_c superconductors disordered by radiation, although much more work is needed to confirm the specific predictions of the theory, as well as further theoretical analysis of microscopic mechanisms of T_c in highly-disordered systems. An important theoretical problem for further studies concerns the role of "statistical fluctuations" in quasi-two-dimensional systems. It was shown in Ref. 17 that these incipient inhomogeneities become important near the Anderson transition and roughly speaking lead to rather substantial smearing of the superconducting transition, as actually is seen in the experiments. We need further theoretical analysis of measurable characteristics such as critical fields, specific heat, etc.

Especially important may be experiments on radiation disordering in isotropic oxides like $Ba_{1-x}K_xBiO_3$, where a different behavior can be expected: these oxides may be more stable to disordering like A-15 or Chevrel phases.

Acknowledgements

The author is grateful to his friends doing experimental research on radiation disordering of superconductors at the Institute for Metal Physics in Sverdlovsk, especially to Prof. B. N. Goshchitskii and Drs. A. V. Mirmelstein, S. A. Davydov and A. E. Karkin with whom all the ideas expressed in this paper were extensively discussed and formulated.

REFERENCES

1. P.W.Anderson. Phys.Rev.109, 1492 (1958).
2. N.F.Mott. Metal-Insulator Transitions.Taylor&Francis,London,1974
3. Anderson Localization. Ed. by Y.Nagaoka & H.Fukuyama. Springer-Verlag, Berlin-NY, 1982.

4. M.V.Sadovskii. Usp.Fiz.Nauk 133, 223 (1981).
5. M.V.Sadovskii. Soviet Sci.Reviews-Phys.Rev. Ed. by I.M.Khalatni kov. Vol.7, p.1, Harwood Academic, NY, 1986.
6. P.A.Lee, T.V.Ramakrishnan, Rev.Mod.Phys. 57, 287 (1985).
7. J.Bardeen, L.N.Cooper, J.R.Schrieffer, Phys.Rev. 106,162 (1957).
8. A.A.Abrikosov, L.P.Gorkov. JETP 35, 1158, (1958); 36, 319 (1959).
9. P.W.Anderson. J.Phys.Chem.Solids, 11, 26 (1959).
10. L.N.Bulaevskii, M.V.Sadovskii, JETP Lett. 39, 524 (1984).
11. L.N.Bulaevskii, M.V.Sadovskii, J.Low Temp.Phys. 59, 89 (1985).
12. A.Kapitulnik, G.Kotliar. Phys.Rev.Lett. 54, 473 (1985).
13. G.Kotliar, A.Kapitulnik. Phys.Rev. B33, 3146 (1986).
14. M.Ma, P.A.Lee. Phys.Rev. B32, 5658 (1985).
15. L.N.Bulaevskii, M.V.Sadovskii, JETP Lett. 43, 76 (1986).
16. L.N.Bulaevskii, A.A.Varlamov, M.V.Sadovskii. Fiz.Tverd.Tela, 28, 1799 (1986).
17. L.N.Bulaevskii, S.V.Panjukov, M.V.Sadovskii. JETP,92,672(1987).
18. B.A.Aleksashin et al. JETP 95, 678 (1989).
19. S.A.Davydov et al. Int.J.Mod.Phys. B33, 87, (1989).
20. B.N.Goshchitskii et al. Proc.Int.Seminar on High-T Superconductivity, Dubna, USSR, World Scientific, Singapore, 1989.
21. B.N.Goshchitskii et al. Proc.M SHTSC,Stanford,USA, Physica C.
22. J.Tateno, N.Masaki, A.Iwase. Phys.Rev.Lett. A138, 313 (1989).
23. V.V.Moshchalkov. Physica C156, 473 (1988).
24. V.N.Prigodin, Yu.A.Firsov. J.Phys.C17, L979 (1984).
25. L.P.Gorkov. JETP 37, 1407 (1959).
26. B.Shapiro, E.Abrahams. Phys.Rev. 24, 4025 (1981).
27. E.A.Kotov, M.V.Sadovskii. Phys.Metals-Metallogr. 60,22(1985).
28. P.W.Anderson, K.A.Muttalib, T.V.Ramakrishnan, Phys.Rev.B28, 117 (1983).
29. N.F.Mott. Phil.Mag. 24, 935 (1971).
30. V.L.Berezinskii, L.P.Gorkov. JETP 77, 2498 (1979).

LOCAL MOMENTS NEAR THE METAL-INSULATOR TRANSITION

Subir Sachdev
Center for Theoretical Physics, P.O. Box 6666
Yale University, New Haven, CT 06511

This paper reviews recent progress in understanding the metal-insulator transition in a system of spin-1/2 interacting electrons in the presence of a non-magnetic random potential. Using results of recent experiments in doped semiconductors, it is argued that the metallic state near the transition can be described by a phenomenological two-fluid model of local moments and itinerant quasiparticles. A mean-field calculation on a disordered Hubbard model which justifies this two fluid model is described. Recent advances in the non-linear sigma model description of the transition are also discussed.

The subject matter of the lecture presented by the author overlapped considerably with previous talks by the author [1] and R.N. Bhatt [2]. We will not repeat here the material covered in these talks which has been published elsewhere [1, 2]; we will concentrate, instead on reviewing some recent theoretical work. Upon combining the present paper with these previous reviews [1, 2], the reader will have a fairly complete record of the talk presented by the author.

The experimental motivation for this investigation comes from experiments in doped semiconductors like $Si : P$ and $Si : B$. The dopant electrons (or holes) in these materials can be described quite accurately by the following Hamiltonian

$$H = -\sum_{i,j} t_{ij} c_{i\sigma}^{\dagger} c_{j\sigma} + U \sum_{i} n_{i\uparrow} n_{i\downarrow} + \sum_{i} (\epsilon_i - \mu) c_{i\sigma}^{\dagger} c_{i\sigma} \tag{1}$$

Here the sites i, j represent the donor orbitals and are placed at positions $\mathbf{R}_i, \mathbf{R}_j$ distributed randomly in a system of volume V. The hopping matrix elements t_{ij} are exponentially decaying functions of $|\mathbf{R}_i - \mathbf{R}_j|$

$$t_{ij} = t_0 \exp(-|\mathbf{R}_i - \mathbf{R}_j|/a_B) \tag{2}$$

where the Bohr-radius a_B is approximately 20 Å. The Hubbard repulsion U acts between two electrons on the same site. The on-site energy ϵ_i is also a random variable. The chemical potential μ is adjusted to control the filling factor of the system. Uncompensated doped semiconductors are at or very close to half-filling. The quantity which is varied experimentally is the concentration n of the electrons. At a critical concentration $n = n_c$ the system is observed to undergo a transition from an insulator ($n < n_c$) to a metal ($n > n_c$).

REVIEW OF EXPERIMENTS

The experimentally observed properties of this transition have been reviewed elsewhere ([1, 2] and references therein). For completeness, we recall here a few important features. The most striking feature of the experimental data [3] is the difference in the properties of the conductivity and spin-susceptibility in the metallic system. The conductivity, which vanishes at $T = 0$ and $n = n_c$ is only weakly temperature dependent, changing by about 15% between $10K$ and $0.1K$. In contrast the spin-susceptibility increases by a factor of 10 between

$T = 10K$ and $T = 0.1K$ for $n = 1.09n_c$. The temperature dependence of the susceptibility is in fact rather similar to that of the insulator, which has an even greater enhancement. This difference leads one to suspect rather different physical effects are responsible for the apparent continuous vanishing of the conductivity at $n = n_c$, and for the enhancement of the spin-susceptibility. In fact, it was found that all of the thermodynamic data on the metal near $n = n_c$ could be described by a phenomenological two-fluid model. The two fluids are (i) an itinerant fermi liquid of quasiparticles and (ii) localized electrons spins interacting with each other via an antiferromagnetic exchange. The spin-susceptibility χ and specific heat C are predicted to have the following temperature dependences

$$\begin{aligned} \frac{C}{T} &= N(E_F) + \left(\frac{T}{T_0}\right)^{-\alpha} \\ \chi &= N(E_F) + \beta(\alpha)\left(\frac{T}{T_0}\right)^{-\alpha} \end{aligned} \tag{3}$$

The two terms represent the contributions of the itinerant fluid and the local moments respectively. The temperature dependence of the local moment susceptibility and specific heat are deduced from the Bhatt-Lee model [4] of the insulating state. The density of states $N(E_F)$ and β are known constants and the only unknown is T_0, which is a measure of the fraction of electrons in localized moments. With this one adjustable parameter, the equations above provide a surprisingly good fit to the experiments [3]. An important *prediction* of these phenomenological equations is that the Wilson ratio $T\chi/C$ remains *finite* at $T = 0$: the experimental data agree well with this prediction. Nuclear magnetic resonance experiments of Alloul and Dellouve [5] have also provided direct experimental evidence for the presence of localized moments at densities $n < 2n_c$.

A major question has been left unresolved in the phenomenological discussion of the experimental results: how can a metallic system in which all states near the Fermi- level are extended, support the existence of localized moments ? Density fluctuations leading to islands of insulating sites cannot be the answer; simulations show that at densities near $n = n_c$, the fluctuations strong enough to create localized states do not occur often enough to account for the experimental results. It is this question which the bulk of the rest of the paper will attempt to address. The discussion is a review of results contained in Ref [6]

SINGLE-IMPURITY

To set the stage let us consider a simplified version of the Hamiltonian $H = H_i$ with a *single* impurity at a site $\mathbf{R}_0$. All the sites are placed on a regular lattice and the hopping matrix elements are given by

$$t_{ij} = \begin{cases} t & i,j \text{ nearest neighbors, } \mathbf{R}_i \neq \mathbf{R}_j \neq \mathbf{R}_0 \\ w & \mathbf{R}_i \neq \mathbf{R}_j \text{ and } \mathbf{R}_i = \mathbf{R}_0 \text{ or } \mathbf{R}_j = \mathbf{R}_0 \\ 0 & \text{otherwise} \end{cases} \tag{4}$$

The on-site energies ϵ_i are all chosen equal and the chemical potential is chosen so that the system is close to half-filling. We assume that U is always small enough, or the temperature T large enough so that the system is in a metallic state and all spin-density wave instabilities can be safely ignored. The model under consideration is closely related to the Anderson-Wolff [7] models of local moment formation in transition metals. By analogy with these models and subsequent work [8] we can understand the properties of H_i.

A mean-field diagonalization of H gives rise to a *resonance* near the Fermi-level. The width, Γ, of this resonance is of order

$$\Gamma \sim \frac{w^2}{t} \tag{5}$$

for small w. The consequences of the existence of this resonance can be elucidated by considering the temperature dependence of χ_0, the local spin susceptibility of the impurity site. At very high temperatures $T \gg U$ we have

$$\frac{\chi_0}{(g\mu_B)^2} = \frac{1}{8kT} \tag{6}$$

where g is gyromagnetic ratio, μ_B is the Bohr magneton. At lower temperatures, two different types of behavior can occur depending upon the value of U/Γ:

(a) Local moment regime For $U/\Gamma > \alpha$ (where α is a numerical constant of order unity) we find

$$\frac{\chi_0}{(g\mu_B)^2} = \frac{1}{4kT} \tag{7}$$

This susceptibility if a factor of 2 larger than the high- temperature result and is simply the Curie susceptibility of a single free spin. At very low temperature $T \sim T_K$, the Kondo effect takes over and the susceptibility is quenched.

(b) Fermi liquid regime For $U/\Gamma < \alpha$ the local moment does not form and the susceptibility becomes temperature independent

$$\frac{\chi_0}{(g\mu_B)^2} \sim \frac{1}{\Gamma} \tag{8}$$

The crucial point for our purposes is that measurements on all of the doped semiconductors are performed at temperatures well above T_K. Thus the enhancement of the susceptibility in the local moment regime is very relevant. We emphasize that no localized state was necessary for the development of the local moment. Instead all that was required was a sufficiently narrow resonance at the Fermi level. Moreover the Anderson-Clogston compensation theorem [7] implies that the *spin density* associated with the local moment is strongly localized at the impurity site.

FULLY DISORDERED SYSTEM

We will now use the insight gained from the single-impurity system to investigate the properties of the fully random H. A suitable mean-field formalism is provided by the effective field method of Feynman and de Gennes [9] This involves finding the best single- particle effective Hamiltonian

$$H_{eff} = -\sum_{ij} t_{ij} c^{\dagger}_{i\sigma} c_{j\sigma} + \sum_{i\sigma} (\tilde{\epsilon}_i - \mu) c^{\dagger}_{i\sigma} c_{i\sigma} + \sum_i \mathbf{h}_i \cdot \mathbf{S}_i \tag{9}$$

where $\mathbf{S}_i$ is the spin of the electron at site i. The on-site energy $\tilde{\epsilon}_i$ and magnetic field $\mathbf{h}_i$ are varitional parameters. A non-zero value of $\mathbf{h}_i$ implies the formation of a local moment at the site i. We shall be particular interested in the spatial distribution of $\mathbf{h}_i$. The variational parameters must be determined by minimizing the following quantity

$$-kT \ln \mathrm{Tr} \exp\left(-\frac{H_{eff}}{kT}\right) + \langle H - H_{eff} \rangle_{eff} \tag{10}$$

where the second expectation value is taken in the canonical ensemble defined by H_{eff}. This problem is too complicated to be solved numerically in the present form. As we are interested in the initial instability to local moment formation, we expand the effective free energy in powers of $\mathbf{h}_i$. For U small enough, $\mathbf{h}_i$ is zero, and the saddle point condition reduces to

$$\tilde{\epsilon}_i = \epsilon_i + U \sum_{\alpha} |\Psi_\alpha(i)|^2 f(\lambda_\alpha) \tag{11}$$

which is simply the familiar Hartree-Fock equation for the Hubbard model. Here $\Psi_\alpha(i)$ is the exact eigenstate of H_{eff} with $\mathbf{h}_i = 0$, and λ_α is the eigenvalue.

$$H_{eff}(\mathbf{h}_i = 0)\Psi_\alpha = \lambda_\alpha \Psi^\alpha \tag{12}$$

The same set of exact eigenstates were introduced by Abrahams *et. al.* [10] in their analysis of interaction effects in disordered systems. Expanding the effective free energy in powers of $\mathbf{h}_i$, we obtain

$$F_{eff}(\mathbf{h}_i) = F_{eff}(\mathbf{h}_i = 0) + \sum_{ijk} \chi_{ij}(\delta_{jk} - U\chi_{jk})(\mathbf{h}_i \cdot \mathbf{h}_k) + \ldots \tag{13}$$

where

$$\chi_{ij} = -\sum_{\alpha,\beta} \Psi_\alpha(i)\Psi^*_\beta(i)\Psi_\beta(j)\Psi^*_\alpha(j)\frac{f(\lambda_\alpha) - f(\lambda_\beta)}{\lambda_\alpha - \lambda_\beta} \tag{14}$$

where χ_{ij} is the susceptibility matrix of free electrons described by $H_{eff}(\mathbf{h_i} = 0)$. The local magnetic field $\mathbf{h}_i$ will acquire a non-zero expectation value when the quadratic form in Eqn (13) first has a negative eigenvalue. It is clear from the expression for F_{eff} that the eigenvectors of the quadratic form are the eigenvectors of the spatial matrix χ_{ij}:

$$\sum_j \chi_{ij} m_a(j) = \kappa_a m_a(i) \tag{15}$$

Expanding the $\mathbf{h}_i$ in term of the complete set $m_a(i)$ we see from Eqn (13) that the system first becomes unstable to local moment formation when the largest susceptibility eigenvalue κ_a exceeds $1/U$. Moreover the spatial distribution of the magnetization will be proportional to the corresponding eigenvalue $m_a(i)$.

This is as far as general considerations can carry us. We now have to use numerical simulations to study the behavior of the system. The results of such a numerical study are presented in Ref. [6] and the reader is referred to that paper for greater details. The mean field equation (11) for a random Hamiltonian H was solved by iteration. The corresponding susceptibility matrix χ_{ij} was diagonalized exactly and the structure of its eigenvectors studied. The eigenvectors of χ_{ij} and $H_{eff}(\mathbf{h}_i = 0)$ were characterized by the inverse participation ratio

$$\begin{aligned} P_H &= \langle |\Psi_\alpha(i)|^4 \rangle \\ P_\chi &= \langle |m_a(i)|^4 \rangle \end{aligned} \tag{16}$$

Here the average is taken over several samples over states near the Fermi level for P_H and over the largest 5 eigenvalues of χ for P_χ. System sizes as large as 400 sites were studied. The results unambiguously showed that the eigenvalues of χ_{ij} - $m_a(i)$ were much more strongly localized that the one-electron eigenstates $\Psi_\alpha(i)$. In particular the $m_a(i)$ were essentially localized on a few sites, while the extent of the $\Psi_\alpha(i)$ was limited by the system size. This localization of the $m_a(i)$ may be viewed as a generalization of the Anderson-Clogston compensation theorem [7].

The physical implication of the results above is clear. Weakly disordered systems contain very local instabilities towards the formation of local moments. These instabilities are associated with the eigenvalues of χ_{ij} and do not require a localized Hartree-Fock eigenstate. The local moments, once formed will lead to a dramatic enhancement of the spin-susceptibility and specific heat of the disordered metal. Moreover since the same spins are responsible for both enhancements, the Wilson ratio $T\chi/C$ is expected to be finite at $T = 0$.

FIELD THEORETIC RESULTS

While the results discussed above do clarify the physical picture they do not make any specific predictions on the critical properties of the metal-insulator transition. The only available method is the field-theoretic non-linear sigma model of interacting electrons in weak disorder [11]. As the method is intrinsically a weak disorder expansion in $d = 2+\epsilon$ dimensions it will be difficult for it to properly capture the local moment instabilities discussed above: these instabilities can occur at weak disorder and require the use of exact eigenstates.

The one-loop renormalization group results of Finkelstein and the physical arguments of Castellani *et. al.* can be encapsulated in the following scaling form for the conductivity σ

$$\sigma(\omega, \xi) = \frac{1}{\xi^{d-2-\theta}} F(\omega \xi^{d+x}) \tag{17}$$

where $\xi \sim (n - n_c)^{-\nu}$ is the correlation length and F is a universal function. At finite temperature we expect that ω can be replaced by T. An attempt to fit the temperature dependent conductivity in $Si : P$ with this scaling form was unsuccesful [12]. The spin diffusivity D_s, χ and C/T have the leading divergences

$$\chi \sim \xi^{\theta+x} \qquad D_s \sim \xi^{-\theta-x} \qquad \frac{C}{T} \sim \xi^{x} \tag{18}$$

Note that the Wilson ratio is predicted to diverge as ξ^{θ}, which is also not in accord with experiments. The one-loop values of the exponents are $\nu = 1/\epsilon$, $\theta = \epsilon$ and $x = 3\epsilon$.

Recently Belitz and Kirkpatrick [13] have analyzed the higher order corrections to the Finkelstein field theory. They found two important results
(i) the metal-insulator fixed point of Finkelstein and Castellani *et. al.* is in fact *suppressed* and does not describe the ultimate critical behavior. The system however does not notice the suppression until it is exponentially close (for small ϵ) to the 'fixed' point. Thus *e.g.* the temperature would have to be of order

$$T \sim \exp\left(-\frac{c}{\epsilon^{3/2}}\right) \tag{19}$$

The ultimate fixed point theory remins inaccessible in the ϵ expansion.
(ii) Because the suppression is exponentially small, the scaling equations (17) and (18) can meaningfully define *effective* exponents. Belitz and Kirkpatrick [13] found that

$$\theta = d - 2 \tag{20}$$

to all orders in ϵ. Comparing with Eqn (17) this indicates that the conductivity σ is *not* driven towards zero. Thus in the scaling region, the spin diffusivity is suppressed, the spin susceptibility enhanced, but the conductivity remains finite. This picture is remarkably similar to that developed in the mean-field calculations discussed above. However it is not clear that the critical region will be large when $\epsilon = 1$, which is necessary for the exponents to have any meaning.

CONCLUSIONS

It is heartening that the two rather different approaches discussed in this review appear to obtain similar results. Neither of the approaches has however so far come up with a theory of the metal insulator transition. It is clear, however, that an important part of the physics of the dirty metallic state is the formation of localized magnetic moments. The length scale associated with the formation of these moments is much smaller that the correlation length ξ. It is clear that the ultimate theory of the transition will have to incorporate this length scale at an initial stage.

I would like to acknowledge contributions of my collaborators M. Milovanovic and R.N. Bhatt.

REFERENCES

[1] S. Sachdev, R.N. Bhatt, and M.A. Paalanen, J. Appl. Phys. **63**, 4285 (1988)

[2] R.N. Bhatt, M.A. Paalanen, and S. Sachdev, Journal de Physique, Colloque C8, 1179 (1988).

[3] M.A. Paalanen, J. Graebner, R.N. Bhatt and S. Sachdev, Phys. Rev. Lett. **61**, 597 (1988).

[4] R.N. Bhatt and P.A. Lee, Phys. Rev. Lett. **48**, 344 (1988).

[5] H. Alloul and P. Dellouve, Phys. Rev. Lett. **59**, 578 (1987).

[6] M. Milovanovic, S. Sachdev and R.N. Bhatt, Phys. Rev. Lett. **63**, 82 (1989).

[7] P.W. Anderson, Phys. Rev. **124**, 41 (1961); P.A. Wolff, Phys. Ref. **124**, 1030 (1961).

[8] F.D.M. Haldane Phys. Rev. Lett. **40**, 489 (1978); H.R. Krishnamurthy, J.W. Wilkins and K.G. Wilson, Phys. Rev. **B 21**, 1003 (1980); **B 21**, 1044 (1980).

[9] R.P. Feynman, *Statistical Physics*, (Benjamin, New York, 1972); P. G. de Gennes, *Superconductivity of Metals and Alloys*, (Benjamin, New York, 1972).

[10] E. Abrahams, P.W. Anderson, P.A. Lee, and T.V. Ramakrishnan, Phys. Rev. **B 24**, 6783 (1981).

[11] A.M. Finkelstein Zh. Eksp. Teor. Fiz. **84**, 166 (1983) [Sov. Phys. JETP **57**, 97 (1983)]; Z. Phys. B **56**, 189 (1984); C. Castellani, B.G. Kotliar, P.A. Lee, Phys. Rev. Lett. **56**, 1179 (1986).

[12] S. Sachdev, unpublished.

[13] D. Belitz and T.R. Kirkpatrick, Nucl. Phys. **B 316**, 509 (1989); T.R. Kirkpatrick and D. Belitz, Phys. Rev. **B 40**, 5227 (1989); D. Belitz and T.R. Kirkpatrick, Phys. Rev. Lett. **63**, 1296 (1989); T.R. Kirkpatrick and D. Belitz, U. of Maryland preprint.

THEORY OF SUPERCONDUCTOR-INSULATOR TRANSITIONS IN DISORDERED FILMS

Matthew P.A. Fisher
IBM Research Division
T. J. Watson Research Center
Yorktown Heights, NY 10598

A scaling theory for the zero temperature $(T = 0)$ superconductor to insulator transition in disordered films is described. Right at the transition, the system is predicted to be "metallic", with a resistance per square having a finite, non-zero value at $T = 0$. This value, moreover, should be universal, independent of all microscopic details. In the presence of an applied magnetic field, an additional $T = 0$ superconductor-insulator transition is accessible at which both resistivities ρ_{xx} and ρ_{xy} should be universal.

INTRODUCTION

In the past decade substantial progress has been made in understanding Anderson localization in electron systems and the metal-insulator transition in dirty interacting Fermion systems[1]. A central conclusion which has emerged, is that in two dimensions (2D) even weak disorder localizes all states: A true 2D metallic phase with non-zero conductivity is not possible at $T = 0$. In this paper, I describe some recent results on related phenomena in bosonic systems[2] In particular, I consider possible superconducting to insulating transitions in disordered systems, the direct bosonic analog to the metal-insulator transition. Attention is focussed on the 2D case, since a number of recent experiments[3] have probed this transition by systematically varying the thickness of amorphous films. In this way the subtle interplay between localization and superconductivity can be examined.

The first part of the paper describes a scaling theory of the $T = 0$ superconductor-insulator transition[2]. Quite remarkedly, as in Anderson localization, two dimensions again emerges as a special dimension. In (and only in) 2D, right at the transition separating superconducting from insulating behavior, the system should be "metallic", with a finite, non-zero resistivity. Thus, perhaps paradoxically, disordered (unpaired) electrons in 2D cannot diffuse at $T = 0$, but Cooper pairs, perched on the brink of superconductivity, can. This 2D metallic state, moreover, is predicted to have a universal resistance per square[4], independent of all microscopic details, depending only on the universality class of the superconductor-insulator transition. These results are consistent with recent experiments on amorphous thin-film superconductors[3].

In the second section, the effects of an applied magnetic field on the low temperature properties of dirty superconducting films are considered. It is argued that an additional $T = 0$ superconductor-to-insulator transition can be accessed, by varying the strength of the applied field. At low fields, vortices introduced by the field are localized and do not dissipate at $T = 0$, so that the film can be superconducting, in a so-called vortex-glass phase[5]. With increasing field, the vortices should de-localize and the film undergo a $T = 0$ transition into an insulator. Right at this transition, in addition to a universal resistivity ρ_{xx}, the Hall resistivity, ρ_{xy}, should be finite and universal.

SCALING THEORY OF TRANSITION

It has recently been argued that the $T = 0$ superconductor-insulator phase transition in thin amorphous[2], films such as those of ref. 3, can be correctly described by a model of charge 2e bosons moving in a 2D random potential[6]. In the superconducting phase the electrons have bound to form Cooper pairs and a description of the low energy physics in terms of charge 2e composite bosons is presumably valid. In the insulating phase, where pairing is destroyed and the individual electrons are presumably localized by the disorder, such a description is inadequate. It seems nonetheless likely that the asymptotic critical properties of the transition are insensitive to the obvious difference between bosonic and fermionic insulating phases, i.e., between the bose glass[7] and fermi glass[1] phases. Indeed, Bose condensation and the superconducting transition in pure systems at finite T belong, e.g., in the same universality class, the difference between the normal bosonic and fermionic phases notwithstanding. Moreover, in 1D the $T=0$ superconductor-insulator transition can be studied directly in terms of a model of electrons with a BCS attractive interaction moving in a random potential. It is found[8] that the critical behavior of this $T = 0$ phase transition is in the same universality class as the superfluid-insulator transition in a model of repulsively interacting bosons, representing the Cooper pairs, moving in a random potential. One therefore expects that the experimentally relevant superconductor-insulator transition in amorphous films can be properly described in terms of charge 2e bosons.

Consider then the Hamiltonian for a system of charged bosons: $H = H_0 + H_1$ with,

$$H_0 = \int d^d x[(\hbar^2/2m) \,|\, \nabla\psi \,|^2 + U(x) \,|\, \psi(x) \,|^2] \tag{1a}$$

$$H_1 = \int d^d x d^d x' V(x - x')[\,|\, \psi(x) \,|^2 - n_0][\,|\, \psi(x') \,|^2 - n_0], \tag{1b}$$

and $\psi(x)$ the usual boson field operator. Here $V(x) = (2e)^2/|x|$ is a Coulomb interaction between the bosons, with n_0 a compensating positive charge background (charge neutrality fixing the boson density at n_0), and $U(x)$ a random potential. Since the 2D screening length is typically macroscopic for thin films (λ^2/d, with d=film thickness), coupling to a fluctuating gauge field in (1) can be ignored.

As the boson density n_0 is increased through some critical density n_c, a $T = 0$ phase transition from a localized bose glass phase[7] to a superconducting phase with $< \psi > \neq 0$ is expected. It is convenient to introduce a parameter $\delta = n_0 - n_c$, which measures the distance to this $T = 0$ transition. In experiments on real amorphous films one could take $\delta = d - d_c$, with d the film thickness, or $\delta = R_N - R_{N,c}$, where R_N is the film's "normal state" resistance per square taken at some convenient reference temperature above the bulk transition temperature. Provided this $T = 0$ transition is continuous, it is characterized by a superconducting correlation length which diverges as $\xi \sim \delta^{-\nu}$, with an exponent satisfying the bound[7] $\nu \geq 2/d$. There is also a characteristic frequency Ω which vanishes at criticality as $\Omega \sim \xi^{-z}$ where z is the dynamical exponent.[7] Near the T=0 critical point all frequencies and the temperature scale[7] with Ω. Thus the Kosterlitz-Thouless[13] transition temperature T_c , at which the (2D) system becomes normal for $\delta > 0$, will scale as

$$T_c \sim \delta^{z\nu}, \tag{2}$$

for $\delta \to 0^+$.

In the superconducting phase the second sound (phonon) mode has a plasmon- like dispersion relation[9,10] $\omega \sim k^{(3-d)/2}$ due to the long-range Coulomb interaction. This mode can be described by an effective imaginary time action[7] which depends only on the phase ϕ of the order parameter $\psi = |\psi| \exp(i\phi)$,

$$S_\phi = (1/2)\int d^dk d\omega[(\rho_s\hbar/2m)k^2 + \hbar\omega^2 |k|^{d-1}/e_R^2] |\phi(k, \omega)|^2. \tag{3}$$

Here ρ_s is the fully renormalized superfluid density and e_R a "fully renormalized" charge, $e_R^2 \equiv \lim_{k \to 0} |k|^{d-1}/C_{nn}(k, \omega = 0)$, with $C_{nn}(k, \omega) \equiv \delta < n(k, \omega) > /\delta\mu(k, \omega)$ the density-density response function. Near the T=0 superconductor-insulator transition[11] ρ_s vanishes as $\rho_s \sim \xi^{-(d+z-2)}$. This essentially follows from power counting in the first term in (3), noting that both S_ϕ and $\phi(x)$ are dimensionless[7]. Likewise, the second term in (3) implies that the charge e_R should scale near the transition as $e_R^2 \sim \xi^{1-z}$. It can be argued[2] that at the superconductor to localized Bose-glass transition, e_R will have a finite value. This implies

$$z = 1, \tag{4}$$

which should hold in all dimensions. Eqn. (4) is the generalization to charged systems of the relation $z = d$, which has been argued to hold[7] at the T=0 superfluid-insulator transition in disordered charge neutral boson systems, such as ^{4}He in porous media.

Scaling of the frequency-dependent conductivity near the superconductor -- insulator transition can be obtained from the relation[12]

$$\sigma(\omega) = (2e)^2\rho_s(-i\omega)/(-im\omega) , \tag{5}$$

where $\rho_s(\omega)$ is a generalized frequency dependent superfluid density defined in terms of a current-current correlation function. Since all frequencies should be scaled by the characteristic frequency Ω near the transition one can write the scaling relation

$$\rho_s(\omega, \xi) = \xi^{-d}(\xi/a)^{2-z}\tilde{\rho}_s(\omega/\Omega), \tag{6}$$

where $\tilde{\rho}_s$ is an appropriate dimensionless scaling function and $\Omega = (\hbar/ma^2)(a/\xi)^z$, with a a short distance cutoff. For $x \equiv \omega/\Omega \to 0$ we must recover the result $\rho_s \sim \xi^{-(d+z-2)}$, so $\tilde{\rho}_s(x)$ must approach a constant. The form as $x\to\infty$ is set by the requirement that at criticality, where both ξ and $\Omega^{-1} \sim \xi^z$ are infinite, $\rho_s(\omega, \xi = \infty)$ is finite:

$\tilde{\rho}_s(x) = c_d x^{(d+z-2)/z}$, with c_d a dimensionless constant. Combining this with (5) implies that at criticality

$$\sigma(\omega, \xi = \infty) = c_d (e^2/\hbar) a^{2-d} (-i\hbar\omega/ma^2)^{(d-2)/z} . \qquad (7)$$

Similarly, at the transition, the finite temperature d.c. conductivity should scale as $\sigma(T, \xi = \infty) \sim T^{(d-2)/z}$.

Eqn. (7) indicates that in 2D the $T = 0$ conductivity at criticality is a finite constant, $c_d e^2/\hbar$, in the d.c. limit. (Logarithmic corrections in ω are not expected in 2D, since $d = 1$ is the lower critical dimension for the transition[7].) Thus, at the superconductor-insulator transition the system exhibits true metallic conduction at T=0, something not possible in 2D normal fermion systems. The Cooper pairs, poised on the brink of becoming superconducting, are capable of ordinary diffusion.

Likewise, the resistance per square at the transition $R^* \equiv 1/\sigma(\omega = 0, \xi = \infty)$, when expressed in units of h/e^2, is a pure number. Since this number is given by the $k = \omega = 0$ limit of a (current-current) response function evaluated at the critical point[12], standard renormalization group (RG) arguments imply that, like critical exponents, it is universal: Its value will depend only on the universality class of the transition, and not on microscopic details.

EFFECTS OF MAGNETIC FIELD

In this section I consider the effects of an applied magnetic field on the low temperature properties of thin amorphous films. Consider a film which in zero field is superconducting below some Kosterlitz-Thouless transition temperature, T_c. In the superconducting state, an applied field will induce vortices in the Cooper pair wave function, all of the same sign. These vortices will interact with one another logarithmically[13] out to the two-dimensional screening length, $\lambda_{2D} = \lambda^2/d$, which in practice is usually macroscopic[14]. In the presence of disorder, which will tend to pin the vortices, the Abrikosov vortex-lattice phase will be destroyed[15]. At finite temperatures the film will then not be a true zero-resistance superconductor: Thermally activated vortex creep will lead to a dissipative (linear) resistance[16]. What happens when the system is cooled to $T = 0$? A classical description of vortex dynamics would predict that all vortex motion ceases in this limit, and the resistance should vanish. This $T = 0$ superconducting phase will exhibit Edwards-Anderson spin-glass type order[17] in the boson field ψ, and as a result is referred to as a "vortex-glass"[18]. But what about quantum fluctuations of the vortices?

Recently a theoretical framework has been developed for treating vortex quantum fluctuations in 2D bosonic systems[19] It is found that the vortices are in fact themselves bosonic objects, with a dynamics not altogether different from that of the Cooper pairs. At zero temperature these vortices can be localized by inhomogeneties, just as regular bosons or fermions can. Thus, the 2D superconducitng vortex-glass phase at $T=0$ should survive quantum fluctuations.

As the applied magnetic field is increased, though, a tantalizing possibility arises. The quantum gas of point vortices, increasing in density with applied field, should eventually "bose" condense, at some critical field H_c. (It turns out that this condensation is only possible at $T=0$.) What properties does a "superfluid" of vortices possess? Since vortex motion causes voltage fluctuations, this phase is, not surprisingly, an insulator with infinite resistance. This phase can alternatively be described in more conventional terms, as a localized fermionic (or Cooper pair) insulator. In any case, it should be possible, by simply varying the strength of the external field, to tune through this (T=0) superconductor-insulator transition.

Films with low (H=0) Kosterlitz-Thouless transition temperatures, will presumably have correspondingly low "critical fields", H_c , above which they are insulating (at $T=0$). The most natural scenario is that as the superconductivity is weakened, by making the film thinner, say, T_c and H_c will vanish together at the (multi-critical) $T=H=0$ superconductor-insulator transition. In this limit, the critical field will vanish as $H_c \sim \xi^{-2}$, where ξ is the superconducting correlation length of the $T=H=0$ transition, introduced before Eqn. 2. Combining this with (2) implies that near this transition

$$H_c \sim T_c^{2/z}. \tag{8}$$

This provides a direct way to measure the dynamical exponent, and should allow for a check on the theoretical prediction, $z=1$, for charged systems.

I now address the expected properties of the system near, and at, the $H \neq 0$ vortex-glass superconductor-to-insulator transition. Most of the scaling results discussed in Sec. 2 apply equally well to this transition. Once again, near the transition one expects a diverging superconducting correlation length, $\tilde{\xi} \sim (H-H_c)^{-\tilde{\nu}}$, and vanishing characteristic frequency, $\tilde{\Omega} \sim \tilde{\xi}^{-\tilde{z}}$, where the tilde's are used to differentiate these quantities from their $H=0$-transition analogs. One expects $\tilde{z}=1$, $\tilde{\nu} \geq 2/d$ (but presumably different from ν), and a universal resistivity at the transition (in 2D). Since the critical fixed point describing this transition is different than the $H=0$- transition fixed-point, different values for the universal resistivities would be expected.

Due to the applied field, a Hall resistivity ρ_{xy} is also expected at the $H \neq 0$ vortex-glass to insulator transition. Like ρ_{xx}, ρ_{xy} should have a universal value at the transition. In the scaling regime close to this transition (ie. $H \to H_c$ and $T \to 0$) both resistivities should satisfy scaling forms,

$$\rho_{x\alpha} = (h/4e^2)\tilde{R}_{x\alpha}[c(H-H_c)/T^{1/\tilde{z}\tilde{\nu}}], \tag{9}$$

with $\alpha = x,y$ and, c a non-universal constant. Here $\tilde{R}_{x\alpha}[X]$ are dimensionless scaling functions, which take constant values at $X \equiv c(H-H_c)/T^{1/\tilde{z}\tilde{\nu}} = 0$, and diverge or vanish exponentially in X (to some power) for large positive and negative X, respectively. Plotting $\tilde{R}_{xx}$ <u>versus</u> $\tilde{R}_{xy}$ would eliminate the (shared) non-universal constant c, giving a unique universal function, $R_{xx}(R_{xy})$. (A similar trick is used in plotting resistivities near the phase transition between plateaus in the integer quantum Hall effect[20].)

Can one estimate the universal resistivities, $\rho^*_{x\alpha} \equiv (h/4e^2)\tilde{R}_{x\alpha}[X=0]$, at this transition? As alluded to above, Cooper pairs and vortices play a dual role near the vortex-glass -to-insulator transition: In the superconducting phase the Cooper pairs have Bose condensed, whereas the vortices are condensed in the insulating phase. It turns out that the $T=0$ vortex-glass to localized Bose-glass transition, in a model system of <u>logarithmically</u> interacting bosons moving in a random potential, is, in fact, <u>self-dual</u>[21]. This can be used to show that the universal resistivities at the transition satisfy,

$$(\rho^*_{xx})^2 + (\rho^*_{xy})^2 = (h/4e^2)^2. \tag{10}$$

Since Cooper pairs <u>do not</u> interact logarithmically, but as $1/r$, (10) will presumably not be identically satisfied in a real physical system. I suspect, though, that a more appropriate model, with $1/r$ interactions, might well give a value not substantially different than in (10) (although presumably not a simple rational times $h/4e^2$). In any event, (10) should serve as a useful guide for experimental investigations of this transition.

The scaling theory, outlined herein, is based on a paper in collaboration with G. Grinstein and S.M. Girvin, and is a generalization of earlier work with D.S. Fisher and P.B. Weichman. I am indebted to them all for their help and support. I also thank D.H. Lee and D.A. Huse for numerous fruitful conversations.

REFERENCES

1. E.g., P.A. Lee and T.V. Ramakrishnan, Rev. Mod. Phys. 57, 287 (1985).
2. Part of this work has been reported elsewhere: M.P.A. Fisher, G. Grinstein and S.M. Girvin, submitted to publication (1989).
3. D.B. Haviland, Y. Liu, and A.M. Goldman, Phys. Rev. Lett. 62 , 2180 (1989); D.B. Haviland, Thesis, University of Minnesota (1989).
4. Our explanation of the finite, universal resistance at the T=0 superconductor -- bose glass transition is based on a scaling treatment of critical fluctuations, and so differs markedly from the explanation based on a single pair of ballistically-moving electrons advanced by T. Pang, Phys. Rev. Lett. 62, 2176 (1989).
5. See M.P.A. Fisher, Phys. Rev. Lett. 62, 1415 (1989) and D.S. Fisher, M.P.A. Fisher and D.A. Huse, in preparation (1989).
6. Earlier experiments on granular films showed "flat resistive tails", in contrast to a sharp T=0 superconductor-insulator transition seen in the amorphous films in Ref. 3. See H.M. Jaeger, D.B. Haviland, B.G. Orr and A.M. Goldman, Phys. Rev. B40, 182 (1989). Charged boson models might be insufficient to explain the features of these granular films.
7. See, e.g., M.P.A. Fisher, P.B. Weichman, G. Grinstein, and D.S. Fisher, Phys. Rev. B40, 546 (1989), and references therein.
8. T. Giamarchi and H.J. Schulz, Europhys. Lett. 3, 1287 (1987), and Phys. Rev. B37, 325 (1988).
9. T.V. Ramakrishnan, Physica Scripta T27, 24 (1989).
10. M.P.A. Fisher and G. Grinstein Phys. Rev. Lett. 60, 208 (1988).
11. M. Ma, B.I. Halperin and P.A. Lee, Phys. Rev. B34, 3136 (1986).

12. See, e.g., A.L. Fetter and J.D. Walecka, Quantum Theory of Many- Particle Systems (McGraw-Hill, NY, 1971), sec. 52.

13. See, J.M. Kosterlitz and D.J. Thouless, J. Phys. C6, 1181 (1973); A.F. Hebard and A.T. Fiory, Phys. Rev. Lett.50, 1603 (1983).

14. M.R. Beasley, J.E. Mooij and T. P. Orlando, Phys. Rev. Lett.42, 1155 (1979).

15. A.I. Larkin and Yu. N. Ovchinnikov, J. Low Temp. Phys. 34 , 409 (1979).

16. P.W. Anderson and Y.B. Kim, Rev. Mod. Phys. 36, 39 (1964).

17. K. Binder and A.P. Young, Rev. Mod. Phys. 58, 801 (1986).

18. In contrast to thin films, in which thermal fluctuations are expected to destroy vortex-glass order at any non-zero temperature, a true finite temperature vortex-glass phase might be possible in bulk 3D systems. See Ref. 5 and R. Koch, V. Foglietti, W.J. Gallagher, G. Koren, A. Gupta and M.P.A. Fisher, Phys. Rev. Lett. 63, 1511 (1989).

19. D.H. Lee and M.P.A. Fisher, Phys. Rev. Lett. 63, 903 (1989); M.P.A. Fisher and D.H. Lee, Phys. Rev. B39, 2756 (1989).

20. H.P. Wei, D.C. Tsui and A.M.M. Pruisken, Phys. Rev.B 33, 1488 (1985).

21. S.M. Girvin and M.P.A. Fisher, unpublished.

NOVEL TREATMENT OF THE HEISENBERG SPIN-1 CHAIN: APPLICATION TO $NI(C_2H_8N_2)_2NO_2(ClO_4)$[a]

A. M. Tsvelik[b]
University of Florida, Gainesville, FL 32611

We suggest a new field theory treatment of the Heisenberg spin-1 chain with a single-ion anisotropy as a theory of three Majorana (real) fermions. We calculate the static and dynamic magnetic susceptibilities and eliminate the reported contradiction between excitation gaps deduced from high-field magnetization measurements and neutron scattering data.

According to the Haldane prediction [1] the one-dimensional Heisenberg antiferromagnetic with integer spin S should have an excitation gap and finite correlation length. This prediction has been confirmed both theoretically (see [2] for a review) and experimentally. The neutron scattering data and measurements of magnetic susceptibility for two spin-1 systems—Cs Ni Cl_3 and $Ni(C_2 H_8 N_2)_2 NO_2(ClO_4)$ (NENP) [3–6] give evidence of an excitation gap. Nevertheless the theory can not provide us the detailed analysis of the experimental data. There are two basic reasons for this. To discuss the first problem we should remind ourselves of some basic ideas of the Haldane's approach. The main idea is to use an effective field theory instead of the initial lattice Hamiltonian. This effective field theory is deduced in the semiclassical approximation $S \gg 1$ and is the O(3) non-linear σ- model. The quantity $1/S$ plays the role of bare coupling constant. At $S = 1$ such description seems to be too crude.

The second problem is that the nonlinear σ-model is actually difficult to use for practical calculations being a highly nontrivial theory. Such calculations are needed, however, for a proper treatment of anisotropy, which is always present in real systems. In particular NENP is found to have a significant single-ion anisotropy. It is well-known that particles of the O(3) non-linear σ-model are triplets: according to general symmetry arguments this triplet should be split in the presence of planar anisotropy into singlet and doublet. From the present knowledge of the non-linear σ-model one cannot provide any more detailed information. But one actually needs extra information to explain the high-field magnetization [4] and the neutron scattering measurements [5,6]. Both types of experiments demonstrate this splitting, but there is an apparent contradiction on the values of gaps deduced from these experiments. The authors of [4] define the excitation gaps as the critical magnetic fields $E^{(H)}_{\perp,\parallel} = g_{\perp,\parallel}\mu_B H_c^{\perp,\parallel}$ for perpendicular and parallel directions to the chain axis. They find $E^{(H)}_{\parallel} = 14.2K$, $E^{(H)}_{\perp} = 19.5K$. In neutron scattering experiments one observes peaks in differential scattering cross-sections of neutrons moving in a given direction. The measurements of scattering in the direction of the chain give $E^{(N)}_{\parallel} = 30K$, and for the perpendicular direction $E^{(N)}_{\perp} = 14K$ [5].

---

[a] PACS # 75. 10. Jm, 75. 50 Ee

[b] On leave from Landau Institute for Theoretical Physics of Academy of Sciences of USSR.

Below we show how to eliminate this apparent contradiction using some alternative approaches. We show that the neutron scattering in the perpendicular direction and the high-field magnetization experiments in the parallel direction measure the mass of doublet: $E_{\perp}^{(N)} = E_{\parallel}^{(H)} = m_1$. The neutron scattering in the parallel direction measures the mass of singlet m_2, and the critical magnetic field in the perpendicular direction is $E_{\perp}^{(H)} = \sqrt{m_1 m_2}$.

The neutron-scattering observations of the one-particle spectrum in NENP [5] show that the ratio of the gap to the band width 2J is not large: $E_{\parallel}/2J \simeq 0.12$. That makes it possible to describe NENP by a continuous field theory.

In the present paper we suggest an alternative field theory for the spin-1 Heisenberg model. It is the theory of three Majorana fermions with the following Lagrangian

$$L = i\bar{\chi}_a \gamma_\mu \partial_\mu \chi_a - m_a \bar{\chi}_a \chi_a + g_a J_\mu^a J_\mu^a \quad (1)$$

where $J_\mu^a = \epsilon^{abc} \bar{\chi}_b \gamma_\mu \chi_c$.

The anisotropy creates the difference between fermionic masses and coupling constants: $m_1 = m_3 < m_2, \quad g_1 = g_3 \neq g_2$.

We have the following arguments for the model (1). At

$$m_a = 0, \;\; g_a = 0 \quad (2)$$

the model (1) has a gapless spectrum and therefore possesses conformal symmetry. It possesses the SU(2) symmetry as well because the Majorana fields may be considered as belonging to the adjoint representation of this group. Consider the spin-1 integrable model with the Hamiltonian

$$H_{int} = J \sum_{n=1}^{N} \left[(\mathbf{S_n S_{n+1}}) - (\mathbf{S_n S_{n+1}})^2 \right] \quad (3)$$

According to [7] this model has the SU(2) symmetry and gapless excitation spectrum . Its central charge $C = 3/2$ [8] is also the same as the central charge of the model (1) in its conformally invariant point (2). Therefore the latter model reproduces the low-energy behavior of the model (3).

The conformal group theory provides us an operator basis for each universality class. Any perturbation may be expressed in terms of this basis. Therefore it is convenient to consider the Heisenberg model with a single-ion anisotropy as a perturbed variant of the model (3):

$$H = H_{int} + gJ \sum_{n=0}^{N} (\mathbf{S_n S_{n+1}})^2 + D \sum_{n=0}^{N} (S_n^z)^2 \quad (4)$$

at $g = 1$.

In the continuous limit one should save only relevant perturbations which break the conformal invariance. It is very important that we are able to enumerate all possible relevant operators as long as we know the universality class of the theory. This universality class is completely specified by the symmetry group and the central charge. It was firstly argued in [8] that the theory (3) belongs to the universality class of the Wess-Zumino-Witten-Novikov (WZWN) theory. The operator content of the WZWN-theory [9] includes in our case only two primary (relevant) fields: the field

$$\Phi^{(1/2)}_{\alpha\beta}(z,\bar{z}),\ \alpha,\beta=1,2$$
$$(z=x+vt, \bar{z}=x-vt,\ v \text{ is a velocity of magnons}) \tag{5}$$

with the scaling dimension $\delta = 3/16$ and the field

$$\Phi^{(1)}_{ab}(z,\bar{z}) = \chi_{+,a}(\bar{z})\chi_{-,b}(z) \tag{6}$$

with the scaling dimension 1/2 (the signs + and − correspond to chiralities of fermions).

There is also one marginal operator with dimension 1: $J^a_+(\bar{z})J^a_-(z)$. Therefore the perturbative treatment of the Hamiltonian (4) may give in the continuous limit only terms included in (1). The field $\Phi^{(1/2)}$ which is non local with respect to χ-field cannot be present because it breaks the time invariance. It is clear because this field enters into the expression for the staggered magnetization of the model (3) (see for example [10]):

$$(-1)^n S^a(x,t) = Tr\left[\sigma^a\Phi^{(1/2)}(x,t)\right] + (\text{ fields with higher dimensions}) \tag{7}$$

We are not sure that the perturbation theory at $g=1$ could provide reliable expressions for masses, and therefore one should consider the parameters of the model (1) as phenomenological. Nevertheless we think that the theory (1) describes the spin-1 case better than the 0(3) non-linear σ-model and have one additional argument for it.

In the integrable limit $g=0$, $D=0$ (3) the model (4) exhibits a linear specific heat at $T \ll J$ [7,8]:

$$C_v/T = \frac{\pi}{6v}C,\ \ C=3/2 \tag{8}$$

The specific heat is linear in T for finite g as well, as long as T is much greater than the masses. But according to Zamolodchikov's theorem [11] the coefficient C in the linear term of specific heat (central charge) can not be larger than the coefficient of the unperturbed theory! Hence the expression (8) gives us the upper bound. Let us take the O(3) non-linear σ-model:

$$L = \frac{1}{2g}\int\left((\partial_\mu\theta)^2 + \sin^2\theta(\partial_\mu\psi)^2\right)\,dx \tag{9}$$

At temperatures higher than mass scale one may substitute $\sin^2\theta$ in (6) by unity which gives us the Lagrangian of two free bosonic fields. This gives for the central charge the value $C=2$, which is higher than the upper bound $C=3/2$.

The additional reason to use model (1) is that it is good for practical purposes of calculations. In the first approximation one can neglect the interaction. It gives only small corrections $\sim g_a \ln(J/m)$. Below we will omit the interaction. After doing this the model (1) is easily quantized. Let us introduce the Fourier-transformed fields

$$\chi_{\pm,a}(k,t) = \int dx e^{ikx} \chi_{\pm,a}(x,t) \tag{10}$$

where the sign $+(-)$ corresponds to the right (left) movers. After this transformation the kinetic part of the Lagrangian (1) becomes

$$L_0 = \sum_{k>0,\tau=\pm} \left[i\chi_{\tau,a}(-k,t)\partial_t \chi_{\tau,a}(k,t) \right] \tag{11}$$

According to the standard procedure of quantization one gets the creation and annihilation operators as follows:

$$a^*_{\pm,b}(k) = \chi_{\pm,b}(-k), a_{\pm,b}(k) = \chi_{\pm,b}(k) \ \ (k > 0). \tag{12}$$

The corresponding Hamiltonian is

$$H_0 = \sum_{k>0,\tau=\pm} \left[k\tau a^*_{\tau,b}(k) a_{\tau,b}(k) + m_b \left(a^*_{\tau,b}(k) a_{\tau,b}(k) \right) \right]. \tag{13}$$

After the canonical transformation

$$\begin{aligned} a_+(k) &= \cos\theta_k b_1(k) + \sin\theta_k b^*_2(k); \\ a_-(k) &= -\sin\theta_k b_1(k) + \cos\theta_k b^*_2(k); \\ \tan 2\theta_k &= -m/k; \end{aligned} \tag{14}$$

one rewrites the Hamiltonian (13) as

$$H_0 = \sum_{k>0,a} \left[\sqrt{m_a^2 + k^2} \left(b^*_{1,a}(k) b_{1,a}(k) + b^*_{2,a}(k) b_{2,a}(k) \right) \right]. \tag{15}$$

To add the magnetic field in the a-direction means to add the integral of motion

$$h_a \int dx \epsilon^{abc} \bar{\chi}_b(x) \gamma_0 \chi_c(x). \tag{16}$$

The quantization of (16) gives

$$H_{mag} = \sum_{k>0,\tau=\pm} h_1 \left(b^*_{\tau,2}(k) b_{\tau,3}(k) + h.c. \right) + (\text{cyclic perm.}). \tag{17}$$

The diagonalization of the Hamiltonian (13) together with (17) gives us:

$$H_0 + H_{mag} = \sum_{k,a=1,2,3} \{\varepsilon_a(k) c_a^*(k) c_a(k)\} \tag{18a}$$

$$c(k>0) = c_+(k), \; c(k<0) = c_-(k)$$

where

$$\varepsilon_a(k) = \varepsilon_a^0(k) - (1-a)\,h \text{ for } h \parallel \hat{z} \tag{18b}$$

$$\varepsilon_{1,2}(k) = \left(\varepsilon_1^0(k) + \varepsilon_2^0(k)\right)/2 \pm \sqrt{\left(\frac{\varepsilon_1^0(k) - \varepsilon_2^0(k)}{2}\right)^2 + h^2} \tag{18c}$$

$$\text{for } h \parallel \hat{x} \text{ or } \hat{y}$$

where

$$\varepsilon_\alpha^0(k) = \sqrt{k^2 + m_\alpha^2} \tag{18d}$$

(Remember that $m_1 = m_3$.)

From this Hamiltonian it is easy to obtain the magnetic moment as a function of the magnetic field:

$$M = \frac{1}{\pi v}\sqrt{(h^2 - m_1^2)} \tag{19a}$$

$$\text{for } h \parallel \hat{z}$$

$$M = \frac{1}{\pi v}\left[-\frac{m_1^2 + m_2^2}{2} + \left[\left(\frac{m_1^2 - m_2^2}{2}\right)^2 + h^4\right]^{1/2}\right]^{1/2} \tag{19b}$$

$$\text{for } h \parallel \hat{x}\hat{y}$$

According to (16) the critical magnetic fields are $E_\parallel^{(H)} = g_\parallel H_c^\parallel = m_1$, $E_\perp^{(H)} = g_\perp H_c^\perp = \sqrt{m_1 m_2}$. Using the values of $E_\perp^{(H)}, E_\parallel^{(H)}$ given in [4] we obtain $m_1 = E_\parallel^{(H)} = 14.2K$, $m_2 = \left[E_\perp^{(H)}\right]^2 / E_\parallel^{(H)} = 26.7K$ which are rather close to the masses observed in neutron scattering experiments [5,6].

The differential cross section of unelastic neutron scattering is directly proportional to the imaginary part of a.c. magnetic susceptibility.Our approach provides us an ability to calculate the a.c. magnetic susceptibility. It follows from (7) that

$$\chi(x,t) = \langle S^+(x,t) S^-(0,0)\rangle$$
$$\sim (-1)^{x/a} \langle Tr\left(\sigma^+ \Phi^{(1/2)}(x,t)\right) Tr\left(\sigma^- \Phi^{1/2}(0,0)\right)\rangle \tag{20}$$

where "a" is a lattice spacing.

It is well known that the 2D Ising model is equivalent to the model of free massive Majorana fermions with mass equal to $m = a(T - T_c)$. The model (1) may be seen as a collection of three Ising models each being at its own temperature, which are directly related to the masses of Majorana fermions. If the current-current interaction was neglected, these models would be independent. According to [9] the following relation between the components of matrix field Φ and the order- and disorder-fields σ and μ takes place at $T = T_c$:

$$Tr\left(\sigma^+\Phi^{(1/2)}\right) = \sigma^1\mu^2\mu^3 + i\mu^1\sigma^2\sigma^3;$$
$$Tr\left(\sigma^-\Phi^{(1/2)}\right) = \sigma^1\mu^2\mu^3 - i\mu^1\sigma^2\sigma^3; \tag{21}$$

In the absence of interaction these relations are still valid beyond the critical point.

To write down the expression for the spin-spin correlation function in the anisotropic case one should specify the correspondence between the labels 1,2,3 in the expression (18) and labels of the fermionic fields in (1). It is important because fermions with different labels have different masses. There are different possibilities to arrange the labels and each arrangement corresponds to a spin susceptibility χ along the appropriate direction of a.c. magnetic field. There are two arrangements giving the same expression for χ:

$$\sigma^1 \Leftrightarrow \chi_1, \sigma^2 \Leftrightarrow \chi_3, \text{(mass } m_1), \ \sigma^3 \Leftrightarrow \chi_2 \text{(mass } m_2); \tag{22}$$

and

$$\sigma^1 \Leftrightarrow \chi_1, \sigma^3 \Leftrightarrow \chi_3, \text{(mass } m_1), \ \sigma^2 \Leftrightarrow \chi_2 \text{(mass } m_2); \tag{23}$$

In these cases one gets from (4) and (18)

$$\chi_\perp(x,t) = < S^+(x,t)S^-(0,0) > = (-1)^{x/a}F_+(m_1r)F_-(m_1r)\times$$
$$\times\left(F_+(m_2r) + F_-(m_2r)\right); \tag{24}$$
$$r = \sqrt{x^2 + v^2t^2}$$

where t is the Matsubara time. The correlation functions $F_+(|\vec{x}|) = <\sigma(\vec{x})\sigma(0)>, F_-(|\vec{x}|) = <\mu(\vec{x})\mu(0)>$ are calculated in [12]. The third arrangement is

$$\sigma^2 \Leftrightarrow \chi_1, \sigma^3 \Leftrightarrow \chi_3; \ \sigma^1 \Leftrightarrow \chi_2 \tag{25}$$

which gives

$$\chi_\parallel(x,t) = (-1)^{x/a}\{F_+(m_2r)F_-^2(m_1r) + F_-(m_2r)F_+^2(m_1r)\} \tag{26}$$

From the symmetry arguments it follows that expression (24) corresponds to the spin-spin correlation functions along the directions transverse to the z-axis and (26) gives the longitudinal susceptibility.

Using the explicit expressions for the F-functions from [12] we get for the imaginary parts of the a.c. susceptibilities as follows :

$$Im\chi_\perp^{(R)}(\omega,k) = f_\perp(s) = \delta\left(s^2 - m_1^2\right) + 2\left[(s-m_2)^2 - m_1^2\right]^{-1/2}$$
$$\times\left[(s+m_2)^2 - m_1^2\right]^{-1/2} + \dots \tag{27}$$
$$Im\chi_\parallel^{(R)}(\omega,k) = f_\parallel(s) = \delta\left(s^2 - m_2^2\right) + 2\mid s\mid^{-1}\left[s^2 - 4m_1^2\right]^{-1/2} + \dots$$
$$s^2 = \omega^2 - v^2(k-\pi/a)^2$$

From (27) one sees that the neutron scattering measurements along the z-direction should exhibit the most strong singularity at the largest mass m_2. One should remember that the high-field magnetization measurements along this axis give the critical field proportional to the smallest mass m_1. Thus the above-mentioned contradiction is removed.

In summary, we have suggested the alternative approach to the 1D spin-1 chain on the basis of model (1). This model is good for practical calculations, and we have managed to calculate the critical magnetic fields and imaginary parts of a.c. magnetic susceptibilities in the present of planar anisotropy.

ACKNOWLEDGEMENTS

I would like to thank Dr. O. de Alcantara Bonfim for a help. This work was supported by DARPA under contract MDA 972–B5–J–1006 and by NSF, DMR–8607941.

REFERENCES

1. F.D.M. Haldane, Phys. Lett. **93A**, 464 (1983).
2. I. Affleck, J. Phys.: Condensed Matter 1,3047 (1989).
3. W.J.L. Buyers, R.M. Morra, R.L. Armstrong, P. Gerlach and Hirakawa, Phys. Rev. Lett. **56**, 371 (1986).
4. Y. Aijro, T. Goto, H. Kikuchi, T. Sakakibara and T. Inami, Phys. Rev. Lett. **63**, 1424 (1989).
5. J.P. Renard, M. Verdaguer, L.P. Regnault, W.A.C. Erkelens, J. Rossat-Mignod and W.G. Stirling, Europhys. Lett. 3,945 (1987).
6. J.P. Renard, M. Verdaguer, L.P. Regnault, W.A.C. Erkelens, J. Rossat-Mignod, J. Ribas, W.G. Stirling and C. Vettier, J. Appl. Phys. **63**, 3538 (1988).
7. H.M. Babujan, Nucl. Phys. **B215**, 317 (1983).
8. .I. Affleck, Phys. Rev. Lett. **56**, 746 (1986).
9. A.B. Zamolodchikov, V.A. Fateev, Sov. J. Nucl. Phys., **43**, 1031 (1986).
10. I. Affleck, F.D.M. Haldane, Phys. Rev. **B36**, 5291 (1987).
11. A.B. Zamolodchikov, JETP Lett. **46**, 161 (1987).
12. T.T. Wu, B. McCoy, C.A. Tracy and E. Barouch, Phys. Rev. **B13**, 316 (1976).

PROBING THE UNIVERSAL CHARACTER OF WAVE PROPAGATION IN RANDOM MEDIA

W. Polkosnik, N. Garcia, A.Z. Genack
Queens College of CUNY, Flushing,NY 11367

We argue that a universal description of the statistical character of wave propagation is given in terms of the level width and level spacing of eigenstates of a random medium and a parameter describing inelastic processes. These parameters are obtained from measurements of the intensity correlation function of microwave radiation as a function of frequency.

Random intensity fluctuations are a common feature of wave propagation in disordered materials. A familiar manifestation of these fluctuations is the speckle pattern in scattered laser light. The value of the spatially varying field depends upon the microscopic structure of the sample and upon the form of the interaction. Because of limitations in the precision to which the structure is known and of calculational uncertainties, the value of the field at a point in multiply scattering samples cannot be calculated even approximately. Nonetheless, the average of dynamical properties and the statistical character of intensity fluctuations are of considerable interest and can in principle be found experimentally. In some respects this situation is analogous to the study of molecular dynamics. Though it is not feasible to specify the velocities and positions of all the molecules in gas, nonetheless, the elastic and inelastic mean free paths can be determined and the form of the velocity distribution can be calculated in many cases.

On length scales greater than the transport mean free path, ℓ, in which the direction of propagation is randomized, the statistical character of transport is not influenced by the microscopic character of the wave interaction with the structure except through the values of various average scattering parameters. This suggests that there may be universal description of all statistical properties in terms of a limited number of parameters.

Any wave can be expressed as a linear superposition of eigenstates of the wave in the medium. We expect, therefore, that appropriate universal parameters are average properties of eigenstates. Two key properties are the typical energy width and spacing of eigenstates. These properties were first used by Thouless[1] to specify the condition for electronic localization.[2] A description of propagation in terms of properties of the underlying eigenstates should apply equally to classical and quantum mechanical waves. This can clearly be done since there is a correspondence between the classical frequency difference $\Delta\nu$ and the quantum mechanical energy difference ΔE, $\Delta\nu = \Delta E/h$. In units of frequency we denote the level width as $\delta\nu$ and the level spacing as $d\nu/dN$. The level spacing is the inverse of the density of states, $dN/d\nu$. The dimensionless level width $\delta = \delta\nu/(d\nu/dN)$ determines the proximity to the localization threshold which is reached when $\delta = 1$. We argue that, in the absence of inelastic processes, a universal description of all statistical properties of propagation are given in terms of $\delta\nu$ and $d\nu/dN$.[3,5] The ultimate proof of this conjecture would be the experimental demonstration that the entire array of propagation phenomena in all regimes are given in terms of these parameters. We begin such a program showing that the intensity correlation function for extended electromagnetic waves is given by a function of $\delta\nu$ and $d\nu/dN$ and an additional parameter describing photon absorption.

Since we seek to study universal aspects of propagation, we are at liberty to choose experimental systems which are particularly amenable to such measurements. An ideal system would have the following properties: (1) the wave can be confined by reflecting surfaces so that $d\nu/dN$, which is inversely proportional to the volume, can be varied. (2) the value of the intensity at a point can be measured and such measurements can be obtained simultaneously at several points (3) availability of tunable monochromatic sources (4) wide range of scattering strengths (5) macroscopic sample so that the sample microstructure and density can be controlled and measurements on length scales smaller than ℓ or the wavelength λ can be made (6) the sample can be static but the constituents can be randomly rearranged to produce a new configuration so that ensemble averaging can be performed (7) quanta of the field are noninteracting bosons so that interaction and correlation effects can be neglected.

One system that satisfies all these requirements is microwave propagation in a random sample inside a container with metallic transverse walls.[5] For K band radiation the wavelength is of the order of 1 cm. We report results for a sample of 1/2 inch polystyrene balls with a volume filling fraction of $f = 0.56$. f is sufficiently large that the spheres fill the sample volume, but loose enough that the configuration can be charged by tumbling the sample. The sample holder is shown in Fig. 1. The sample is contained in a cylindrical 7.3cm diameter copper tube. The sample length L is set by the position of a plunger which is attached to a computer controlled translation stage. Intensity fluctuations as a function of incident frequency are detected by a Schottky diode on the surface of the plunger. Absorbing material is placed behind the plane of the diode so that the wave is not reflected back into the sample. Intensity measurements are local since the diode junction is much smaller than λ. When the frequency scan is completed, the sample holder is rotated about its axis for a few seconds to produce a new configuration in preparation for the next scan.

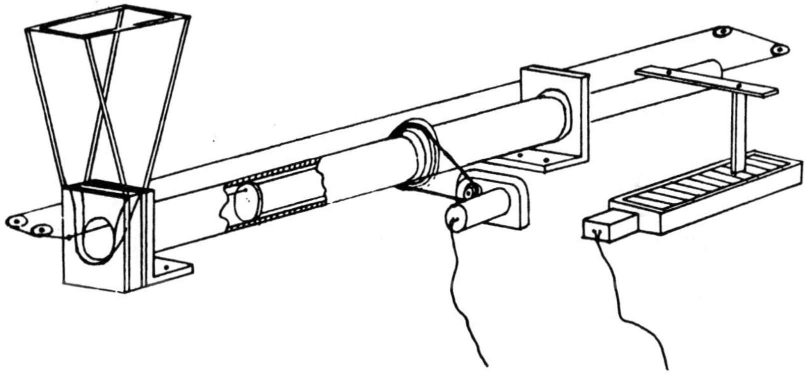

Figure 1 Experimental setup to measure the scale dependence of the transmitted microwave intensity and the intensity fluctuations as a function of frequency.

The transmission versus thickness is obtained by averaging the intensity at each sample thickness as the sample is continuously tumbled. A fixed number of spheres is added through the hopper shown in Fig. 1 for each length increment. The results for ν = 22 GHz are shown as the points in Fig. 2. The solid curve drawn through the points represents a fit of the data to the prediction of the photon diffusion model, using a normalization constant. The model gives,

$$T(L) = \sinh(\alpha a)/\sinh(\alpha L) \qquad (1)$$

where $\alpha = (D\tau_a)^{-\frac{1}{2}}$ is the absorption coefficient for $L > L_a = \alpha^{-1}$, D is the diffusion coefficient, τ_a is photon absorption time and $a = 5\ell/3$ for the case of normal incident radiation. This fit gives the absorption length L_a = 25cm. This characterizes inelastic processes in this sample. The agreement with the classical particle diffusion model demonstrates that interference effects that occur when $\ell > \lambda$ are not significant and that the wave is extended.

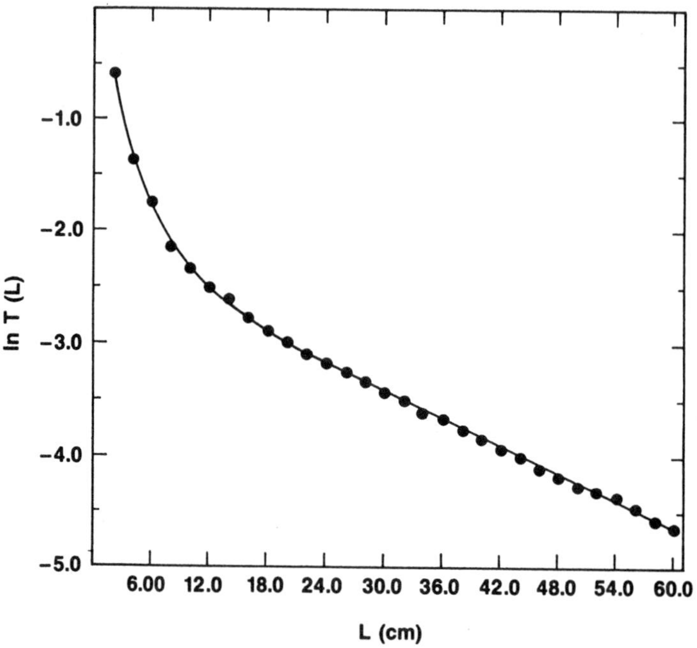

Figure 2 The dots represent the scale dependence of the transmitted intensity of microwave radiation (ν = 22GHz) through a sample of 1/2" polystyrene spheres. The solid line is a fit of the photon diffusion model to the data.

The parameters $\delta\nu$ and $\delta\nu/dN$ can be determined from measurements of intensity correlation in the medium. The degrees of long range intensity correlation is inversely proportional to the number of independent parameters needed to represent a wave.[3] For monochromatic excitation, only eigenstates within the level width contribute significantly to the field. The amplitudes of these eigenstates are the parameters which specify the wave. The number of these parameters is $N_{ind} = (dN/d\nu)\delta\nu = \delta$.[3] Thus we expect that the degree of intensity cross correlation is proportional to $1/\delta$.[3]

We consider the intensity correlation function as a function of frequency.[5,6] In order to eliminate effects of the frequency dependence of the instrumental response, all spectra are normalized by the ensemble average spectrum. We refer to the normalized intensity at frequency ν for a specific configuration as $I(\nu)$ Since the ensemble average of $I(\nu)$, $\langle I(\nu)\rangle$ is unity, the cumulant intensity correlation function with frequency can be written as

$$C(\Delta\nu) = \langle (I(\nu)-1)\ (I(\nu+\Delta\nu)-1)\rangle = \langle I(\nu)\ I(\nu+\Delta\nu)\rangle -1 \quad (2)$$

In the limit $\delta >> 1$, this may be expressed as an expansion in δ^{-1} which is proportional to the degree of long range correlation in the sample, $C = C_1 + C_2 + \ldots$.[4,7] The first term is the intensity correlation function in the field factorization approximation is

$$C_1(\Delta\nu) \sim \langle |E(\nu)\, E^*(\nu + \Delta\nu)|^2 \rangle = |\tilde{G}^E(\Delta\nu)|^2, \tag{3}$$

where E is the complex field and $\tilde{G}^E$ is the field-field correlation function.[6,8,9] This term neglects correlation in the amplitude of partial waves reaching a point in a specific configuration. The field-field correlation function is the Fourier transform of the time of flight distribution,[9]

$$G^E(\Delta\nu) = |\int_0^\infty I(t)\, e^{i2\pi\Delta\nu t}\, dt|, \tag{4}$$

where I(t) is the ensemble average of the time of flight distribution. In the weak scattering limit I(t) is obtained by solving the diffusion equation. Using the diffusion solution for I(t) and performing the integral in eq. 4 gives [3,5,9]

$$G^E(\Delta\nu) = |[\sinh(q_o a)/\sinh(q_o L)]|/[\sinh(\alpha a)/\sinh(\alpha L)] \tag{5}$$

where q_o is the root with negative imaginary part of

$$q^2 = 1/L^2 + (i2\pi\Delta\nu)/D \tag{6}$$

The level width is the frequency over which the incident wave must be tuned in order to change the field configuration at the output. This is essentially the half width of $G^E(\Delta\nu)$. Thus in the limit $\delta >> 1$, $\delta\nu$ is a function of D and L_a. For stronger scattering it is also a function of δ. From eqs. 3 and 5, the leading term in the expansion for the cumulant intensity correlation function with frequency in the presence of absorption for $\delta > 1$ is,

$$C_1(\Delta\nu) = A_1|\sinh^2(q_o a)/\sinh^2(q_o L)|/[\sinh^2(\alpha a)/\sinh^2(\alpha L)] \tag{7}$$

where $A_1 = 1$ for $\delta >> 1$ but in general is a function of δ.

The first order correction [7,10] due to the presence of long range intensity correlation is,

$$C_2(\Delta\nu) = A_2\delta^{-1}F_2(q_o L) \tag{8}$$

where A_2 is of order unity for $\delta > 1$, and,

$$F_2(q_o L)=|(\alpha/q_o)[\coth(q_o L)-q_o L/\sinh^2(q_o L)]/[\coth(\alpha L)-\alpha L/\sinh^2(\alpha L)]$$

This form of F_2 is in agreement with our recent measurements of the intensity cross correlation function with frequency at two different points in the medium. Equation 9 is similar to that calculated by Feng et al. for the first order correction due to correlation of the intensity correlation function versus the angle of rotation of the sample.[7]

The cumulant intensity correlation function for L = 30 cm averaged over spectra from 500 configurations is shown as the points in Fig. 3. The fit of the data obtained using eq. 7, with the value of L_a obtained from the measurement of T(L) and with D as a fitting parameter is shown as the solid line in the figure. The fit gives $D = 2.9 \pm 0.2 \times 10^{10}$ cm^2/s. Clearly long range intensity correlation is negligible in this case.

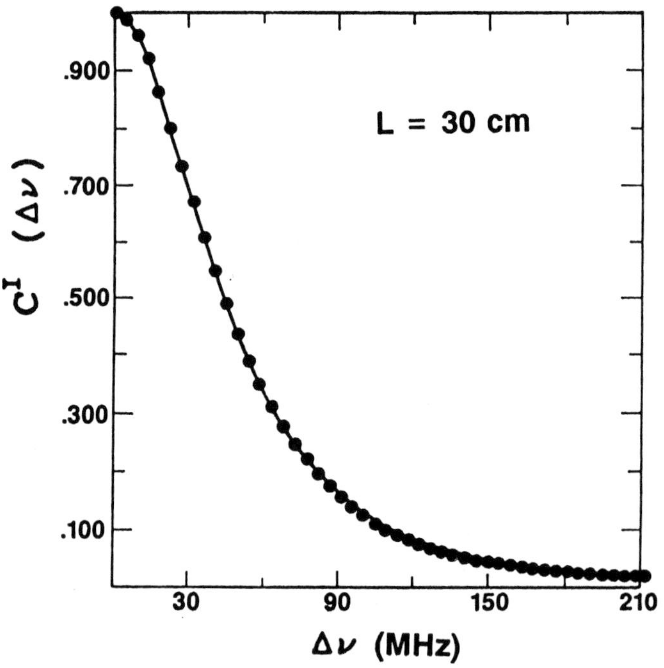

Figure 3 Ensemble average of the autocorrelation function for 500 sample configurations of 1/2" polystyrene spheres. The sample thickness L = 30 cm.

The cumulant intensity correlation function for L = 140 cm averaged over 5000 configurations is shown as the points in Fig. 4. The tail of $C(\Delta\nu)$ falls as $\Delta\nu^{-\frac{1}{2}}$. This is a characteristic of F_2 and indicates the influence of long range intensity correlation within a given configuration. The fit of

$C_1 + C_2$ to the data is shown as the curve in Fig. 4. The fit gives $D = 3.0 \times 10^{10}$ cm^2/s.

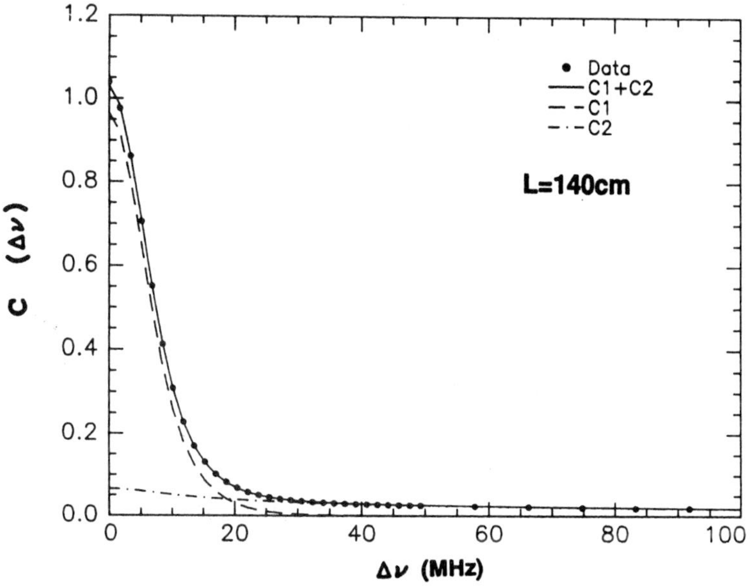

Figure 4 Ensemble average of the autocorrelation function for 500 sample configurations of 1/2" polystyrene spheres. The sample thickness L = 140 cm.

Using the values of D and L_a in eqs. 5 and 6 we find for the half width of the field correlation function $\delta\nu = 3.5$ MHz. The density of states in the sample is

$$dN/d\nu = (2k^2/\pi v)AL \tag{10}$$

where A is the cross sectional area of the sample and $k = 2\pi/\lambda$. For our sample $dN/d\nu = 5.9 \times 10^{-6}$. This gives $\delta = \delta\nu(dN/d\nu) = 20.6$. This value determined from C_1 is in excellent agreement with the value of δ obtained from recent measurements of the variation of the cross correlation function with separation between the points in the medium.

In conclusion we have shown that the intensity correlation function of microwave radiation can be expressed as a function of two universal parameters, $\delta\nu$ and δ, in the limit $\delta \rangle 1$. These parameters are average properties of eigenstates of the medium. These results demonstrate that the configuration average of microscopic statistical quantities can be obtained for microwave radiation in regimes in which the influence of long range intensity correlation is significant.

We wish to thank E. Kuhner for his expert technical assistance. This work has been supported by grants from the PSC-BHE of CUNY and from the Exxon Research Foundation.

REFERENCES

1. D.J. Thouless, Phys. Rev. Lett. **39**, 1167 (1977).
2. E. Abrahams, P.W. Anderson, D.C. Licciardello and T.V. Ramakrishnan,Phys. Rev. Lett. **42**, 673 (1979).
3. A.Z. Genack in "Classical Wave Localization" ed. P. Sheng (World Scientific Singapore, 1990);
4. N. Garcia, W. Polkosnik and A.Z. Genack submitted for publication.
5. N. Garcia and A.Z. Genack, Phys. Rev. Lett. **63**, 1678 (1989).
6. A.Z. Genack, Phys. Rev. Lett. **58**, 2043, (1987).
7. S. Feng, C. Kane, P.A. Lee and A.D. Stone, Phys. Rev. Lett. **61**, 834 (1988).
8. B. Shapiro, Phys. Rev. Lett. **57**, 2168 (1986).
9. A.Z. Genack and J.M. Drake, Europhys. Lett., to be published (1990).
10. M.J. Stephen and G. Cwilich, Phys. Rev. Lett. **59**, 285 (1987).

SINGULARITY IN THE POSITIVE HALL COEFFICIENT NEAR PRE-ONSET TEMPERATURES IN HIGH-T_C SUPERCONDUCTORS

G.C. Vezzoli[1,2,5], M.F. Chen[1], and F. Craver[1], B.M. Moon[2], and A. Safari[2]
T. Burke[3], W. Stanley[4]

Hall measurements using continuous extremely slow cooling and reheating rates as well as employing eqiulibrium point-by-point conventional techniques reveals a clear anomally in R_H at pre-onset temperatures near T_c in polycrystalline samples of $Y_1 Ba_2 Cu_3 O_{7-}$ and $Bi_2Sr_2Ca_2Cu_3O_{10}$. The anomaly has the appearance of a singularity or Dirac-delta function which parallels earlier work on $La_{1-x}Sr_xCuO_4$ [1]. Recent single crystal work on the Bi-containing high-T_c superconductor is in accord with a clearcut anomaly [2]. The singularity is tentatively interpreted to be associated (upon cooling) with initially the removal of positive holes from the hopping conduction system of the normal state such as from the increased concentration of bound or virtual excitons due to increased exciton and hole lifetimes at low temperature. Subsequently the formation of Cooper pairs by mediation from these centers (bound-holes and/or bound excitons) may cause an ionization of these bound or virtual excitons thereby re-introducing holes and electrons into the conduction system at T_c.

1. Introduction

Hall effect studies [1] of the first high-T_c superconductor [3] $La_{1-x} Sr_x Cu O_4$ reveal a Dirac-delta like singularity in the positive Hall coefficient at a temperature slightly higher than T_c, as reproduced in Fig 1. Subsequent Hall effect studies [4] of the high-T_c superconductor $Y_1Ba_2Cu_3O_{7-}$[5] also revealed a singularity beginning (upon cooling) at the pre-onset temperature corresponding to the temperature at which the resistance (R) vs temperature (T) data first deviates from linearity (reproduced in Fig 2). In both of these studies there is indication that at the low-temperature terminus of the singularity the Hall coefficient becomes slightly negative, and then at slightly lower temperatures becomes zero (in the superconducting state). There is also indication of unstable, fluctuating, or oscillatory behavior in the Hall coefficient as temperature is decreased to the value where $R = R_H = 0$ (see Ref 4).

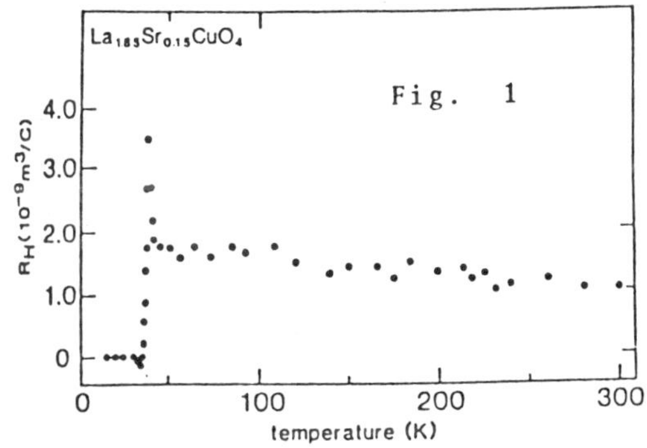

HALL CONSTANT IN $La_{1.85}Sr_{0.15}CuO_4$
VERSUS TEMPERATURE

[1] U.S. Army Materials Technology Laboratory, Materials Science Branch, Watertown, Massachusetts 02172
[2] Rutgers University, Departments of Electrical Engineering and Ceramic Science, Busch Campus, Piscataway, New Jersey 08854
[3] Pulsed Power Center, US Army Electronics Technology and Devices Laboratory, Fort Monmouth, New Jersey 07703
[4] Decision Software, Cambridge, Massachusetts 02139
[5] United States Military Academy, Physics Department, West Point, New York 10996

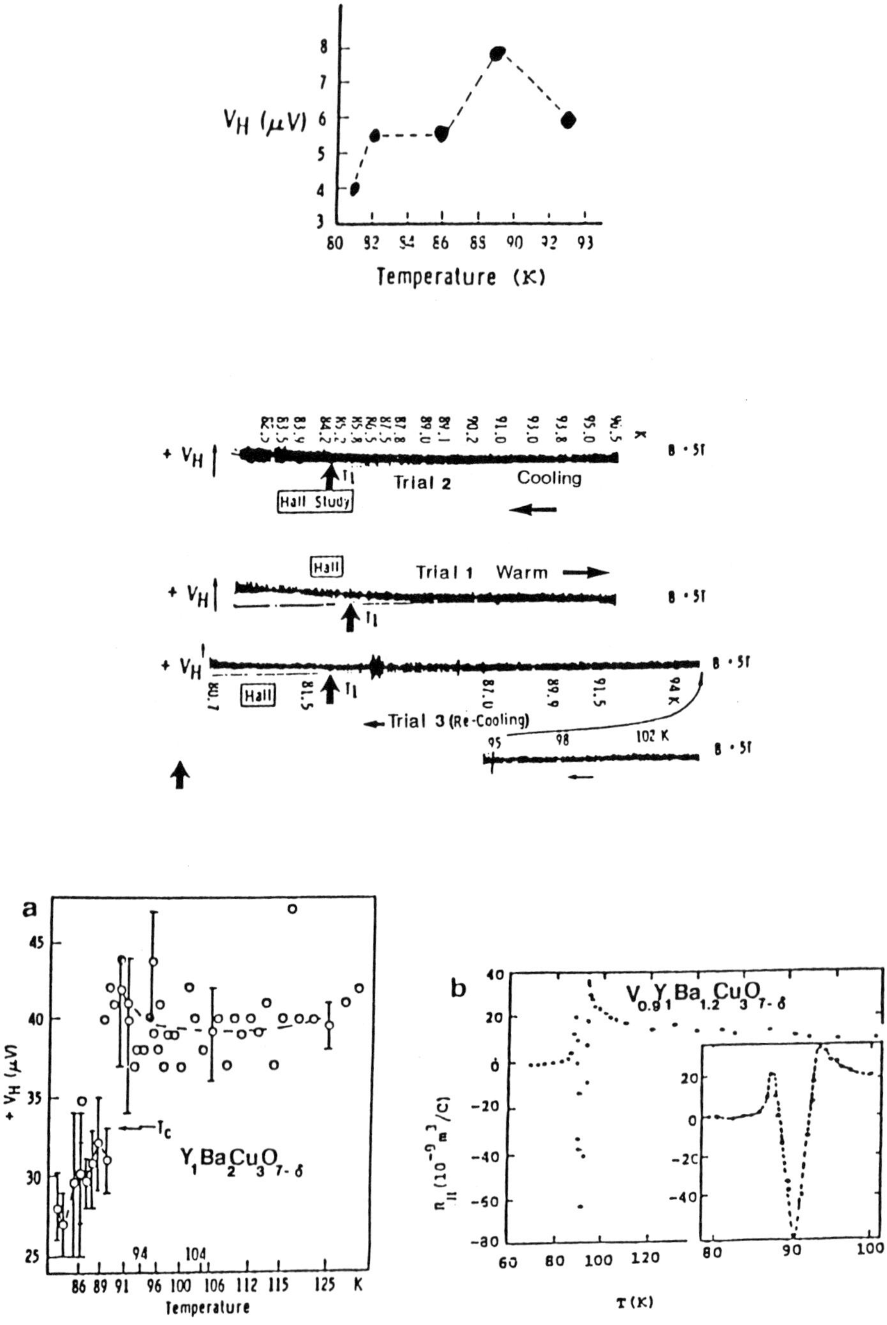

V_H (μV)
Temperature (K)
+ V_H
Trial 2
Cooling
Hall Study
Hall
Trial 1 Warm
B = 5T
Trial 3 (Re-Cooling)
a
+ V_H (μV)
Temperature
Y1Ba2Cu3O7-δ
b
V0.91Y1Ba1.2Cu3O7-δ
T (K)

Fig. 2

2. Experimental Data

Figs 3 and 4 give digital and analogue Hall data for Bi-containing high-T_c superconductor samples containing a substantial proportion of the $Bi_2Sr_2Ca_2Cu_3O_{10}$ phase. These data show a major singularity at T = 126K as well as indication of two other singularity-type anomalies at lower temperatures (suggesting three coexisting phases). At a temperature of about 114K the Hall coefficient of the 2-2-2-3 phase decreasees abruptly suggesting that 114K is T_c for this phase.

Fig 5 shows data (from the sample described in Fig 3) giving the resistance during the transition to the superconducting state as a function of inverse temperature and applied magnetic field. These data show that the magnetic field causes a divergence in the R vs 1000/T data. These linear R vs 1000/T dependencies at B from 2 to 14 tesla back-extrapolate to intersection at T = 114K, confirming this value for T_c. Thus T_c represents the temperature for the formation of a sufficient number of Cooper pairs such that the B field has the effect of elevating resistance (by breaking Cooper pairs) at constant temperature in the transitional state.

3. Interpretation and Discussion

Simultaneous theoretical studies in our laboratory [6] show that a structure of the high-T_c type (such as a derivative of the K_2MnF_4 structure or a stacked quasi Perovskite) which contains two types of polyhedral building blocks such as CuO_5 Coulombic pyramids and Cu-O-Cu chains (or planes) gives rise to high values of internal electric fields. The field is highest at or near apical ions in the CuO_5 pyramid such as 04 in $Y_1Ba_2Cu_3O_7$ and at ions associated with fluctuations from spin antiferromagnetism. These high internal fields contribute to polarizations or charge separation yielding electron-hole pairs such as associated with virtual excitons at the multivalence Bi and Cu sites and bound excitons on the oxygens. These localized centers (electron-hole pairs) have a characteristic lifetime that varies inversely with temperature. Therefore, the concentration of these bound centers increases with decreasing temperature. At a value of temperature where the lifetime of these centers has risen to the value about equal to the dielectric relaxation time of the normal state, the charge separation or polarizability should peak. In such circumstances positive holes will be dramatically removed from conduction roles and utilized as constituents of these centers. The concentration of these electron-hole unrecombined centers (or their bound holes) at this temperature is large enough to interact Coulombically or quantum mechanically with conduction electrons and then act as Cooper-pairing mediators. (We have calculated this interaction in Ref 6 and show it leads to coherence lengths of the order of 12 to 15A). The abrupt removal of holes from the conduction system will be associated with a steeply rising positive Hall coefficient as we observe in Figs 3 and 4. At such a condition there should exist strong competition between the Cooper-pairing process and the electron-hole pair process. The weak binding energy of the electron-hole pairs (excitons) should succumb to the Cooper-pairing energy (12 to 14 mev) and cause ionization of the excitons. Such an effect will deliver holes and electrons into the conduction system and cause the magnitude of R_H to decrease sharply. If in this process the free electron concentration overtakes the conducting hole concentration then the Hall coefficient can with further slight decrease in temperature become transiently negative prior to becoming zero in the superconducting state.

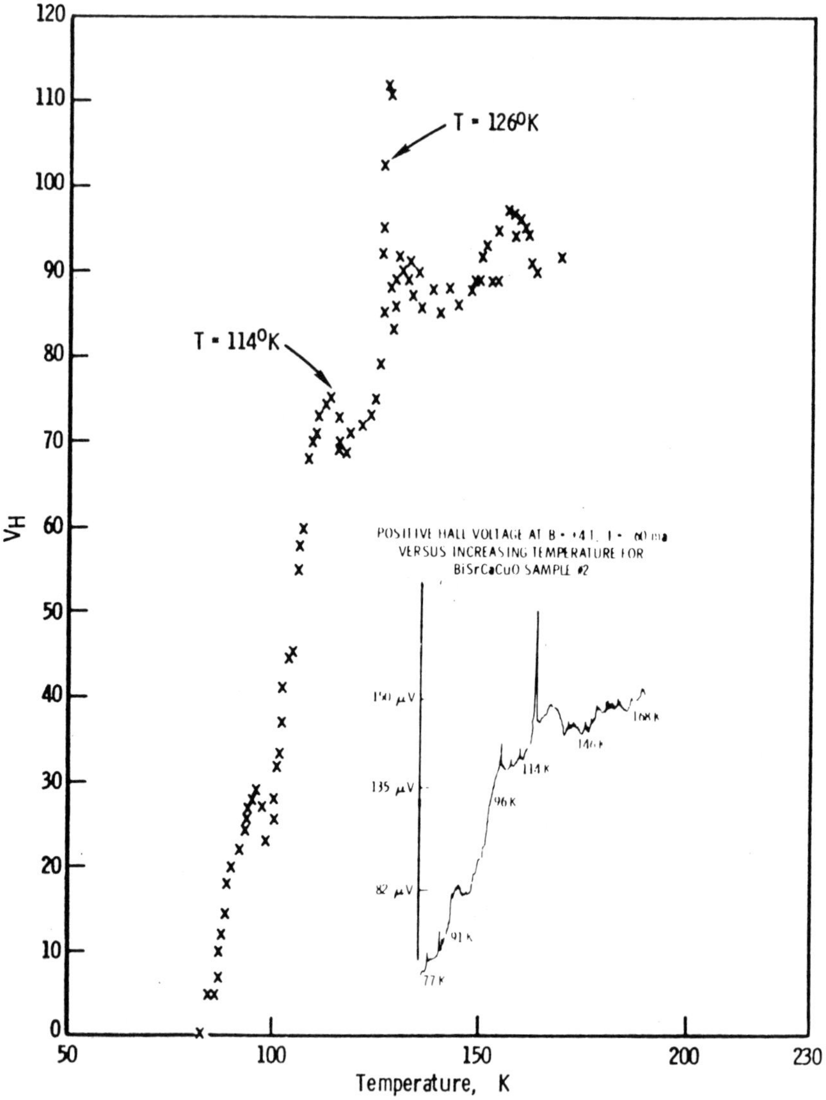

HALL EFFECT VERSUS TEMPERATURE °K AT B = 4T FOR BiSrCaCuO
(HEATING CYCLE ~ 0.1°K/min)

Fig. 3. The Hall Effect vs Temperature for BiSrCaCuO giving digital and analogue (inset) data showing singularity at 126K and rapid decrease to zero at 114K. Sample fabricated at U.S. Army Materials Technology Laboratory.

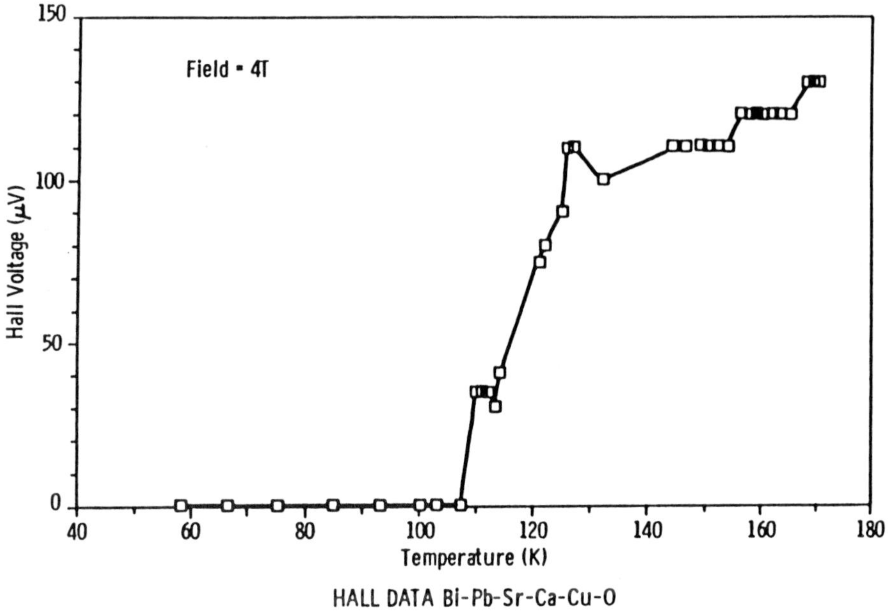

Fig. 4

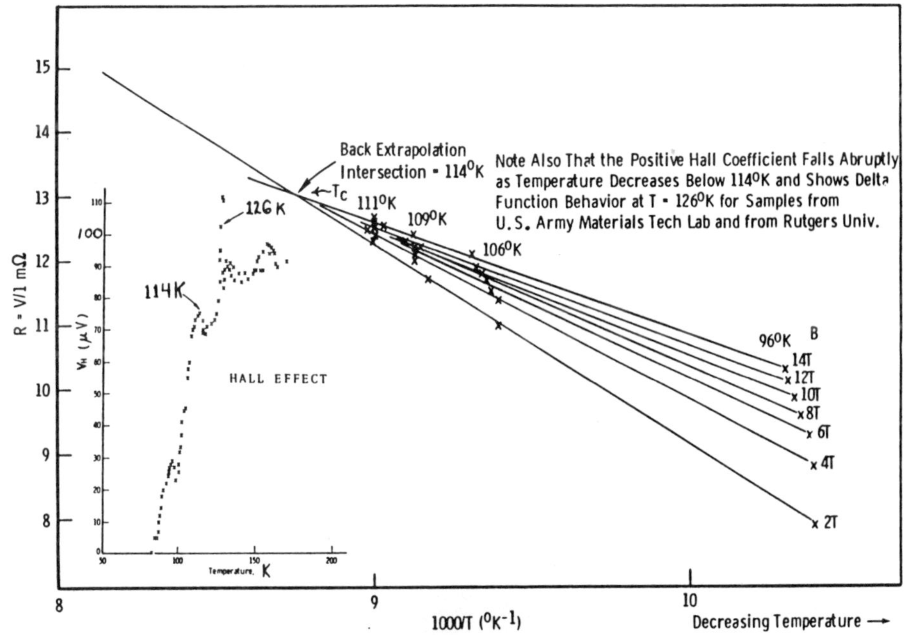

ELECTRICAL RESISTANCE VERSUS RECIPROCAL TEMPERATURE FOR $Bi_YCa_ZSr_WCu_RO_X$ AS A FUNCTION OF APPLIED MAGNETIC FIELD DURING TRANSITION TO SUPERCONDUCTING STATE

Fig. 5

References

1. M.F. Hundley, A. Zettl, A. Stacy, and M.L. Cohen, Phys. Rev. B35, 8800 (1987).

2. A. Zettl, Phys Rev (in press).

3. J.G. Bednorz and R.A. Mueller, Z. Phys B64, 189 (1986).

4. G.C. Vezzoli, T. Burke, B.M. Moon, B. Lalevic, A. Safari, H.G.K. Sundar, R. Bonometti, C. Alexander, C. Rau, and K. Waters, J. Mag and Mag Mat'ls 79, 146 (1989).

5. C.W. Chu, P.H. Hor, R.L. Meng, L. Gao, Z.J. Huang, and Y.Q. Wang, Phys Rev. Lett. 58, 405 (1987).

6. G.C. Vezzoli, J. Mag and Mag Mat'ls (in press); also Progr in High Temperature Superconductivity Vol 17, p 116, World Publishers, London 1989. (Proc of High-T_c Superconductors: Magnetic Interations, Oct 11-13, 1988, Gaithersburg, Md).

FREQUENCY DEPENDENCE OF RESONANT TUNNELING

Carlo Jacoboni
IBM T.J. Watson Research Center, Yorktown Heights, New York 10598
and
Dipartimento di Fisica, Modena University, 41100 Modena, Italy
Peter J. Price
IBM T.J. Watson Research Center, Yorktown Heights, New York 10598

The "Ohmic" diode conductance versus frequency is computed for resonant tunneling electrons, by means of a Kubo-like formula. When the Fermi level is at the transmission peak energy, then for a symmetrical double-barrier structure it falls off at the resonance width divided by Planck's constant, while for the non-symmetrical case it rises from the smaller zero-frequency conductance to a peak at about this frequency, before falling off like the symmetrical case. For a resonance peak of a "quantum wire" with a smoothly varying disordered potential, the result is quite similar to the non-symmetrical double barrier case.

Coherent transmission of electrons can occur in a number of solid-state structures. Semiconductor instances are a double-barrier parallel-plane heterostructure [1], and a quasi-one dimensional "quantum wire" with a randomly varying potential [2]. The familiar Lorentzian peaks in the transmission probability, $T(E) = T_0\,\Lambda(E)$ where T_0 is the maximum value of T and $\Lambda(E) = 1/[1 + ((E - E_0)/\Delta E)^2\,]$, can have very narrow half widths ΔE. For a substantially asymmetrical structure in the double-barrier case, and in general for the disordered quasi-one dimensional case, $T(E_0) = T_0$ is $<< 1$ (whereas for a symmetrical structure $T_0 = 1$). It is natural to associate a time scale for coherent electron motion with $\hbar/\Delta E \equiv t_0$. This t_0 is in fact equal to the formal transit time, $t_{\text{trans}}(E) = \hbar\, d(\text{phase})/dE$, at $E = E_0$; but the wave packet which would normally be considered to reify this measure can not in fact be transmitted without distortion, in the resonance range of E. More physical is the decay time of the quasi-level at E_0, which is equal to $\hbar/2\Delta E = t_0/2$. In the observable category is the conductance of the structure considered as a "diode" embedded in conductors on each side, with electron densities given by their Fermi energies. Computational simulation of the electrical behavior of this system when a bias voltage is applied, by means of the Wigner transform of the one-electron density matrix, is an obviously attractive approach. It has recently had some success applied to double-barrier heterostructures [3, 4]. We report here another computational approach, applied to both double-barrier and quantum-wire systems, which has provided unexpected results [5].

Assume a bias potential applied over the range $x_L < x < x_R$ of the "diode" structure, in the form $V(x, t) = V_0 s(x) \sin \omega t$ where $s(x)$ is equal to $-1/2$ for $x < x_L$ and to $+1/2$ for $x > x_R$, and increases monotonically with x in (x_L, x_R). Then the real part σ of the corresponding Ohmic conductance can be shown to be given by

$$\sigma(\omega) = 2\sum_m \sum_n |\bar{J}_{mn}|^2 \frac{f(E_m) - f(E_n)}{E_n - E_m} \delta(E_m - E_n + \hbar\omega) \tag{1}$$

where $f(E)$ is the Fermi function for the surrounding conductors and $\bar{J}_{mn}$ is the average over (x_L, x_R), with weighting function ds/dx, of the matrix element, $J_{mn}(x)$, of the current operator $J(x)$. That is, $\bar{J}_{mn} = \int J_{mn}(x)\,(ds/dx)\,dx$. (The actual distribution of the applied potential, represented by $s(x)$, is determined self-consistently by the space charge, as part of the conduction phenomenon, which can include neighboring "image" charge in the surrounding conductors. It can be shown that in general space charge effects are substantial at frequencies below $\sim \Delta E/h$.) This formula (1) is of course to be applied in the limit where the "normalization lengths" on the two sides, which determine the energy eigenstates $|n)$, tend to infinity, so that the spectrum of levels becomes infinitely dense. There are two independent series of these states, corresponding to the existence of two independent real wave functions in (x_L, x_R) for any given E. The wave functions obtained by integrating the Schrödinger equation over this interval must be transformed to the correct pair of linear combinations [5], for substitution in (1). In the low temperature limit ($kT << \Delta E$), the Fermi function becomes a step function and $\sigma(\omega)$ is then proportional to

$$\frac{1}{\hbar\,\omega}\int_{\zeta}^{\zeta+\hbar\omega} F(E-\hbar\omega\,,E)\,dE \tag{2}$$

where $F(E_1\,,E_2)$ is the sum of the four $|\,\overline{J}\,|^2$ terms connecting the pair of states at E_1 to the pair of states at E_2 (the density of states being effectively constant over this relatively small E range).

For a symmetrical structure (in particular, a symmetrical "quantum well" enclosed by two square or mirror-image barriers) it can be shown that near a resonance $F(E_1\,,E_2)$ is proportional to $\Lambda(E_1)+\Lambda(E_2)$. Consequently

$$\frac{\sigma(\omega)}{\sigma_0}=\frac{\Delta E}{2\,\hbar\omega}\left[\arctan\frac{\hbar\omega+\zeta'}{\Delta E}+\arctan\frac{\hbar\omega-\zeta'}{\Delta E}\right] \tag{3}$$

where $\zeta'=\zeta-E_0$ is the Fermi energy relative to the resonance value of E, and σ_0 is the zero-frequency conductance, $\sigma(0)$, for $\zeta'=0$. Figure 1 displays this function. It shows $Y(X)\equiv\sigma/\sigma_0$ versus $X\equiv\hbar\omega/\Delta E$, for the indicated values of $Z\equiv|\,\zeta-E_0\,|\,/\Delta E$. For $X=0$, we recover the Lorentzian $Y=1/(1+Z^2)$. For $Z\neq 0$, $Y(X)$ has a maximum which, for large Z, occurs at an X value a little greater than Z, with maximum value $\sim 1/Z$. This strong peak thus represents "excitations" from the Fermi level to the resonance quasi-level. For the high frequency limit $X>>Z$, we have $\sigma/\sigma_0\rightarrow\pi\,\Delta E/2\hbar\omega$.

For the three-dimensional double barrier diode, the conductance is given by an integral $\int d^2\mathbf{K}\,\sigma_{\mathbf{K}}(\omega)$ where $\mathbf{K}$ is the lateral two-dimensional wavevector and $\sigma_{\mathbf{K}}$ is the foregoing one-dimensional conductance. Assuming the usual separability and parabolic energy function, the integration is equivalent to const $\int d\zeta\,\sigma_{\mathbf{K}}$, taken over a range $\zeta_1<E_0$ to $\zeta_2>E_0$ with $E_0-\zeta_1$ and ζ_2-E_0 large compared to ΔE. For $\sigma_{\mathbf{K}}$ given by (3), the result is a σ independent of frequency, on the scale $\hbar\omega/\Delta E$. (This follows more directly from the relationship $\sigma(\omega)=$ const $\int dE\,F(E-\hbar\omega\,,E)$, with equivalent limits, applied to the present case.) Thus the lateral dimensions are important not only in allowing lateral scattering processes but also, in this example, transforming the frequency dependence.

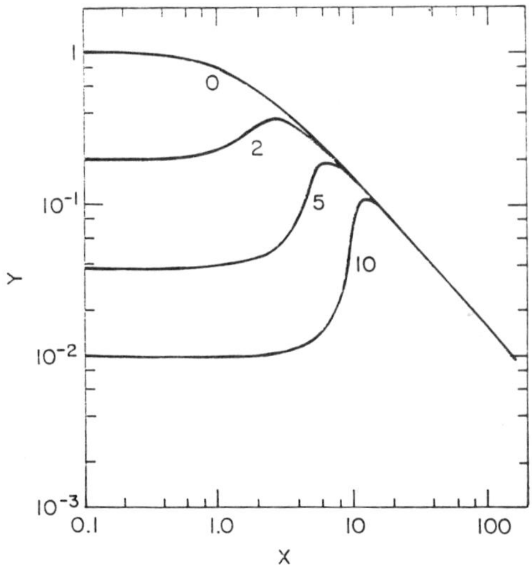

Figure 1. Relative conductance $Y = \sigma(\omega)/\sigma_0$ versus $X = \hbar\omega/\Delta E$, for the indicated values of $Z = \mid \zeta - E_0 \mid /\Delta E$.

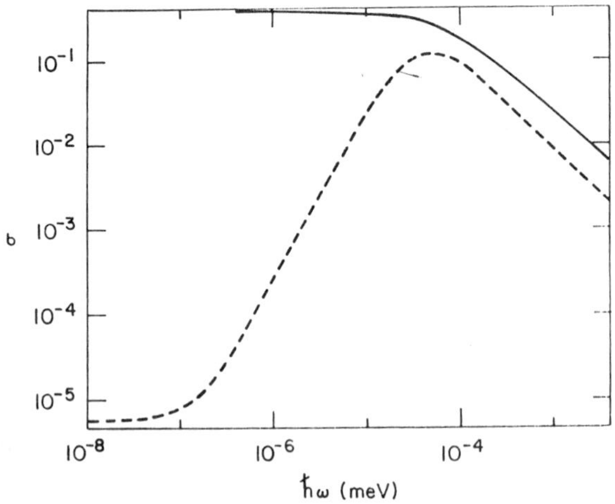

Figure 2. Conductance, in arbitrary units, versus $\hbar\omega$, for the double-barrier structures specified in the text. Symmetrical case: full line. Asymmetrical case: dashed line.

For non-symmetrical structures, one can evaluate the integral (2) numerically, from the computed wave functions and hence $\overline{J}$ elements [5], in particular assuming that $s(x)$ increases linearly $(ds/dx = 1/(x_R - x_L))$ in (x_L, x_R). A comparison was made of symmetrical ("S") and asymmetrical ("A") double-barrier structures, as follows: The electron mass was taken as 0.067 times the free mass (the Γ GaAs value); the well width was 40 nm; the two potential barriers both had a height of 100 meV. For S, both barrier widths were 20 nm; for A, one was 20 nm and the other was 40 nm. Then the two had virtually identical quasi-levels E_0, with the A value of ΔE equal to half the S value. The resonance investigated was at $E_0 = 44.31$ meV, with $\Delta E = 4.45 \times 10^{-5}$ meV for the S structure, and $T_0 = 1.50 \times 10^{-5}$ for A. Figure 2 compares conductance $\sigma(\omega)$, computed for $\zeta = E_0$, in the two structures. The full line (S) corresponds to the $Z = 0$ curve in Fig. 1. The dashed line resembles the $Z \neq 0$ lines, but with a much steeper rise by a factor $\sim 1/T_0$ (the ratio of the zero-frequency conductances). Again at high frequencies σ varies as $1/\omega$. The source of this behavior in the form of the wave functions ψ_n has been discussed elsewhere [5].

A similar phenomenon was found for $\sigma(\omega)$ in a "quantum wire" with a disordered potential, taken as a smoothly varying $V(x)$. Such a potential form can be generated, using pseudo-random numbers, for a specified autocorrelation function $V_2(z)$, where $V_2(z) = < V_1(x)\, V_1(x+z) >$ with $V_1(x) = V(x) - < V >$. For a Gaussian autocorrelation it is found [5, 6] to have $T(E)$ functions similar to those for the disordered $V(x)$ which is modelled by an array of delta functions [2]: Narrow resonance peaks are superposed on a monotonic "base line" curve. When eqs.(1) and (2) are used to compute $\sigma(\omega)$, for low temperatures with ζ equal to a resonance energy E_0, by means of the same algorithms as for the double-barrier case, the result [6] is a curve strikingly similar to the "A" curve of Fig. 2. With this much larger conductance at a frequency $\sim \Delta E/h$ than at zero frequency, phonon-induced transmission, which can contribute substantially to conduction at zero frequency, could be of less relative importance.

Thus the intuitively expected "roll off" of resonant conductance at a frequency $\sim \Delta E/h$ is verified, but in general with a more complicated dependence including a peak, at non-zero frequency, which can be large compared to the value at zero frequency. It should be noted, however, that the form of $s(x)$, which gives the distribution of the externally applied voltage over the tunneling structure and which will depend on the electronic space charge, can be expected to be frequency dependent, so that the actual frequency dependent $\sigma(\omega)$ will include this feature not given by the present calculation. The latter is readily included, however, in the Wigner function modeling [3, 4]. The latter can be, and has been, used to investigate the time and frequency dependence in the specially interesting bias range, for double barrier diodes, where the differential conductance is negative. It would evidently be of interest to apply it to the "quantum wire" case.

REFERENCES

1. L.L. Chang, L. Esaki and R. Tsu, Appl. Phys. Lett. **24**, 593 (1974).
2. M. Ya. Azbel and P. Soven, Phys. Rev. B**27**, 831 (1983).
M. Ya. Azbel and D. DiVincenzo, Phys. Rev. B**30**, 6877 (1984).
3. W.R. Frensley, Superlatt. and Microstruct. **4**, 497 (1988).
4. N.C. Kluksdahl, A.M. Kriman, D.K. Ferry and C. Ringhofer, Phys. Rev. B**39**, 7720 (1989).
5. C. Jacoboni and P.J. Price, preprint.
6. C. Jacoboni and P.J. Price, to be published.

Index of Contributors

Subject Index